T0264928

Modern Magnetooptics and Magnetooptical Materials

Studies in Condensed Matter Physics

Modern Magnetooptics and Magnetooptical Materials

A K Zvezdin and V A Kotov

General Physics Institute (IOF RAN)
Moscow

Published in 1997 by
Taylor & Francis Group
270 Madison Avenue
New York, NY 10016

Published in Great Britain by
Taylor & Francis Group
2 Park Square
Milton Park, Abingdon
Oxon OX14 4RN

International Standard Book Number-10: 0-7503-0362-X (Hardcover)
International Standard Book Number-13: 978-0-7503-0362-X (Hardcover)
Series Editors: **J M D Coey**, Trinity College, Dublin and **D R Tilley**, Universiti Sains Malaysia

Library of Congress Cataloging-in-Publication Data

Catalog record is available from the Library of Congress

Taylor & Francis Group
is the Academic Division of Informa plc.

**Visit the Taylor & Francis Web site at
http://www.taylorandfrancis.com**

Contents

Preface

In 1845, Faraday found that when plane-polarized light was transmitted through glass in a direction parallel to that of an applied magnetic field, the plane of polarization was rotated. This effect constituted the first demonstration of the connection between magnetism and light, and the origin of magnetooptics. Since Faraday's original discovery, magnetooptics has become a highly competitive and fascinating field of research, which is of great importance from both the basic science and applications points of view.

Several recent trends in science and technology highlight a definite need for a book devoted entirely to magnetooptics. Also, many college-trained research workers and engineers nowadays find themselves working on projects that are closely connected with magnetooptics. Such people need a concise, understandable, and fundamental discussion of magnetooptical phenomena, materials, and their applications. While many books of a more general nature contain some discussion of magnetooptical phenomena, in most cases the treatment is superficial and inadequate. On the other hand, the specialist papers and review articles tend to lack sufficient introduction to the subject for the average research worker to grasp the underlying physics of the—often quite complicated—formalism.

Our main concern in this book is to clarify the concepts in magnetism and optics that are most relevant to a deeper understanding of magnetooptics, magnetooptical materials, and applications. The book also provides the necessary theoretical details in simple form. With a mature reader in mind, we have attempted to make each chapter reasonably independent, so that most topics can be studied with only occasional reference to other material in the book. This may be particularly helpful to the phenomenologically inclined reader, who might want to bypass the more lengthy mathematical discussions.

We did not attempt to discuss every aspect of magnetooptics on the same footing. Some topics are discussed in depth; others are discussed superficially. Many topics could not be included, and others are treated so briefly as to merely whet the reader's interest. In the latter case the reader is directed to relevant books or recent papers. Duplication of existing texts has been mostly avoided, but sufficient background material is given to make the book useful to nonspecialized readers. Each chapter starts at a rather elementary level, and then proceeds step by step to more difficult matters.

The book is divided into an introduction and three parts. The introduction gives a broad view of the milestones of magnetooptics, and the subject matter is almost free of mathematical derivations. In part 1, the emphasis is on fundamental principles. An effort is made to bridge the gap between theory and practice through the use of graphs and real numerical data. The most important characteristics of magnetooptical materials are summarized in part 2. Part 3 deals with applications.

A wide range of topics is included, as may be seen from the contents listing. In writing this book, we have assumed that the reader has been introduced to Maxwell's equations in an intermediate course on electricity and magnetism.

This book should be useful to students and research engineers in the fields of solid-state physics, magnetism, magnetooptics, optoelectronics, magnetic memories, electromagnetic communications, etc.

Over the past nearly 20 years we have greatly profited from the cooperation of and discussions with numerous scientists, and we now take the opportunity to thank all of them. We wish to mention in particular Ron Atkinson, G Barbero, J A Bland, A D Boardman, M V Chetkin, A V Druzhinin, H Fujimori, G S Krinchik, N F Kubrakov, V Lavruk, V M Mayevskii, V Nekvasil, R V Pisarev, A I Popov, R and H Szymczak, Yu A Uspenskii and Š Višňovský.

Professor G S Krinchik read the first Russian version of the book, and offered a good deal of useful advice and remarks.

This book was initiated by Professor J M D Coey, whom we wish to thank cordially for his support.

A K Zvezdin and V A Kotov
May 1997

Acknowledgments

We gratefully acknowledge permission to reproduce illustrations from the following publishers.

The American Institute of Physics

Figure 8.1: Wood D L, Remeika J P and Kolb E D 1970 *J. Appl. Phys.* **41** 5315–22

Figure 8.7, data from: Krumme J-P, Doorman V, Hansen P, Baumgart H, Petruzzello J and Viegers M P A 1989 *J. Appl. Phys.* **66** 4393–407

Figure 8.7, data from: Nakamura H, Ohmi F, Kaneko Y, Sawada Y, Watada A and Machida H 1987 *J. Appl. Phys.* **61** 3346–8

Figure 8.8, data from: Moser F, Ahrenkiel R K, Carnall E, Coburn T, Lyu S L, Lee T H, Martin T and Pearlman D 1971 *J. Appl. Phys.* **42** 1449–51

Figure 8.9: Ahrenkiel R K and Coburn T J 1973 *Appl. Phys. Lett.* **22** 340–1

Figure 8.13, data from: Kurtzig A J and Guggenheim H J 1970 *Appl. Phys. Lett.* **16** 43

Figure 8.13, data from: Kurtzig A J, Wolfe R, LeCraw R C and Nielsen J W 1969 *Appl. Phys. Lett.* **14** 350

Figures 9.1, 9.2, 9.7, data from: Wood D L and Remeika J P 1967 *J. Appl. Phys.* **38** 1038

Figure 9.3, data from: Wemple S H, Dillon J F Jr, van Uitert L G and Grodkiewicz W H 1973 *Appl. Phys. Lett.* **22** 331

Figure 9.6, data from: Thuy N P, Višňovský Š, Prosser V, Krishnan R and Vien T K 1981 *J. Appl. Phys.* **52** 2292–4

Figure 9.18 and table 9.3, data from: Scott G B and Page J P 1977a *J. Appl. Phys.* **48** 1342

Figure 9.18, data from: Larsen P K and Robertson J M 1974 *J. Appl. Phys.* **45** 2867

Figure 10.2, data from: Sawatzky E 1971 *J. Appl. Phys.* **42** 1706–7

Figure 10.2, data from: Sawatzky E and Street G B 1973 *J. Appl. Phys.* **44** 1789–92

Figure 10.2, data from: Katsui A 1976a *J. Appl. Phys.* **47** 3609–11

Figure 10.6, data from: Reim W, Husser O E, Schoenes J, Kaldis E and Wachter P 1984 *J. Appl. Phys.* **55** 2155–6

Figure 11.2, data from: Dillon J F Jr, Furdyna J K, Debska U and Mycielski A 1990 *J. Appl. Phys.* **67** 4917–9

Figure 12.1: Reim W and Weller D 1988 *Appl. Phys. Lett.* **53** 2453

Figure 12.3: Kudo T, Johbetto H and Ichiji K 1990 *J. Appl. Phys.* **67** 4778–80

Figure 13.3, data from: Shirasaki M, Targaki N and Obokata T 1981 *Appl. Phys. Lett.* **38** 833

Figure 13.3, data from: Umegaki S, Inoue H and Yoshino T 1981 *Appl. Phys. Lett.* **38** 752–4

Figure 13.10: Yan X and Xiao S 1989 *J. Appl. Phys.* **65** 1664–5

Figure 13.23: Whitcomb E C and Henry R D 1978 *J. Appl. Phys.* **49** 1803

Figure 14.3: Kryder M H 1985 *J. Appl. Phys.* **57** 3913–8

Figure 15.3, data from: Tamada H and Saitoh M 1990 *J. Appl. Phys.* **67** 949–54

Figures 15.4 and 15.6, data from: Ando K, Takeda N, Koshizuka N and Okuda T 1985a *J. Appl. Phys.* **57** 718

Figure 16.1: Tien P K, Martin R J, Blank S L, Wemple S H and Varnerin L J 1972a *Appl. Phys. Lett.* **21** 207–9

Figure 16.2: Henry R D 1975 *Appl. Phys. Lett.* **26** 408–11

Table 8.3: Wolfe R, Kurtzig A J and LeCraw R C 1970 *J. Appl. Phys.* **41** 1218–24

The American Physical Society

Figure 5.10: Oppeneer P M, Maurer T, Sticht J and Kubler J 1992 *Phys. Rev.* B **45** 10924–33

Figure 6.1: Stähler S, Schütz G and Ebert H 1993 *Phys. Rev.* B **47** 818

Figure 6.3: Idzerda Y U, Tjeng L H, Lin H J, Gutierrez C J, Meigs G and Chen C T 1993 *Phys. Rev.* B **48** 4144

Figure 8.2: Kahn F J, Pershan P S and Remeika J P 1969 *Phys. Rev.* **186** 891–918

Figure 9.3, data from: Wemple S H, Blank S L and Seman J A 1974 *Phys. Rev.* B **9** 2134

Figure 9.3 and tables 9.1, 9.2, data from: Scott G B, Lacklison D E and Page J L 1974 *Phys. Rev.* B **10** 971–86 and Scott G B, Lacklison D E, Ralph H I and Page J L 1975 *Phys. Rev.* B **12** 2562–71

Figure 9.15: Wittekoek S, Popma T J A, Robertson J M and Bongers P F 1975 *Phys. Rev.* **12** 2777

Figure 9.18: Hansen P, Witter K and Tolksdorf W 1983 *Phys. Rev.* B **27** 6608

The Czech Academy of Science

Figure 8.6: Višňovský Š, Siroky P and Krishnan R 1986 *Czech. J. Phys.* B **36** 1434

Figure 9.6: Višňovský Š 1986 *Czech. J. Phys.* B **36** 625

Elsevier Science Ltd

Figure 8.3: Popma T J A and Kamminga M G J 1975 *Solid State Commun.* **17** 1073–5

Figure 8.8: Carnall E Jr, Pearlman D, Coburn T J, Moser F and Martin T W 1972 *Mater. Res. Bull.* **7** 1361–8

Figure 8.11: Dillon J F Jr, Kamimura H and Remeika J P 1966 *J. Phys. Chem. Solids* **27** 1531–49

Figure 11.1: Gaj J A, Gałązka R R and Nawrocki M 1978 *Solid State Commun.* **25** 193

Figure 12.2: Zhou S M, Lu M, Zhai H R, Miao Y Z, Tian P B, Wang H and Xu Y B 1990 *Solid State Commun.* **76** 1305–7

The Institute of Electrical and Electronics Engineers, Inc.

Figure 8.5: Ahrenkiel R K and Coburn T J 1975 *IEEE Trans. Magn.* **11** 1103–8

Figure 8.10: Brandle H, Schoenes J, Hulliger F and Reim W 1990 *IEEE Trans. Magn.* **26** 2795–7

Figure 9.12: Daval J, Ferrand B, Milani E and Paroli P 1987 *IEEE Trans. Magn.* **23** 3488–90

Figure 10.2, data from: Chen D, Otto G N and Schmit F M 1973 *IEEE Trans. Magn.* **9** 66–83

Figure 10.2, data from: Sawatzky E and Street G B 1971 *IEEE Trans. Magn.* **7** 377–80

Figure 11.1: Koyanagi T, Watanabe T and Matsubara K 1987 *IEEE Trans. Magn.* **23** 3214–6

Figure 13.2: Scott G B and Lacklison D E 1976 *IEEE Trans. Magn.* **12** 292

Figure 13.8, data from: Nakano T, Yuri H and Kihara U 1984 *IEEE Trans. Magn.* **20** 986–8

Figure 13.10: Kobayashi K and Seki M 1980 *IEEE J. Quantum Electron.* **16** 11–22

Figure 13.13, data from: Numata T, Ihbuchi Y and Sakurai Y 1980 *IEEE Trans. Magn.* **16** 1197

Figure 13.15: Krawczak J A and Torok E J 1980 *IEEE Trans. Magn.* **16** 1200

Figure 13.18: Lacklison D E, Scott G B, Pearson R F and Page J L 1975 *IEEE Trans. Magn.* **11** 1118

Figure 13.19: MacNeal B E, Pulliam G R, Fernandez de Castro J J and Warren D M 1983 *IEEE Trans. Magn.* **19** 1766

Figure 13.20: Nomura T 1985 *IEEE Trans. Magn.* **21** 1544–5

Figure 14.3: Meiklejohn W H 1986 *Proc. IEEE* **74** 1570–81

Figure 15.1: Daval J, Ferrand B, Geynet J, Challeton D, Peuzin J C, Leclert A and Monerie M 1975 *IEEE Trans. Magn.* **11** 1115–7

Figures 16.3 and 16.4: Wolfe R, Hegarty J, Dillon J F Jr, Luther L C, Celler G K and Trimble L E 1985 *IEEE Trans. Magn.* **21** 1647–50

Table 14.1, data from: Dancygier M 1987 *IEEE Trans. Magn.* **23** 2608–10

The Society of Photo-Optical Instrumentation Engineers

Figure 8.4: Martens J W D and Peeters W L 1983 *Proc. SPIE* **420** 231–5

Taylor & Francis Ltd

Figure 9.4, data from: Mee C D 1967 *Contemp. Phys.* **8** 385

Chapter 1

Introduction

1.1 Faraday and Voigt geometries; longitudinal and transverse magneto-optical effects

Magnetooptics deals with phenomena arising as a result of interaction between light and matter when the latter is subjected to a magnetic field. In the case of magnetically ordered matter (ferromagnets, ferrimagnets, etc) magnetooptical effects may appear in the absence of an external magnetic field as well. The presence of a magnetic field changes the dispersion curves of the absorption coefficient and leads to the appearance or variation of optical anisotropy. A great number of magnetooptical phenomena are the direct or indirect outcome of the splitting of system energy levels in an external magnetic field. This splitting is the Zeeman effect[1]. Essentially, all other magnetooptical effects are consequences of the Zeeman effect.

Basic magnetooptical phenomena can be classified according to the relative orientation of the wave vector of the light emission k and the magnetic field H. Two basic geometries are distinguished (figure 1.1):

(1) Faraday geometry: the light travels along the field direction ($k \parallel H$); and

(2) Voigt geometry: the light travels perpendicularly to the field direction ($k \perp H$).

In both cases the Zeeman effect is observed—longitudinal in the Faraday geometry and transverse in the Voigt geometry.

In an absorption spectrum the optical anisotropy of a magnetized medium manifests itself as dichroism, i.e. the difference between the absorption coefficients for the two orthogonal polarizations. Dichroism is defined as the difference between the absorptions of the circularly polarized components

---

[1] In 1896, P Zeeman, Dutch physicist and Nobel prize winner, discovered that, when a sodium flame is placed between the poles of an electromagnet, the two yellow lines are considerably broadened. Shortly afterward, Lorentz presented a simple theory for these observations, based upon the electron theory of matter.

$(k_+ - k_-)$ in the case of the Faraday geometry (so-called magnetic circular dichroism (MCD)), and in the Voigt geometry it is determined from the difference between the absorptions of components polarized parallel and perpendicular to the magnetic field (the magnetic linear dichroism (MLD)). These effects are analogues of the transverse and longitudinal Zeeman effects.

From the Kramers–Kronig relations it follows that the splitting of the dispersion curves of the absorption coefficient (i.e. MCD or MLD) is connected with the splitting of the dispersion curves of the refractive index. In this case it shows up as a difference between the refractive indices for the two circularly polarized components (the magnetic circular birefringence or Faraday effect) in the Faraday geometry and for the two linearly polarized components (the magnetic linear birefringence) in the Faraday geometry.

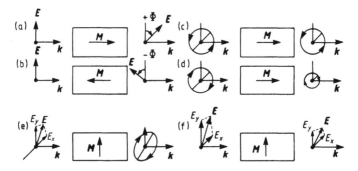

Figure 1.1. Magnetooptical effects resulting from the interaction of optical radiation with the medium: magnetic circular birefringence ((a), (b)), magnetic circular dichroism ((c), (d)), magnetic linear birefringence (e), and magnetic linear dichroism (f).

1.2   The Faraday effect

In 1845 Michael Faraday discovered that when a block of glass is subjected to a magnetic field, it becomes optically active. When plane-polarized light passes through glass in a direction parallel to the applied field, the plane of polarization is rotated.

The angle of rotation of the plane of polarization, θ, is in the simplest case proportional to the magnitude of the magnetic field H and the distance L travelled by light in a medium along the direction of the field:

$$\theta = VHL. \tag{1.1}$$

The constant V, called the Verdet constant, is defined as the rotation per unit path, per unit field strength. The Verdet constant depends upon the properties of the medium, the frequency of light, and the temperature T.

The sign of the angle θ depends on the sign of H. Therefore if the light travels twice through a field, first along the field direction, and then after normal reflection from a mirror in the opposite direction, the value of θ is doubled. This is a phenomenological distinction of the Faraday effect from the effect of natural optical activity. In the latter case, when light returns, $\theta = 0$.

The circular anisotropy of the magnetized medium is explained by the fact that magnetic field splitting of the energy levels of the ground and excited states creates energy non-equivalence of quantum states with different eigenvalues of the angular momentum. From a phenomenological point of view, the Faraday effect is explained as has been pointed out above by the fact that the refractive indexes, n_+ and n_- for the left-hand and right-hand circularly polarized light in the case of an originally inactive substance, become different when the latter is placed in a magnetic field. Linearly polarized light with a given plane of polarization can be represented as a superposition of the right-hand and the left-hand circularly polarized waves with a definite phase difference. As a consequence of the difference between n_+ and n_-, the right-hand and the left-hand circularly polarized waves propagate with different velocities, c/n_+ and c/n_-. The plane of polarization of the linearly polarized light therefore rotates through the angle

$$\theta = \frac{\omega}{2c}(n_+ - n_-)L \tag{1.2}$$

where ω is the angular frequency, c is the velocity of light, and L is the path length of an optical beam in the medium. The magnetic-field-induced Larmor precession of electron orbits is the simplest mechanism for the Faraday effect. Instead of one eigenfrequency of the electron (ω), two (ω_+ and ω_-) arise, corresponding to the right-hand and the left-hand circular oscillations when there is a magnetic field in the medium. Theory shows that

$$\omega_\pm = \omega_0 \pm \frac{1}{2}\frac{e}{m}\frac{H}{c} \tag{1.3}$$

where e and m are the electron charge and mass.

The appearance of two resonance frequencies (ω_- and ω_+) in a medium placed in a magnetic field gives rise to the splitting of the absorption line. The difference between the resonance frequencies in the Zeeman doublet, ω_+ and ω_-, results in a displacement of the curves $n_+(\omega)$ and $n_-(\omega)$ relative to each other on the frequency scale:

$$n_\pm(\omega) \approx n(\omega) \pm \frac{dn}{d\omega}\frac{eH}{2mc} \tag{1.4}$$

where $n(\omega)$ is the refractive index in the absence of the field H. Substituting (1.4) and (1.2) into (1.1) yields the well-known Becquerel formula

$$V = \frac{e}{2mc^2}\lambda\frac{dn}{d\lambda} \tag{1.5}$$

where $\lambda = 2\pi c/\omega$ is the wavelength of light. Formula (1.5) is consistent with experimental data for diamagnetic media. Complete analysis of the phenomenon has to be carried out on the basis of quantum mechanics (chapter 5). Figure 1.2 shows that formula (1.1) describes with high accuracy the Faraday rotation in paramagnetic glasses in magnetic fields up to 7 MOe (Pavlovskii *et al* 1980).

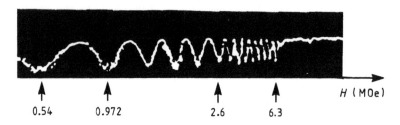

Figure 1.2. The Faraday effect for the glass TF-5 ($V = 0.0495'$ cm^{-1} Oe, $l = 5$ mm). The intervals between minima are 180°. The numbers correspond to the magnetic field measured in MOe. After Pavlovskii *et al* (1980).

Giant values of rotation ~ 1–$10°$ cm^{-1} Oe^{-1} were observed for the magnetic semiconductor $Cd_{0.95}Mn_{0.05}Te$ (Gaj *et al* 1993).

1.3 The Cotton–Mouton or Voigt effect

In 1902 Voigt discovered that, when a magnetic field is applied to a vapour through which light is passing perpendicularly to the field, birefringence takes place. A similar but stronger effect was discovered in 1907 by Cotton and Mouton for a liquid. In liquids like nitrobenzene, very strong double refraction is observed. This birefringence is attributed to a lining up of the magnetically and optically anisotropic molecules in the applied-field direction.

The magnetic linear birefringence effect is even in the magnetic field. Usually it depends quadratically on the magnetic field intensity, and results in a change of ellipticity of circularly polarized light propagating through a medium. The magnetic linear birefringence observed in magnetic materials is in many cases stronger than the linear magnetooptical effects (the Faraday effect, etc).

Interesting and strong magnetooptical effects are observed in liquid crystals, composed as a rule from diamagnetic molecules and possessing strong anisotropy of the magnetic susceptibility and electrical polarization. Although the magnetic susceptibility of liquid crystals is rather weak, the energy of a magnetic field may appear to be sufficient to change the orientational order in a liquid crystal because of the cooperative character of its response to an external magnetic field. In turn, a change of the liquid crystal's orientational structure by virtue of strong optical anisotropy of the molecules appears in the magnetically induced change of the birefringence. Very strong magnetooptical effects of the same character are observed in magnetic liquids.

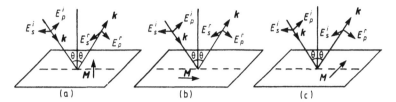

Figure 1.3. Magnetooptical Kerr effects taking place when light is reflected from the surface of a magnetized material: polar (a); longitudinal (meridional) (b); and transverse (equatorial) (c) effects (the subscripts p and s in E_p and E_s indicate p and s polarization of the wave).

1.4 The magnetooptical Kerr effect (MOKE)

Optical anisotropy of a magnetized medium manifests itself also in the reflection of light from its surface. Phenomena arising here are generally referred to as the magnetooptical Kerr effect. It consists in an influence of the magnetization of the medium on the reflected light[2].

Depending on the orientation of the magnetization vector relative to the reflective surface and the plane of incidence of the light beam, three types of magnetooptical effect in reflection are distinguished: the polar, longitudinal (meridional), and transverse (equatorial) Kerr effects. The magnetization vector M is oriented perpendicularly to the reflective surface and parallel to the plane of incidence in the polar effect (figure 1.3(a)). The geometry of the longitudinal Kerr effect is shown in figure 1.3(b). The influence of magnetization in both of these effects is reduced to a rotation of the plane of polarization and the appearance of ellipticity of the reflected linearly polarized light.

The general feature of the polar and longitudinal effects is the presence of a non-zero projection of the wave vector k of the electromagnetic wave on the magnetization direction M. The polar Kerr effect is of great importance for optical data storage, since it is the basis for reading information from magnetooptical disks.

The geometry of the transverse (equatorial) Kerr effect is depicted in figure 1.3(c). It is observed when the magnetization vector is oriented perpendicularly to the plane of incidence of the light, and it is revealed in the change of intensity and phase of the linearly polarized light reflected by the magnetized medium.

This effect corresponds to the transverse magnetooptical effect since the projection of the light wave vector in the magnetization direction is here equal to zero. The transverse Kerr effect, unlike the Cotton–Mouton effect, is odd in

---

[2] The magnetooptical Kerr effect was discovered by the Scottish physicist John Kerr in 1888. He observed that when plane-polarized light is reflected at normal incidence from the polished pole of an electromagnet, it becomes elliptically polarized with the major axis of the ellipse rotated with respect to the plane of polarization of the incident beam.

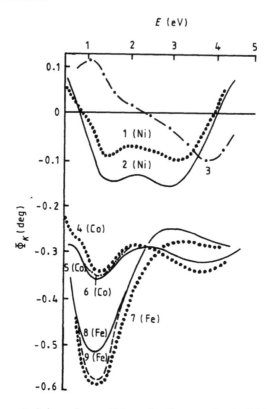

Figure 1.4. The spectral dependence of the polar Kerr rotation in Ni (dotted curve (1): Krinchik and Artem'ev (1968); solid curve (2): van Engen (1983)) and the ellipticity of $Ni_{0.92}Au_{0.08}$ alloy (chain curve (3): Buschow (1988)); of the polar Kerr rotation in Co (dotted curve (4): Krinchik and Artem'ev (1968); solid curve (5): van Engen (1983); chain curve (6): Carey *et al* (1983)); and of the polar Kerr rotation in Fe (dotted curve (7): Krinchik and Artem'ev (1968); solid curve (8): van Engen (1983); dashed curve (9): Carey *et al* (1983)).

the magnetization. Therefore it is widely employed in magnetic investigations, particularly for observations of magnetic domains at the surface of a magnetized sample. Also, this effect allows the design of transversely magnetized mirrors as non-reciprocal optical devices.

In general the Kerr effects are small; for example, the polar Kerr angles for perpendicular incidence for iron, cobalt, and nickel are typically less than a degree in the visible range of the spectrum, and the transverse effect is 0.01 or less (figure 1.4).

Transition metal compounds containing Mn and Pt are known to show a giant magnetooptical Kerr effect. Van Engen *et al* (1983b) discovered the giant Kerr effect in the Heusler compound PtMnSb, which has a maximum Kerr rotation of 1.2° at 1.7 eV photon energy. There is a huge Kerr rotation of 1.18° in MnPt$_3$ (Kato *et al* 1983b).

1.5 The permittivity tensor and the sign convention

The formal description of magnetooptical effects is based on consideration of the influence of a magnetic field on the dielectrical permittivity tensor of the medium ϵ_{ij}. If ϵ_{ij} is symmetric ($\epsilon_{ik} = \epsilon_{ki}$) at $H = 0$, it becomes non-symmetric ($\epsilon_{ik}(H) = \epsilon_{ki}(-H)$) in the presence of a magnetic field H. In this case the requirement of reality of the ϵ_{ik}-tensor for a transparent medium is replaced with the requirement of its hermiticity:

$$\epsilon_{ik} = \epsilon_{ki}^{*}.$$

From the present considerations it follows that the symmetric part of the tensor ϵ_{ik} of a transparent magnetized medium is real, and the antisymmetric part is imaginary. Besides which, the real components of ϵ_{ik} have to be even functions of the magnetic field.

A second and most important point which must be made in relation to magnetooptical calculations is as regards the choice of sign convention, for the form of the permittivity tensor and also the sense of magnetooptically induced rotations and ellipticities.

There are advantages and disadvantages to all of the basic sign schemes and, in the final analysis, that adopted may be a question of personal taste or may have been chosen for historical reasons. However, whatever choice is made with regard to the sign convention, the scheme must be clearly stated and, where possible, justified as being an appropriate and reasonable choice.

To begin with, assume that the plane-wave solutions to the Maxwell equations are taken to have a negative time exponent $e^{-i\omega t}$, and take the form of the skew-symmetric permittivity tensor as

$$[\varepsilon] = n^2 \begin{bmatrix} 1 & -iQ & 0 \\ +iQ & 1 & 0 \\ 0 & 0 & 1 \end{bmatrix} \tag{1.6}$$

where the complex index of refraction n and the so-called magnetooptical parameter Q are

$$n = n' + in'' \tag{1.7}$$

$$Q = Q' + iQ'' \tag{1.8}$$

and the magnetization lies along the z-direction.

The form of the permittivity given in equation (1.6) ensures that the signs of the Kerr and Faraday rotations and ellipticities are positive for a clockwise rotation of the axes of the polarization ellipse and a clockwise rotation of the E-vector respectively, when viewed along the positive z-direction (see section 2.2).

It is perhaps worth making a number of points in relation to other schemes, so that the reader familiar with an alternative convention can easily convert from one scheme to another (Atkinson and Lissberger 1992). First, it is equally valid in the plane-wave solution to the Maxwell equations to take the positive time exponent as $e^{+i\omega t}$. The consequence of doing this is that the complex refractive index must now be written as $n' - in''$, again to ensure an exponential decay of the wave amplitude with increasing propagation distance. On choosing the same form for the magnetooptical parameter, namely $Q' - iQ''$, all other transformations between this scheme and the preferred one are simply obtained by changing the sign associated with all of the imaginary quantities. This applies throughout the whole of any magnetooptical analysis, including, for example, the Kerr or Faraday ellipticities which may be regarded as imaginary components of the complex Kerr of Faraday rotations.

Second, a commonly used form for the permittivity tensor has the signs of the off-diagonal elements reversed compared to those shown in equation (1.6). The transposition from one scheme to the other is effected quite simply by multiplying any terms in Q by minus one.

1.6   Other physical aspects

1.6.1   Time-reversal symmetry, and circular anisotropy in a rotating medium

The origin of circular birefringence and dichroism is evident from a symmetry viewpoint. Magnetic field, which is an axial vector, has the symmetry of a circular current set out in a plane perpendicular to the vector of the magnetic field. In a medium placed in a magnetic field, the rotation directions in a plane oriented perpendicularly to the magnetic field are non-equivalent. Therefore the optical properties of a magnetized medium for light propagating along the field direction with right-hand and left-hand circular polarization have to be different. By analogy we can assert that a mechanically rotating medium ought to reveal circular anisotropy along its rotation axis. This follows from the fact that angular velocity, like magnetic field, is an axial vector. The effect of *rotation of the plane of polarization on a rotating medium* predicted by J J Thomson in 1885 was recently discovered experimentally (see Jones 1976)[3].

The Faraday rotation and the Kerr effect can be assigned as non-reciprocal effects. They are characterized by different phase velocities and/or attenuations of light waves travelling along the same optical path but in opposite directions. Non-reciprocal optical effects are usually observed in magnetized media. A magnetic moment can be induced by a magnetic field in paramagnets and diamagnets or can arise spontaneously, as in ferromagnets or ferrimagnets. In

---

[3] This effect is a close analogue of the Barnette effect, i.e. magnetization appears in a rotating medium. The latter is in its turn related to the Einstein–de Haas effect consisting in the sample twisting during its magnetization in a reverse sense.

all of these the time-reversal symmetry (T) is broken.

A number of attempts have been made to detect the non-reciprocal optical effects due to time-parity violation in high-T_c superconductors predicted within some theoretical models (see, e.g., Spielman *et al* 1990). All of these attempts failed, but meanwhile non-reciprocal optical rotation was observed in a conventional antiferromagnet (Krichevtsov *et al* 1993).

1.6.2 Light-induced magnetooptical effects; the angular momentum of a light beam

Effects in which emission not only probes the magnetic state of a medium but also changes it in an active way are referred to as magnetooptical effects. It is well known that an elliptically polarized wave has angular momentum. Therefore, circularly (and elliptically) polarized light giving rise to electronic transitions of the atoms induces orientation of their moments relative to the light beam direction. These effects are most conspicuous in magnetic semiconductors.

From this point of view it is interesting to consider the photo-induced change in magnetization of solid or gas-like paramagnets effected by circularly polarized emission. This phenomenon was discovered in 1952 by A Kastler and it is known as *optical orientation*. Atomic absorption of circularly polarized photons results in a change of magnetization of the medium. In principle, optical orientation can appear as a result of emission of sufficiently small intensity if the magnetization relaxation times are sufficiently large.

It should be noted that angular momentum exchange between circularly polarized emission and a medium may also produce macroscopic angular momentum in the medium, i.e. a twist of the sample (the Sadovskii effect predicted in 1899). This effect is rather small, but was observed experimentally (see Rosenberg 1950, Santamato *et al* 1986).

1.6.3 Non-linear magnetooptical effects

Different effects arise in transparent media. It is known from electrodynamics that a rotating electric field acts as an effective magnetic field, whence it immediately follows that a medium becomes magnetized under the action of an intense circularly polarized wave. The magnetization direction is defined by the sign of the circularly polarized light. This effect is called the *inverse Faraday effect*. This non-linear optical effect can be observed if the amplitude of the electromagnetic field is fairly large. The inverse Faraday effect has been observed in crystals doped with paramagnetic centres and in metal vapours.

Non-linear magnetooptics—in particular, *magnetism-induced second-harmonic generation (MSHG)*—is of general interest, and is becoming an important tool in the investigation of fundamental problems of magnetic films, multilayers, and surfaces. MSHG derives its sensitivity from the breaking of the symmetry at the surface or at interfaces. Because the optical penetration length is of order 50–100 nm, not only surfaces but also buried interfaces can be studied by this method. There is a large enhancement of the non-linear MOKE (Aktsipetrov *et al* 1990, Reif *et al* 1991, Spierings *et al* 1993, Pustogowa *et al* 1994, Vollmer *et al* 1995) in comparison with the linear MOKE.

Magnetism-induced MSHG was also observed in the magnetoelectric materials Cr_2O_3 (Fiebig *et al* 1994) and $BiFeO_3$ (Agal'tsov *et al* 1989).

1.6.4 Magnetooptical luminescence

The circular anisotropy of luminescence can be related to magnetooptics. The picture of emission transitions between excited and ground states is distinguished only by the direction of the transitions from the similar picture for the absorption spectra. Therefore the difference spectrum displaying the difference between circularly polarized luminescence components:

$$p = \frac{I_+ - I_-}{I_+ + I_-}$$

can be assigned as a spectrum of the magnetic circular luminescence.

The luminescence spectrum carries information about the magnetization of the excited ions (Zapasskii and Feofilov 1975).

1.6.5 Magnetooptical effects in the x-ray region

The transmission and reflection magnetooptical effects considered above constitute the foundation of established methods for studying the magnetic materials in the visible, infra-red and ultra-violet regions. There are analogous magnetooptical effects in the x-ray region. Although the x-ray polarizability of materials is very slight (see chapter 6), there is great scientific interest and progress in the use of x-ray magnetooptical effects as a probe of magnetic materials. This has become possible due to both theoretical advances in describing near-x-ray-absorption-edge phenomena and great experimental progress achieved with the use of synchrotron radiation (SR). Broad-spectrum, high-brightness sources, which provide intense monochromatic beams of 10^{12}– 10^{13} photons per second, and the unique polarization properties of SR permit the realization of experimental methods involving the production, monitoring, and detection of polarized x-rays.

Magnetic x-ray dichroism (MXD) was predicted by Erskine and Stern (1975) and Thole *et al* (1985a, b) and observed using both linearly (van der Laan *et al* 1986) and circularly (Schütz *et al* 1987) polarized synchrotron radiation. Magnetic circular x-ray dichroism (MCXD) is mainly used as a tool at the x-ray absorption edges of 3d or 4f elements, where large asymmetries in the white lines upon reversal of either the sample magnetization or the photon spin are observed. MXD has opened up the possibility for element-specific analysis of magnetic materials.

MXD is also lattice site specific, because the final electron states are disturbed by the crystal field, and this disturbance is different when the same ions occupy different lattice sites. This method supplements neutron diffraction because it does not require a large volume of material (it could determine for example the orientation of the moment on impurities). In addition it can determine the orientation of the magnetic moments absolutely. MXD has all the versatility of x-ray absorption spectroscopy: one can study bulk spectra by measuring the fluorescence of the core hole, or just the first few atomic layers by measuring the intensity of the Auger electrons. This makes MXD especially suitable for studying the magnetism of surfaces and thin films. Moreover, the sum rule for MCXD spectra allows a quantitative determination of the orbital contribution to the total magnetization. This direct measurement of orbital angular momentum is highly desirable for probing spin–orbit interaction and macroscopic phenomena originating from it (e.g. magnetocrystalline anisotropy).

In general, MXD attracts great interest as a tool for studying phenomena (spin–orbit coupling, exchange interaction, hybridization, and crystal-field splitting) affecting the electronic structure of a sample. Many alloy surfaces and multilayers exhibiting unusual magnetic properties have been probed by MCXD.

Some other phenomena connected with the interaction of x-rays and magnetic structure such as the Kerr effect (Bonarski and Karp 1989) and Faraday rotation (Siddons *et al* 1990) have also been observed using SR.

1.6.6 Dichroic photoemission

The observation of large MCXD effects has stimulated the search for similar effects in photoemission. The well-known merits of photoemission are high surface sensitivity and a sampling depth that can be tuned by varying the kinetic energy of the photoelectrons. There is a strong interaction between the core holes and the valence electrons which can lead to dichroic photoemission. This phenomenon differs from MXD, where the core electron is excited into the unoccupied part of the localized levels. In x-ray emission the electron is excited into continuum states which have no magnetic structure far above the Fermi energy, and the dichroism is induced by core–valence interaction. The first study of dichroic photoemission was made for the 2p core level of Fe, with the $2p_{1/2}$ and $2p_{3/2}$ electrons being emitted into the continuum spectrum (Baumgarten *et al* 1990). Large MCD effects and a high-resolution photoemission spectrum were observed for the localized 4f states in rare earths (Navas *et al* 1993).

The use of circularly polarized SR in photoemission experiments opens up new perspectives in surface and thin-film magnetism. It allows measurement, in particular, of the magnitudes and orientations of individual magnetic sublattices, as well as the study of spin–orbit interaction, and it is very attractive also for domain-imaging applications.

1.7 Magnetooptical materials

During the last 20–30 years a wide class of magnetic materials has been studied. Existing magnetooptical materials can be divided into two groups. The first group includes metals and alloys which are partly transparent only at film thicknesses lower than 100 nm. For these materials the polar Kerr effect is the most important magnetooptical effect because it involves the reflection of light from the surfaces of magnetized media. Rare-earth–transition metal alloys are a most important group of materials, because thin films of such alloys have been widely used in magnetooptical disk memory systems.

Another important group of magnetooptical materials includes the dielectric and semimagnetic materials, which look very promising as regards applications in spatial and time light modulators, non-reciprocal devices, and integrated magnetooptics. Magnetic dielectrics, such as ferrimagnetic garnets, orthoferrites, spinel ferrites, and another oxides, are rather transparent media. The ferrimagnetic garnets are more useful for practical purposes; for example, bismuth-substituted iron garnets are widely used in devices for optical signal processing.

1.7.1 Dielectrics

In chapter 8 the magnetooptical properties of different classes of magnetic dielectric, such as orthoferrites, manganites, spinel ferrites, magnetoplumbite, barium hexaferrite, chalcogenide spinels, chromium trihalides, europium oxide and monochalcogenides, ferric borate, and ferric fluoride, are described. For most of these compounds the spectral dependencies of the Faraday and Kerr effects are presented, and the nature of the magnetooptical activity of magnetic dielectrics is discussed.

1.7.2 Ferrimagnetic garnets

In chapter 9 the results of investigations of optical and magnetooptical properties of different iron garnets are presented, with most attention being paid to the optics and magnetooptics of bismuth-substituted iron garnets and the relationships of these properties to the technology of such crystals and films. The influence of cerium, lead, and cobalt substitution on the optical and magnetooptical properties of iron garnets are also reported.

The ferrimagnetic garnets are more useful for practical purposes; for example, bismuth-substituted iron garnets are widely used in developing different magnetooptical devices for optical signal-processing systems and optical communication systems (optical isolators and circulators, switches, time and spatial modulators, non-reciprocal devices, sensors, magnetooptical heads, and so on).

1.7.3 Metals and alloys

In chapter 10 the experimental results obtained on ferromagnetic metals and compounds are presented. Included in this section will also be a discussion of magnetooptical properties of alloys based on simple metals, binary intermetallics including CoPt, MnBi and pseudo-binary compounds, ternary compounds including PtMnSb and Heusler alloys, amorphous films of rare-earth–transition metal alloys used in magnetooptical disk memories, and low-temperature uranium- and rare-earth-containing compounds with outstanding magnetooptical properties including U(Sb, Te), CeSb and others.

1.7.4 Semimagnetic semiconductors

Semimagnetic semiconductors are a new class of materials combining the properties of ordinary and magnetic semiconductors. These materials include semiconductor crystals doped with 3d ions of transition metals, and solid solutions containing a magnetic component.

1.7.5 Bilayers, multilayers, superlattices, and granular materials

Chapter 12 deals with new artificial magnetooptical media including bilayer, trilayer, and quadrilayer structures, granular transition metal–dielectric thin films, compositionally modulated structures, and a new class of phase-separated magnetic materials in which two phases couple antiferromagnetically across a phase boundary.

Artificially structured magnetic multilayers are an interesting new class of magnetooptical materials. When the layer thicknesses approach the nanometre region there is clearly an opportunity to control the short-range order, and hence the anisotropy and magnetooptical properties. Interfacial atomic interactions become increasingly significant, and new thin-film properties, some with technological and practical importance, are likely to arise.

Another approach for improving the magnetooptical properties of thin films has been to use multilayers of magnetic elements with metals that have a plasma edge in the wavelength range of interest—for example, the Co/Au system.

1.8 Applied magnetooptics

1.8.1 Magnetooptical devices

In chapter 13 various thin-film magnetooptical devices—magnetooptical modulators, bistable optical switches, non-reciprocal devices (optical isolators), magnetooptical circulators, deflectors, transparencies and displays, read heads, and low-insertion-loss magnetooptical elements for laser gyroscopes—are described. Very interesting aspects of the spatial filtering of optical signals now widely used in practice are also discussed there. Magnetooptical modulators are the devices that control the intensity of optical radiation in optical communication and data-processing systems by applying a magnetic field. Low-frequency modulators based on domain wall movement and high-frequency modulators based on the rotation of the magnetization vector of a saturated sample are described. Development of optical fibres with extremely low losses has stimulated development of non-reciprocal elements (optical isolators), which are used in communication systems for the suppression of reflected signals, which appear in fibre-optic connectors and other contact elements of functional optical devices and which have adverse effects on the optical performance of laser sources.

1.8.2 Magnetooptical memory

Chapter 14 deals with magnetooptical memories, disks and tapes. Information on these devices is recorded using the thermomagnetic method: laser radiation is focused onto a spot about one micrometre in diameter. When an area of the storage medium is heated locally, it remagnetizes as a result of the simultaneous effect of the demagnetizing field and a constant or pulsed bias field. In magnetooptical random-access data storage with a fixed storage medium, the information is recorded and read by a laser beam using a special light deflector.

The computer industry has long dreamed of data-storage systems which would combine the major advantages of optical disk recording (gigabyte capacity and non-contact technology) and magnetic recording (erasability, longevity). Magnetooptical disks using rare-earth–transition metal alloy films exhibiting large perpendicular magnetic anisotropy offer high-density recording (densities near 10^8 bits cm^{-2})—ten times that of high-performance disk drives and up to 100 times the density of low-end disk drives, lifetimes longer than ten years, and data transfer rates up to 1 Mbyte s^{-1} with typical access times of approximately 40 ms.

The disks are removable, like flexible disks, but are rigid, and may be rotated at speeds as high as for conventional winchester disks. The write and erase operations can be performed by laser light modulation or magnetic field modulation at the GaAs laser wavelength. Magnetooptical tape provides a further increase in the information capacity; wide practical use of this tape will be possible after the development of an integrated optical head which would provide for multichannel reading and writing of information.

1.8.3 Integrated magnetooptics

One of the main objectives of integrated magnetooptics—a novel intensively developing branch of applied magnetooptics—is the transmission and processing of optical signals propagating in thin-film magnetooptical waveguides, in which the waveguiding and controlling topological elements and structures are formed using modern methods of thin-film deposition, lithography, and laser annealing.

The thin-film magnetooptical isotropic and anisotropic waveguides, and the problems of optical losses, induced optical anisotropy and mode conversion are examined in chapter 15.

Integrated magnetooptical waveguide modulators, and non-reciprocal elements and structures are also described in chapter 15. Special attention is paid to a new branch of integrated magnetooptics at the interface of microwave technology and integrated optics, which is based on waveguide mode interaction with magnetostatic waves.

PART 1

PHYSICS

Chapter 2

Polarized light and gyrotropic media

In most magnetooptical experiments and devices the incident light is polarized. The main effect of the magnetic medium on the transmitted or the reflected waves is to change their states of polarization. In order to analyse complex optical systems involving polarized waves, a number of formal algebras and approximate methods have been developed. We present here briefly the main equations of the electrodynamics of magnetooptical media, and the most common matrix methods. A more complete description of these methods can be found in many textbooks (see, e.g., Robson 1974, Gerrard and Burch 1975).

2.1 Material equations and boundary conditions

The macroscopic theory of magnetooptical phenomena is based on the Maxwell equations:

$$\operatorname{rot} E = -c^{-1}\frac{\partial B}{\partial t} \qquad\qquad \operatorname{div} B = 0 \qquad\qquad (2.1)$$

$$\operatorname{rot} B = c^{-1}4\pi j + c^{-1}\frac{\partial E}{\partial t} \qquad \operatorname{div} E = 4\pi\rho \qquad (2.2)$$

where ρ and j are respectively the densities of the induced charges and the current, which are interrelated by the continuity equation:

$$\frac{\partial\rho}{\partial t} + \operatorname{div} j = 0.$$

Instead of ρ and j, the polarization P and magnetization M are usually used, where

$$\rho = -\operatorname{div} P \qquad\qquad (2.3)$$

$$j = \frac{\partial P}{\partial t} + c\operatorname{rot} M. \qquad\qquad (2.4)$$

The set of the equations (2.1), (2.2) must be transposed using the material equations, which express P and M in terms of the fields E and B.

Given the intensity of the magnetic field H and the electric induction D, as

$$H = B - 4\pi M \qquad \text{and} \qquad D = E + 4\pi P$$

the Maxwell equations for a region of the medium can be rewritten as

$$\text{rot } E = -c^{-1} \frac{\partial B}{\partial t} \qquad \text{div } B = 0$$

$$\text{rot } H = c^{-1} \frac{\partial D}{\partial t} \qquad \text{div } D = 0.$$

In linear electrodynamics it is assumed that the vectors P and E, as well as M and H, are linearly interrelated (in the Fourier representation); that is, these vectors are related by means of material equations, which are derived on the basis of certain microscopic concepts of the electronic and atomic structure of the medium:

$$P_i(\omega) = \widehat{\alpha}_{ij}(\omega) E_j(\omega) \qquad M_i(\omega) = \widehat{\chi}_{ij}(\omega) H_j(\omega)$$

where $\widehat{\alpha}_{ij}(\omega)$ and $\widehat{\chi}_{ij}(\omega)$ are the tensors of the electric and magnetic polarization; the fields E and D, as well as B and H, are related via the expressions

$$D_i(\omega) = \varepsilon_{ij}(\omega) E_j(\omega) \qquad H_i(\omega) = (\mu^{-1}(\omega))_{ij} B_j(\omega)$$

where $\varepsilon_{ij}(\omega)$ and $\mu_{ij}(\omega)$ are respectively the tensors of the electric and magnetic permeability of the medium. The relation between D and E, and that between H and B are assumed to be local; that is, it is assumed that there is no spatial dispersion.

At the interfaces of the different media the following conditions are valid:

$$E_t^1 = E_t^2 \qquad D_n^1 = D_n^2 \qquad (2.5)$$
$$B_n^1 = B_n^2 \qquad H_t^1 = H_t^2 \qquad (2.6)$$

where the indices 1 and 2 correspond to the media, while the subscripts n and t indicate the normal and tangential components of the fields at the interface.

The formulation of the electrodynamics of the material medium in terms of these four vectors (E, D, B, and H) is not unique, because the definition of the induced charges and currents in terms of the vectors P and M is ambiguous (Silin and Rukhadze 1961, Landau *et al* 1984). Because the densities ρ and j are governed by the continuity equation, in general only one additional vector of a field is required for defining the induced charges and currents. For example, it is possible to define the current density in terms of a generalized conductance (in the Fourier representation):

$$j_i(\omega, k) = \sigma_{ij}(\omega, k) E_j(\omega, k)$$

where $\sigma_{ij}(\omega, k)$ is the generalized electrical conductance tensor. Hence

$$\rho = -\omega^{-1} k_i \sigma_{ij}(\omega, k) E_j(\omega, k)$$

and the generalized D-vector is

$$\widetilde{D}_i(\omega, k) = \widetilde{\varepsilon}_{ij} E_j(\omega, k)$$

where

$$\widetilde{\varepsilon}_{ij}(\omega, k) = \delta_{ij} - \omega^{-1} \, \mathrm{i} \, 4\pi \sigma_{ij}(\omega, k).$$

This scheme does not require the introduction of the field H. The magnetic effects are described by the tensor $\widetilde{\varepsilon}_{ij}$.

This three-vector electrodynamics has some advantages from the fundamental physics point of view, but it is not always convenient in practice. For example, in order to describe electromagnetic properties of materials having non-unity 'ordinary' magnetic susceptibility μ, it is necessary to take into account spatial dispersion in the tensor $\varepsilon_{ij}(\omega, k)$, or, in other words, the non-local relationship between $D(\omega)$ and $E(\omega)$:

$$D(\omega) = \widehat{\varepsilon}(\omega) E(\omega) + (c/\omega)^2 \, \mathrm{rot}((\widehat{\mu}^{-1} - 1) \, \mathrm{rot}\, E).$$

In addition, the boundary conditions are also complicated; it is as difficult to define them as it is to obtain the material equations, because it is necessary to apply microscopical theory. The first pair of equations, (2.5), remain valid, while the second pair, (2.6), change. Induced surface currents and charges should be taken into account here (for more details see Agranovich and Ginzburg 1984). In magnetooptics, the Maxwell equations with two pairs of vectors, E, D and B, H, are usually used[1]. This approach is worth following.

2.2 The tensors $\widehat{\varepsilon}$ and $\widehat{\mu}$, and the gyration vector g

Within the macroscopical theory of magnetooptical phenomena, the particular properties of a medium are defined by the form of the $\widehat{\varepsilon}$- and $\widehat{\mu}$-tensors. For a magnetically ordered state some general properties of these tensors can be established phenomenologically. It is enough to consider just the $\widehat{\varepsilon}$-tensor, because the properties that we describe below are similar for $\widehat{\varepsilon}$- and $\widehat{\mu}$-tensors. In the magnetically ordered state the $\widehat{\varepsilon}$-tensor depends on the order parameter. In ferromagnets the order parameter is the magnetization M, in antiferromagnets it is the sublattice magnetization, and so on. We shall confine our discussion mainly to ferromagnets.

---

[1] Expression (2.4) may actually be considered as the first two-term expansion of the current in the multipole moments, provided that the expansion parameter obeys the condition $a/\lambda \ll 1$, where a is a specific length—for example, the lattice constant of a dielectric or the electron mean free path of metals—and λ is the wavelength of light. If this constraint is obeyed, four-vector electrodynamics is quite adequate.

Let us consider the simplest case first: that of an optically isotropic ferromagnet. The presence of magnetization reduces the symmetry to the single-axis one. The $\hat{\varepsilon}$-tensor can be represented as a sum of symmetric and asymmetric tensors, which in the coordinate system with the z-axis directed along M takes the following form:

$$\hat{\varepsilon} = \begin{pmatrix} \varepsilon_1 & 0 & 0 \\ 0 & \varepsilon_1 & 0 \\ 0 & 0 & \varepsilon_0 \end{pmatrix} + \begin{pmatrix} 0 & -ig & 0 \\ ig & 0 & 0 \\ 0 & 0 & 0 \end{pmatrix}. \tag{2.7}$$

The D-vector can be shown to be

$$D = \varepsilon_0 E + i[g \times E] + b(E - m(m \cdot E)) \tag{2.8}$$

where $m = M/M$, $b(M) = \varepsilon_1 - \varepsilon_0$, ε_0 is the dielectric permeability of the medium at $M = 0$, and g is the gyration vector. In an isotropic medium where $g = g(M)m$, normally $g(M) = aM$. If there is absorption, then

$$\varepsilon_0 = \varepsilon_0' + i\varepsilon_0'' \qquad g = g' + ig'' \qquad b = b' + ib''$$

are complex functions of the frequency. The second terms in the formulae (2.7) and (2.8) describe the gyrotropic effects: the magnetic gyrotropic birefringence and the magnetic circular dichroism. The last terms describe the optical magnetic anisotropy: the magnetic linear birefringence and the magnetic linear dichroism. The constants g and b become zero when $M \to 0$.

The magnetooptical (Voigt) parameter Q is defined as

$$Q = Q' + iQ'' = g/\varepsilon_1.$$

Normally $|Q| \ll 1$. The gyrotropy of the magnetic permeability tensor is described in terms of the Q_M-parameter. If both Q and Q_M need to be taken into consideration, the magnetooptical medium is called bi-gyrotropic.

In crystals, the dependence of the $\hat{\varepsilon}$-tensor on M is more complicated, namely

$$\varepsilon_{ik} = \varepsilon_{ik}^0 - ie_{ikl}g_l + \delta_{iklm}M_lM_m \tag{2.9}$$

where $g_l = a_{lq}M_q$, e_{ikl} is the antisymmetric 3D-order pseudotensor (the Levi–Civita tensor). The polar tensors ε_{ik}^0, a_{lq}, δ_{iklm} are defined by the crystallographic symmetry (crystallographic class).

Let us consider, for instance, materials with cubic symmetry. In the coordinate system with the axes directed along the crystallographic axes [100], [010], and [001], the $\hat{\varepsilon}$-tensor (2.9) is given by

$$\varepsilon_{ik} = (\varepsilon_0 + b_2M^2 - (b_1 + b_2)M_i^2)\delta_{ik} + b_3M_iM_k - iae_{ikl}M_l. \tag{2.10}$$

The gyration vector of a magnet with many sublattices can be presented as $\sum_n a_{ik}^n M_k^{(n)}$, where $M^{(n)}$ is the magnetization of sublattice n.

There is a special class of magnetic media—magnetoelectrics—in which both the time parity and the space parity are destroyed. Theoretical analysis (Hornreich and Shtrikman 1968, Lyubimov 1969) of the propagation of light in magnetoelectrics showed that new magnetooptical phenomena should be found when the spatial dispersion in (2.7) is taken into account. Some of these effects have been observed in the 'classical' magnetoelectrics Cr_2O_3 (Krichevtsov *et al* 1986, 1993). These new effects are much smaller than the corresponding magnetooptical effects in ferromagnets.

2.3   Normal modes and Fresnel equations

In magnetooptics the normal modes of an electromagnetic field are of particular importance. They are solutions of the Maxwell equations (2.1) and (2.2) that have a harmonic dependence on the time and the coordinates:

$$\exp[-i(\omega t - kr)]. \tag{2.11}$$

Taking into account (2.11), the main wave equation

$$-\nabla^2 E + \operatorname{grad} \operatorname{div} E = -c^{-2}\frac{\partial^2 D}{\partial t^2} \tag{2.12}$$

can be derived from equations (2.1)–(2.4).

For the normal modes, equation (2.12) becomes

$$n^2 E - n(n \cdot E) = \widehat{\varepsilon} E \tag{2.13}$$

where $n = (c/\omega)k$ is the refraction vector. Equations (2.13) have a non-trivial solution when the determinant of the coefficients vanishes:

$$\det \|n^2 \delta_{ik} - n_i n_k - \varepsilon_{ik}(\omega)\| = 0. \tag{2.14}$$

Equation (2.14) is called the Fresnel equation. It defines the refraction vectors $n = n(\omega)$ of the normal modes. Substituting the refraction vector into equation (2.11) one can determine the eigenvectors of each mode. In the general case the modes are elliptically polarized. In a gyrotropic medium ($g \neq 0$) the refraction vector depends on the tracing direction of the ellipse. This right–left anisotropy, i.e., $n_+ \neq n_-$, is usually called the gyrotropy of the medium.

Sometimes it is more convenient to deal with the vector D rather than with the intensity vector E, because due to the condition div $D = 0$, the D-vector always lies in the plane that is perpendicular to the direction of propagation of the wave. The equation for the refraction vector in this case takes the following form:

$$\det \|n^2(\varepsilon^{-1})_{ik} - n_i n_l (\varepsilon^{-1})_{lk} - \delta_{ik}\| = 0.$$

It is not difficult to extend equation (2.14) to the bi-gyrotropic media. In this case (2.14) can be reduced to a fourth-order equation in n for a definite wave propagation direction. The general solution of this equation is rather unwieldy. Two particular orientations of the vectors n and g—namely, longitudinal $(n \parallel g)$ and transverse $(n \perp g)$—often arise.

For longitudinal orientation

$$n^2 = \varepsilon_0 \mu_0 [1 \pm (Q + Q_M)] \tag{2.15}$$

where Q and Q_M are the Voigt parameters of the tensors ε and μ. The signs $\pm$ correspond to the right-handed and left-handed elliptical polarizations (see section 2.4).

For the transverse orientation the normal modes are an s wave (TE)

$$H_x \neq 0 \qquad H_y \neq 0 \qquad H_z = 0$$

and a p wave (TM)

$$H_x = 0 \qquad H_y = 0 \qquad H_z \neq 0.$$

In this case

$$
\begin{aligned}
n_s^2 &= \varepsilon_0 \mu_0 (1 - Q_M^2) \\
n_p^2 &= \varepsilon_0 \mu_0 (1 - Q^2).
\end{aligned}
\tag{2.16}
$$

This difference between the dependences of the eigenvectors n on the magnetooptical parameters Q and Q_M for the longitudinal and transverse orientations makes it possible to separate the gyroelectric and gyromagnetic contributions in a bi-gyrotropic medium.

2.4   Polarization of light

We now consider the polarization of light in more detail. Let us describe the electromagnetic wave as before in the form

$$
\begin{aligned}
E(r, t) &= E_0 \exp[-i(\omega t - k \cdot r)] \\
B(r, t) &= B_0 \exp[-i(\omega t - k \cdot r)]
\end{aligned}
$$

where $k = (\omega/c)n$ is the wavevector, which defines the propagation direction of the wave, and n is the refraction vector. As E and B are perpendicular to each other, defining just one of them, for instance E, is enough to characterize the polarization completely.

The plane that the B- and k-vectors belong to is called the plane of polarization. The simplest type of polarization is the linear one. In an electromagnetic wave of this type the plane of polarization does not change during propagation. Linearly polarized light can be obtained by passing natural light through a polarizer.

In a more general case the light is elliptically polarized. In this type of polarization the **E**-vector of a propagating electromagnetic wave describes an ellipse in the plane that is perpendicular to the direction of propagation. The ellipse is defined directly by the equation for the electromagnetic wave

$$E_x = a \operatorname{Re}\{\exp[i\varphi(r, t)]\}$$
$$E_y = b \operatorname{Re}\{\exp[i\varphi(r, t) - \delta]\}$$

(2.17)

where a and b are the real amplitudes of the wave, and $\varphi(r, t) = kr - \omega t$. This ellipse is called the polarization diagram or polarization ellipse of the electromagnetic wave (figure 2.1).

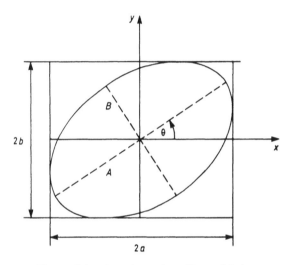

Figure 2.1. The polarization ellipse of light.

The tracing direction of the ellipse is important. If we face the light beam (we look in the direction opposite to **k**), the **E**-vector can trace the ellipse either clockwise or anticlockwise. These waves are called right and left polarized respectively. The difference between them is especially important for magnetooptics, because they interact differently with magnetized media and have different phase velocities.

The following parameters are commonly used for the description of a polarization diagram. The angle θ, which defines the orientation of the major axis, is measured from the x-axis. This parameter is called the orientation angle of the polarization ellipse. The ellipticity angle ψ is defined as the angle between the major axis of the ellipse and the diagonal of the rectangle that circumscribes the ellipse. Ellipticity is also defined via the coefficient e, which is the ratio of the minor and major axes of the polarization ellipse ($e = \tan \psi$). The angle ψ is measured in the direction in which the **E**-vector rotates in accordance with formulae (2.17), its sign being $-$ for the left-polarized light and $+$ for the right-polarized wave. Evidently $-1 \leq e \leq 1$ and $-\pi/4 \leq \psi \leq \pi/4$.

If $\psi = 0$, we have linearly polarized light. If $\psi = \pm\pi$, then the light is circularly polarized and $a = b$.

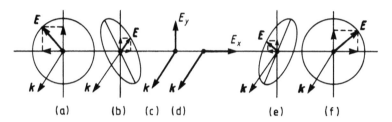

Figure 2.2. Different kinds of light polarization: left circular (a) and right circular (f); left elliptic (b) and right elliptic (e); and linear ((c), (d)).

2.5 Jones matrices; the spin and helicity of a photon

When light interacts with a substance or with an optical device, its polarization can be changed. This fact has two important consequences. First, changing the light polarization is widely and successfully used in solid-state physics to study energy spectra and elementary interactions of light with matter. Second, the change in light polarization that occurs as a result of the interaction with the medium provides an effective way of controlling the light polarization, and therefore is of great interest for technological applications.

We shall describe the polarization state of the electromagnetic wave (2.16) by the column vector (Jones 1941a, b)

$$\Psi_J = \begin{pmatrix} E_x \\ E_y \end{pmatrix} = e^{i\varphi} \begin{pmatrix} a \\ b e^{i\delta} \end{pmatrix}. \tag{2.18}$$

The factor $e^{i\varphi}$ may be omitted here[2], since the polarization state does not depend on φ. When light passes through an optical device or a medium, the E-vector changes. From the mathematical point of view, the effect of the device or the medium on the E-vector can be presented as an operator that converts the incident (input) electromagnetic wave into the output one:

$$\begin{pmatrix} E_x \\ E_y \end{pmatrix}_{\text{out}} = \begin{pmatrix} M_{11} & M_{12} \\ M_{21} & M_{22} \end{pmatrix} \begin{pmatrix} E_x \\ E_y \end{pmatrix}_{\text{in}}. \tag{2.19}$$

The matrix **M** is called the propagation matrix or Jones matrix. Jones matrices are convenient for calculations of complex optical systems that consist of different elements. The result for light transmission—the E-vector at the output—can be calculated by multiplication of the Jones matrices of all of the optical devices in

---

[2] In fact the components a and b of the Jones vector can be considered as the probability amplitudes of the corresponding quantum state.

the order in which the light passes through them in the system. For illustration, we now present Jones matrices for some important optical elements.

An *ideal polarizer* leaves the x-component of an electric field unchanged, and absorbs the y-component. Therefore here

$$\mathbf{M} = \begin{pmatrix} 1 & 0 \\ 0 & 0 \end{pmatrix}. \tag{2.20}$$

A *rotation plate* splits the light beam into ordinary and extraordinary beams, which propagate in the plate with different phase velocities:

$$v_o = c/n_0 \quad \text{and} \quad v_e = c/n_e$$

where n_o and n_e are the refraction indexes for the ordinary and extraordinary waves. The phase delays of the beams, within the plate, are

$$\varphi_0 = \omega n_0 d/c \quad \text{and} \quad \varphi_e = \omega d n_e/c$$

where d is the thickness of the plate. The Jones matrix here is

$$\mathbf{R}(\delta) = \exp(i\bar{\varphi}) \begin{pmatrix} e^{i\delta/2} & 0 \\ 0 & e^{-i\delta/2} \end{pmatrix} \tag{2.21}$$

where $\bar{\varphi} = (\varphi_0 + \varphi_e)/2$, and

$$\delta = (n_e - n_0)\frac{\omega}{c}d = 2\pi(n_e - n_0)\frac{d}{\lambda}. \tag{2.22}$$

The phase factor $\exp(i\bar{\varphi})$ is inessential, as long as we are not concerned with the effects of interference between different beams. As is well known, the quarter-wave plate is a retardation plate, with $\delta = \pi/2$. The Jones matrix of the quarter-wave plate is

$$\mathbf{R}(\lambda/4) = e^{i\bar{\varphi}} \begin{pmatrix} i & 0 \\ 0 & -i \end{pmatrix}.$$

The Jones matrix of the half-wave plate is

$$\mathbf{R}(\lambda/2) = e^{i\bar{\varphi}} \begin{pmatrix} 1 & 0 \\ 0 & -1 \end{pmatrix}.$$

In all cases it is assumed that the optical axis of the plate (the so-called fast axis), which coincides with the direction of the E-vector of the extraordinary wave, is oriented along the x-axis, and reflections at the plate surfaces have been neglected.

We now consider the *transformation of linear polarization to right- and left-handed circular polarization.* Let us consider the coordinate system x', y', z

rotated by an angle α around the z-axis. In this case the vectors $\mathbf{\Psi}$ and $\mathbf{\Psi}'$ and the matrices $\mathbf{M}$ and $\mathbf{M}'$ are related by the equations

$$\mathbf{\Psi} = \mathbf{T}\mathbf{\Psi}'$$

and

$$\mathbf{M} = \mathbf{T}\mathbf{M}'\mathbf{T}^{-1}$$

where

$$\mathbf{T} = \begin{pmatrix} \cos\alpha & -\sin\alpha \\ \sin\alpha & \cos\alpha \end{pmatrix} \tag{2.23}$$

$$\mathbf{T}^{-1} = \begin{pmatrix} \cos\alpha & \sin\alpha \\ -\sin\alpha & \cos\alpha \end{pmatrix}. \tag{2.24}$$

If $\alpha = \pi/4$, we obtain

$$\mathbf{R} = \frac{1}{\sqrt{2}} \begin{pmatrix} 1 & -i \\ i & -1 \end{pmatrix}$$

for a quarter-wave plane.

This Jones matrix transforms linearly polarized waves to right or left circularly polarized waves. Acting on the Jones vectors

$$\begin{pmatrix} 1 \\ 0 \end{pmatrix} \qquad \text{and} \qquad \begin{pmatrix} 0 \\ 1 \end{pmatrix}$$

representing linear polarization along the x- and y-axes, respectively, the matrix $\mathbf{T}\mathbf{R}(\lambda/4)\mathbf{T}^{-1}$ gives the vectors

$$\frac{1}{\sqrt{2}} \begin{pmatrix} 1 \\ +i \end{pmatrix} \qquad \text{and} \qquad \frac{1}{\sqrt{2}} \begin{pmatrix} 1 \\ -i \end{pmatrix} \tag{2.25}$$

which represent left and right circularly polarized light, respectively. If we select an appropriate value of α in the matrix $\mathbf{T}\mathbf{R}(\lambda/4)\mathbf{T}^{-1}$ we can produce a similar operation for elliptical polarization.

We now look at the *general case: elliptical polarization.* First let us consider the case where the angle θ is equal to 0. An elliptically polarized wave can be considered as a superposition of the right and left circularly polarized waves given by (2.25):

$$\mathbf{\Psi}_J = \begin{pmatrix} E_x \\ E_y \end{pmatrix} = E_0 e^{i\varphi} \begin{pmatrix} \cos\psi \\ i\sin\psi \end{pmatrix} \tag{2.26}$$

where the angle ψ is by definition the ellipticity angle. The angle θ is assumed here to be equal to 0.

In the general case where $\theta \neq 0$ (see figure 2.1) we need to rotate (2.26) as above by an angle θ—i.e. the Jones vector can be represented, according to (2.23), by the formula (Jones 1941a, b)

$$\begin{pmatrix} E_x \\ E_y \end{pmatrix} = E_0 e^{i\varphi} \begin{pmatrix} \cos\theta & -\sin\theta \\ \sin\theta & \cos\theta \end{pmatrix} \begin{pmatrix} \cos\psi \\ i\sin\psi \end{pmatrix} \qquad (2.27)$$

whence it immediately follows that

$$\chi \equiv \frac{E_y}{E_x} = \frac{\tan\theta + i\tan\psi}{1 - i\tan\theta\tan\psi} \qquad (2.28)$$

and, conversely,

$$\tan 2\theta = \frac{2\,\mathrm{Re}\,\chi}{1 - |\chi|^2} \qquad (2.29)$$

$$\sin 2\psi = \frac{2\,\mathrm{Im}\,\chi}{1 + |\chi|^2}. \qquad (2.30)$$

In the practically important case where $|\chi| \ll 1$, from equations (2.29) and (2.30) it follows that

$$\theta \approx \mathrm{Re}\,\chi \qquad \psi \approx \mathrm{Im}\,\chi. \qquad (2.31)$$

There is another way to obtain formulae for the state of polarization of light. Let the Jones vector (2.27) be decomposed into the right and left circularly polarized waves given by (2.25); then

$$\tilde{\chi} \equiv \frac{E_+}{E_-} = e^{2i\theta} \frac{\cos\psi - \sin\psi}{\cos\psi + \sin\psi} \qquad (2.32)$$

whence it readily follows that (Azzam 1978)

$$\theta = \frac{1}{2}\arg\tilde{\chi} \qquad (2.33)$$

$$\psi = \tan^{-1}\frac{1 - |\tilde{\chi}|}{1 + |\tilde{\chi}|}. \qquad (2.34)$$

2.5.1 Spin and helicity of photons

From the point of view of quantum electrodynamics, the concept of polarization is associated with the existence of the spin momentum of a photon. In the simplest case, the angular momentum of a photon is $j = 1$; therefore, its projection onto the direction of motion (the wave vector) has three possible values: $m = \pm 1, 0$. For photons—particles with zero rest mass—the state with $m = 0$ cannot be realized, because of the well-known gauge invariance of the Maxwell equations. To put this another way, because electromagnetic waves are transverse, for waves travelling along the z-axis we have $E_z = 0$, i.e. there is no state for which $m = 0$. So, photons, as particles with zero mass, can exist in two states with moments of magnitude $\pm\hbar$, directed along the direction of the momentum of the photon. The states with $m = \pm 1$ (the helicities) describe photons having left (right) circular polarization. They can be represented by the two basis vectors

$$\frac{1}{\sqrt{2}} \begin{pmatrix} 1 \\ +i \end{pmatrix} \quad \text{and} \quad \frac{1}{\sqrt{2}} \begin{pmatrix} 1 \\ -i \end{pmatrix}$$

which are immediately recognizable as the Jones representations for left-handed and right-handed circularly polarized light, respectively. Elliptically polarized light is a superposition of the states with $m = \pm 1$; the particular case of linear polarization is also a superposition of these states.

2.6   The Stokes vector; the Poincaré sphere

Let us consider briefly one more method, which is elegant and in some cases convenient, for mathematically describing light polarization in a transparent medium.

When we consider formula (2.17), which represents polarization of an electromagnetic wave, it strikes us that it is analogous to the quantum-mechanical description of a two-level system (i.e., a system with a quasi-spin $S_{qu} = 1/2$); this is not surprising, because, as we have mentioned above, a photon can be described in terms of the helicity states $m = \pm 1$. This analogy becomes clearer if we rewrite (2.17) in other terms:

$$E = E_0 \begin{pmatrix} \cos\alpha/2 & \exp(-i\beta/2) \\ \sin\alpha/2 & \exp(i\beta/2) \end{pmatrix}.$$

This column vector (spinor), when considered as a wave function of a two-level system, describes the state in which the direction of S_{qu} in space is defined by polar (α) and azimuthal (β) angles. This means that each polarization state can be associated with a direction of the quasi-spin. In other words, the set of polarization states is unambiguously mapped onto the set of directions which are defined by the angles α and β, or, equivalently, onto a set of points that belong to some sphere, which is called the Poincaré sphere.

The transformation of the polarization of light passing through a medium corresponds to a turn of the quasi-spin S_{qu} in a space around some axis or to a displacement of the corresponding point on the Poincaré sphere.

Traditionally, a slightly different method is used in optics for the description of the polarization state using the Poincaré sphere. The polarization state of the wave may be characterized by a four-dimensional real Stokes vector, introduced

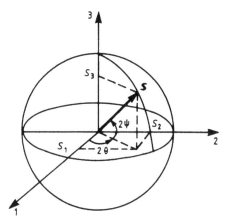

Figure 2.3. The Poincaré sphere representation of the Stokes vector $S = (S_0, S_1, S_2, S_3)$.

by G Stokes in 1852 (for details, see Robson 1974, Gerrard and Burch 1975):

$$\psi_s = \begin{pmatrix} S_0 \\ S_1 \\ S_2 \\ S_3 \end{pmatrix}$$

where

$$S_0 = E_x E_x^* + E_y E_y^* = (S_1^2 + S_2^2 + S_3^2)^{1/2}$$
$$S_1 = E_x E_x^* - E_y E_y^* = a^2 - b^2 = S_0 \cos 2\psi \cos 2\theta$$
$$S_2 = E_x E_y^* + E_x^* E_y = 2ab \cos \delta = S_0 \cos 2\psi \sin 2\theta$$
$$S_3 = i(E_x E_y^* - E_x^* E_y) = 2ab \sin \delta = S_0 \sin 2\psi.$$

There is a simple connection between the Stokes and Jones vectors:

$$S_i = \Psi^+ \sigma_i \Psi \qquad i = 0, 1, 2, 3$$

where Ψ^+ is the Hermitian conjugate of the Jones vector Ψ, i.e. the row matrix $(E_x^* \ E_y^*)$ and

$$\sigma_0 = \begin{pmatrix} 1 & 0 \\ 0 & 1 \end{pmatrix} \qquad \sigma_1 = \begin{pmatrix} 1 & 0 \\ 0 & -1 \end{pmatrix}$$
$$\sigma_2 = \begin{pmatrix} 0 & 1 \\ 1 & 0 \end{pmatrix} \qquad \sigma_3 = \begin{pmatrix} 0 & -i \\ i & 0 \end{pmatrix}.$$

This means that in the quantum approach to optics the Stokes parameters $S_i = \Psi^+ \sigma_i \Psi$ can be considered as the expectation values of σ_i.

The Poincaré sphere Σ' is shown in figure 2.3. Its north and south poles correspond to the circular polarization, the points of intersection of Σ' with the plane (1, 2) correspond to linear polarization, and all of the other points correspond to elliptic polarization. The radius of the sphere is usually associated with the intensity of the wave. Representation of polarization using the Stokes vector or the Poincaré sphere is an explicit and effective technique in the analysis of complex optical systems.

On occasion, it is convenient to use the three-dimensional vector

$$S_B = (S_1, S_2, S_3)$$

where S_i are the components of the Stokes vector. This vector is the mean of the quasi-spins $S_{qu} = (\sigma_1, \sigma_2, \sigma_3)$. In quantum electronics a similar vector is called the Bloch vector. It is known that the Bloch vector is proportional to the angular momentum of a system. An arbitrary polarization state may be represented by a point on the Bloch sphere, which is an analogue of the Poincaré sphere. The propagation (Jones) matrices are from this point of view completely analogous to the unimodular matrices, which are well known in quantum mechanics and are used for the description of the rotation of spin systems.

It is interesting to note in this context the paper by Bland *et al* (1993) in which the authors have highlighted the similarities between neutrons reflected from ferromagnetic media and the reflection of light from media rendered gyrotropic by the application of a magnetic field. They have shown that this analogy can be used to give valuable magnetometric information for ultrathin films and multilayers.

Chapter 3

Magnetooptical effects

3.1 The Faraday effect, and magnetic circular dichroism

3.1.1 Magnetic circular birefringence—the Faraday effect

In a magnetized medium the refractive indexes for right- and left-handed circularly polarized light are different. This effect manifests itself in a rotation of the plane of polarization of linearly polarized light.

Let us derive the main formulae for the Faraday effect for the simple example of the above-considered 'isotropic' medium; its optical properties are defined by the formulae (2.7) and (2.8). Let the z-axis of the coordinate system be directed along the direction of the magnetization and let the light propagation direction coincide with the z-axis; then $n = (0, 0, n)$ and

$$E = E_0 \exp[-i\omega(t - c^{-1}nz)].$$

Since $\operatorname{div} E = 0$ in this case, the wave equations (2.12) and (2.13) can be rewritten as follows:

$$(n^2 - \varepsilon_1)E_{0x} + ig E_{0y} = 0 \tag{3.1}$$

$$- ig E_{0x} + (n^2 - \varepsilon_1)E_{0y} = 0. \tag{3.2}$$

A non-trivial solution of this system exists if (the Fresnel equation (2.14))

$$n_\pm^2 = \varepsilon_1 \pm g = \varepsilon_1(1 \pm Q). \tag{3.3}$$

Substituting $n_\pm^2$ into (3.1) and (3.2) we obtain two modes with right (e_+) and left (e_-) circular polarization (2.25):

$$e_+ = \frac{1}{\sqrt{2}} \begin{pmatrix} 1 \\ -i \end{pmatrix} \exp[-i\omega(t - c^{-1}n_+ z)] \tag{3.4}$$

$$e_- = \frac{1}{\sqrt{2}} \begin{pmatrix} 1 \\ +i \end{pmatrix} \exp[-i\omega(t - c^{-1}n_- z)] \tag{3.5}$$

which have different refractive indexes $n_\pm$. This phenomenon is the Faraday effect, or the effect of magnetic circular birefringence.

Let us consider the effect of magnetic circular birefringence on the transmission of an electromagnetic wave through a gyrotropic medium. Let a linearly polarized (along the x-axis) electromagnetic wave with the amplitude E_0 enter into the medium ($z = 0$). The wave excites two modes with the eigenvectors e_+ and e_-:

$$E = c_1 e_+ + c_2 e_-. \tag{3.6}$$

The constants c_1 and c_2 can be found using the boundary conditions at $z = 0$, $E_x = E_0$, $E_y = 0$:

$$c_1 + c_2 = \sqrt{2} E_0 \qquad i(c_1 - c_2) = 0$$

which yields $c_1 = c_2 = E_0/\sqrt{2}$. For $z > 0$ we have

$$
\begin{aligned}
E(z) &= \frac{1}{\sqrt{2}} E_0 e_+ + \frac{1}{\sqrt{2}} E_0 e_- \\
&= E_0 \begin{pmatrix} \cos(c^{-1}\omega\,\Delta n\,z) \\ -\sin(c^{-1}\omega\,\Delta n\,z) \end{pmatrix} \exp[-i\omega(t - c^{-1}n_0 z)]
\end{aligned} \tag{3.7}
$$

where $\Delta n = (n_+ - n_-)/2$ and $n_0 = (n_+ + n_-)/2$.

If $n_0^2 = \varepsilon_1$, then $\Delta n = (1/2)n_0 Q$. If ε_1 and Q are real, then the wave in the gyrotropic medium remains linearly polarized, i.e. $\psi = 0$, since by (2.30) $\mathrm{Im}\,\chi = 0$. The angle of rotation of the plane of polarization can be found from (2.29):

$$\tan 2\theta = -\frac{2\tan(c^{-1}\omega\,\Delta n\,z)}{1 - \tan^2(c^{-1}\omega\,\Delta n\,z)} = \tan 2(-c^{-1}\omega\,\Delta n\,z)$$

whence it follows that

$$\theta = \Phi_F z = -c^{-1}\omega\,\Delta n\,z \approx -\frac{\omega n_0}{2c} Qz \tag{3.8}$$

where $\Phi_F = -\pi n_0 Q/\lambda$ is the specific Faraday rotation (the rotation of the plane of polarization of the wave per unit length of the sample).

It immediately follows from (2.14) and (3.7) that the Jones matrix for the layer of thickness z can be presented in the form

$$\mathbf{T}_F = \begin{pmatrix} \cos(c^{-1}\omega\,\Delta n\,z) & \sin(c^{-1}\omega\,\Delta n\,z) \\ -\sin(c^{-1}\omega\,\Delta n\,z) & \cos(c^{-1}\omega\,\Delta n\,z) \end{pmatrix} \tag{3.9}$$

provided that the wave propagates along the direction of the magnetization.

3.1.2 Magnetic circular dichroism (MCD)

If a medium exhibits absorption, then the absorption coefficients of the right- and left-handed circularly polarized light are different. If light is transmitted through a medium in which there is MCD, its polarization is changed from linear to elliptical. Let $|\varepsilon'| \gg \max(|\varepsilon''|, |g'|, |g''|)$; then as defined in (3.3)

$$n_\pm = n_0 \pm \frac{1}{2} n_0^{-1} g' + i\left(k_e \pm \frac{1}{2} n_0^{-1} g''\right) \tag{3.10}$$

where $n_0^2 = \varepsilon_1$ and k_e is the extinction coefficient, which is equal to $k_e = \frac{1}{2} n_0^{-1} \varepsilon''$ in this approximation. To understand the effect under consideration clearly, studying just a simplified case for which $|c^{-1}\omega \Delta n z| \ll 1$ will be sufficient. Inserting $\Delta n = \frac{1}{2} n_0^{-1}(g' + g'')$ from (3.10) into (3.7) shows that

$$E = E_0 \begin{pmatrix} 1 \\ -(2cn_0)^{-1}(g' + ig'')\omega z \end{pmatrix} \exp[-i\omega(t - c^{-1}n_0 z) - c^{-1}\omega k_e z]. \tag{3.11}$$

According to formulae (2.27)–(2.30), this is the equation for an elliptically polarized wave with the major axis of the ellipse oriented at the angle

$$\theta = -\frac{g'}{2cn_0}\omega z = -\frac{\omega n_0}{2c} Q' z$$

and with the ellipticity

$$\psi = -\frac{g''}{2cn_0}\omega z = -\frac{\omega n_0}{2c} Q'' z.$$

The specific magnetic circular birefringence and the magnetic circular dichroism can be unified into one general concept—the complex specific Faraday rotation:

$$\tilde{\Phi}_F = \Phi_F + i\Psi_F = -\frac{\omega g}{2cn_0} = -\frac{\pi n_0}{\lambda} Q \tag{3.12}$$

where $\lambda = 2\pi c/\omega$ is the wavelength in vacuum.

Although the main formulae for the Faraday effect and MCD (e.g. equation (3.12)) have been derived for an 'isotropic ferromagnet', they are also valid for the more general case of a material with cubic symmetry or a crystal with a symmetry axis (the tetrahedral, hexahedral, and trihedral classes), provided that the wave vector and magnetization vector are directed along this axis. If the light propagates in other directions in these crystals, or in crystals of other symmetry classes, the situation is more complicated because of the effects of natural birefringence.

Magnetic circular birefringence and dichroism are H-odd effects, i.e. they change sign when the magnetic field changes sign. However, rotation of the plane of polarization, and circular dichroism quadratic in the magnetic field were observed for the two-sublattice tetragonal antiferromagnetic crystals CoF_2 and FeF_2 (Kharchenko *et al* 1985). This effect is caused by specific features of the magnetic symmetry of these antiferromagnets.

It was found that an electric field induces a linear variation of the magnetooptical Faraday rotation (the non-reciprocal rotation due to an electric field) in Cr_2O_3 (Pisarev *et al* 1991) and in magnetic garnet films (Krichevtsov *et al* 1989a, b). This effect is to some extent similar to the magnetoelectric effect. As it is related to the spatial dispersion, the non-reciprocal rotation induced by an electric field is a small effect. Over a wide range of temperatures $T < 250$ K, the non-reciprocal rotation in Cr_2O_3 is characterized by a value of $-40''$ kV^{-1} (Krichevtsov *et al* 1988), which can be detected by modern polarimetric methods.

3.2 Magnetic linear birefringence and magnetic linear dichroism

3.2.1 Magnetic linear birefringence (the Cotton–Mouton or Voigt effect)

This effect results from the difference between the refractive indexes of the two components of light radiation linearly polarized parallel and perpendicular to the direction of magnetization, when the light propagates in a transversely magnetized medium (the wave vector is perpendicular to the magnetization). Linearly polarized light that has its polarization plane oriented at an angle to the magnetization direction becomes elliptically polarized on propagation through the medium.

If the light propagates in the direction that is perpendicular to the gyration vector (magnetic field, magnetization, etc)—say, along the x-axis—the vectors D and E are not parallel to each other. In this case the E-vector may, by (2.12), have all three of its components—E_x, E_y, and E_z—non-zero, while the D-vector has only two non-zero components—D_z and D_y. Therefore, for determination of the modes, it is better to use equation (2.15), which at $n = (n, 0, 0)$ defines two linearly polarized modes:

$$d_{\parallel} = \begin{pmatrix} 0 \\ 1 \end{pmatrix} \exp[-i\omega(t - c^{-1}n_{\parallel}x)]$$

$$d_{\perp} = \begin{pmatrix} 1 \\ 0 \end{pmatrix} \exp[-i\omega(t - c^{-1}n_{\perp}x)]$$

where

$$n_{\parallel} = (\varepsilon_0)^{1/2} \qquad n_{\perp} = (\varepsilon_1 - \varepsilon_0^{-1}g^2)^{1/2} \approx (\varepsilon_0)^{1/2}(1 - Q^2) \qquad (3.13)$$

and the vectors $d_{\parallel}$ and $d_{\perp}$ are defined in the $D_z D_y$-plane.

It is precisely the difference $\Delta n_{CM} = |n_{\parallel} - n_{\perp}|$, arising for the $d_{\parallel}$- and $d_{\perp}$-modes, that causes the Cotton–Mouton or Voigt effect.

This linear birefringence is magnetic, because $b(M)$, $g(M)$, and—by (2.8)—the difference $|n_{\parallel} - n_{\perp}|$ fall to zero if $M \rightarrow 0$.

The effect is often revealed in experiments as a relative phase shift of the two polarization components per unit length of a sample:

$$B_{C-M} = (\omega/c)|\text{Re}(n_{\parallel} - n_{\perp})|. \qquad (3.14)$$

Let an electromagnetic wave with $D_y = D_0 \cos \theta_0$ and $D_z = D_0 \sin \theta_0$ (the inessential phase factor is omitted here) be incident on a gyrotropic medium. It excites two modes in the medium:

$$D = c_1 d_\| + c_2 d_\perp$$

where c_1 and c_2 can be found from the above boundary conditions at $x = 0$. They are equal to $D_0 \cos \theta_0$ and $D_0 \sin \theta_0$ respectively. Therefore

$$D = D_0 \begin{pmatrix} \cos \theta_0 \exp(c^{-1} i \omega n_\perp x) \\ \sin \theta_0 \exp(c^{-1} i \omega n_\| x) \end{pmatrix} \exp(-i\omega t)$$

$$= D_0 \begin{pmatrix} \cos \theta_0 \\ \sin \theta_0 \exp i\delta \end{pmatrix} \exp(-i\omega t + i c^{-1} \omega n_\perp x) \qquad (3.15)$$

where

$$\delta = c^{-1} \omega (n_\| - n_\perp) x \equiv B_{CM} x. \qquad (3.16)$$

Equation (3.15) defines an elliptically polarized wave. Its polarization parameters θ and ψ can be calculated using formulae (2.28)–(2.30), whence it follows that

$$\tan 2\theta = \tan 2\theta_0 \cos \delta$$
$$\sin 2\psi = - \sin 2\theta_0 \sin \delta.$$

In the practically important case where $|\delta| \ll 1$ and $\theta_0 = \pi/4$, in a transparent medium the orientation angle θ of the ellipse is constant, $\theta = \theta_0 = \pi/4$, and the ellipticity angle is $\psi = -\delta/2$.

The Jones matrix for a layer of thickness x in this geometry has the form

$$T_{C-M} = \begin{pmatrix} \exp(c^{-1} i\omega n_\perp x) & 0 \\ 0 & \exp(c^{-1} i\omega n_\| x) \end{pmatrix}. \qquad (3.17)$$

3.2.2 Magnetic linear dichroism (MLD)

For an absorbing transversely magnetized medium this effect is due to the difference

$$\Delta \alpha = \alpha_\| - \alpha_\perp = 2 \operatorname{Im}(\delta/x) = \frac{4\pi}{\lambda} \operatorname{Im}(n_\| - n_\perp) \qquad (3.18)$$

between the absorption coefficients $\alpha_\|$ and $\alpha_\perp$ of the two linearly polarized waves, where δ is defined by (3.16).

The presence of the MLD results in a rotation of the orientation angle θ of the ellipse during the wave propagation. In this case, by (2.28) and (3.15),

$$\chi = (D_z/D_y) = \tan \theta_0 \exp(-\operatorname{Im} \delta + i \operatorname{Re} \delta)$$

and

$$\tan 2\theta = \frac{2 \tan \theta_0 \, e^{-\mathrm{Im}\,\delta} \cos(\mathrm{Re}\,\delta)}{1 - \tan^2 \theta_0 \, e^{-2\,\mathrm{Im}\,\delta}}$$

$$\sin 2\psi = -\frac{2 \tan \theta_0 \, e^{-\mathrm{Im}\,\delta} \sin(\mathrm{Re}\,\delta)}{1 + \tan^2 \theta_0 \, e^{-2\,\mathrm{Im}\,\delta}}$$

whence, in the event that $\theta_0 = \pi/4$, $|\delta| \ll 1$, it immediately follows that

$$\Delta\theta = \theta - \pi/4 = -\mathrm{Im}\,\delta/2 = \Delta\alpha \, x/4 \qquad \psi = -\mathrm{Re}\,\delta/2 \qquad (3.19)$$

where δ is given by formulae (3.16), (2.13), and (3.14).

The specific magnetic linear birefringence $\Psi_{C-M} = \psi/x$ and the magnetic linear dichroism $\Phi_{C-M} = \phi/x$ may be unified into the generalized Cotton–Mouton effect via

$$\widetilde{\Phi}_{C-M} = \Phi_{C-M} + i\Psi_{C-M} = -\frac{i\pi}{\lambda}(n_\parallel - n_\perp)$$

where $\lambda = 2\pi c/\omega$.

In the presence of magnetooptical anisotropy—for example, when the $\widehat{\varepsilon}$-tensor is described by (2.17) and the magnetization m is directed arbitrarily with respect to the crystal's axes—the main directions and axes of the optical indicatrix depend on the direction of m. In this case a determination of the magnetic linear birefringence via the difference $n_\parallel - n_\perp$ and of the magnetic linear dichroism via $\alpha_\parallel - \alpha_\perp$ may not always be valid.

The linear magnetic birefringence and dichroism are H-even effects. However, the specific magnetic symmetry of some antiferromagnetic crystals allows a linear dependence of these effects on the external magnetic field (Ostrovskii and Loktev 1977). This effect has been observed for the antiferromagnets Fe_2O_3 (Leycuras *et al* 1977, Merkulov *et al* 1981), $DyFeO_3$ (Kharchenko and Gnatchenko 1981), and CoF_2 (Kharchenko *et al* 1978).

3.3 The Faraday effect in the presence of natural birefringence; gyro-anisotropic media

Natural birefringence restricts the possibilities for achieving significant angles of rotation of the plane of polarization in crystals with low symmetry. Let the dielectric permeability tensor of a gyrotropic rhombic crystal be expressed as follows:

$$\widehat{\varepsilon} = \begin{pmatrix} \varepsilon_{11} & -ig & 0 \\ ig & \varepsilon_{22} & 0 \\ 0 & 0 & \varepsilon_{33} \end{pmatrix}$$

where $\varepsilon_{11} \neq \varepsilon_{22} \neq \varepsilon_{22}$. Then equations (2.11) and (2.12) readily define the following elliptically polarized normal modes, propagating along the z-axis of the crystal:

$$\begin{pmatrix} E_{x1} \\ E_{y1} \end{pmatrix} = A_1 \begin{pmatrix} 1 \\ i\eta^{-1} \end{pmatrix} \exp[-i\omega(t - c^{-1}n_+ z)]$$

$$\begin{pmatrix} E_{x2} \\ E_{y2} \end{pmatrix} = A_2 \begin{pmatrix} 1 \\ -i\eta \end{pmatrix} \exp[-i\omega(t - c^{-1}n_- z)]$$

where

$$n_{\pm}^2 = \frac{1}{2}\{\varepsilon_{11} + \varepsilon_{22} \pm [(\varepsilon_{22} - \varepsilon_{11})^2 + 4g^2]^{1/2}\}$$

$$\eta = 2g\{\varepsilon_{11} - \varepsilon_{22} - [(\varepsilon_{11} - \varepsilon_{22})^2 + 4g^2]^{1/2}\}^{-1}.$$

By solving the boundary-value problem with $E_x = E_0$, $E_y = 0$ at $z = 0$, one can as previously express the output wave at the point $z = h$ in terms of the wave at the input point $z = 0$ as follows (Tabor and Chen 1969):

$$\begin{pmatrix} E_x \\ E_y \end{pmatrix}_{z=h} = \begin{pmatrix} \cos\left(\dfrac{\kappa}{2}\right) + i\cos\tau\sin\left(\dfrac{\kappa}{2}\right) & -\sin\tau\sin\left(\dfrac{\kappa}{2}\right) \\ \sin\tau\sin\left(\dfrac{\kappa}{2}\right) & \cos\left(\dfrac{\kappa}{2}\right) - i\cos\tau\sin\left(\dfrac{\kappa}{2}\right) \end{pmatrix}$$

$$\times \begin{pmatrix} E_x \\ E_y \end{pmatrix}_{z=0} \tag{3.20}$$

where

$$\cos\tau = (1 + \eta^2)^{-1}(1 - \eta^2) \qquad \sin\tau = 2\eta(1 + \eta^2)^{-1} \kappa = (n_+ - n_-)\frac{\omega}{c}h.$$

If $|\varepsilon_{11} - \varepsilon_{22}| \gg |g|$, we have $\kappa = \omega c^{-1}\,\Delta n\,h$.

If the incident wave is polarized along the x-axis, the propagating wave at the point $z = h$ is given by

$$E_x = E_0(\cos(\kappa/2) + i\cos\tau\sin(\kappa/2))$$
$$E_y = E_0\sin\tau\sin(\kappa/2) \tag{3.21}$$

whence it follows that the amplitude E_y cannot exceed the value $E_0\sin\tau$, where $\sin\tau \approx -2g/(\varepsilon_{11} - \varepsilon_{22})$, which is quite small, if the condition $|g| \ll |\varepsilon_{11} - \varepsilon_{22}|$ is satisfied. If $\sin(\kappa/2) = 0$, there is no E_y-component.

In the case of an isotropic medium $\varepsilon_{11} = \varepsilon_{22}$, so $\sin\tau = 1$. Thus, the presence of the natural birefringence can substantially adversely affect the conditions, rendering them unsuitable for observation of the domain structure, unless the crystal geometry is chosen in such a way that the light propagates along the optical axis of the crystal.

The important feature of the magnetooptical effects in crystals with natural birefringence is an oscillating dependence of the ellipticity and the major-axis

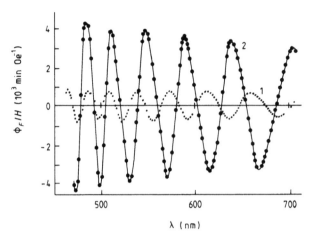

Figure 3.1. The spectral dependence of the rotation angle of the major axes of the polarization ellipse in a TbFeO₃ plate which is cut perpendicular to the [100] axis, for values of T of (curve 1) 290 K and (curve 2) 90 K (Valiev *et al* 1987a).

rotation angle of the polarization ellipse on the wavelength and on the sample thickness. In the event that $|\tau| \ll 1$, according to the formulae (2.28)–(2.30) and (3.20),

$$\chi = E_y/E_x \approx \tau/2(\sin\kappa - 2\mathrm{i}\sin^2\kappa/2)$$
$$\tan 2\theta = \tau\sin\kappa \qquad \sin 2\psi = -2\tau\sin^2\kappa/2$$

whence it readily follows that

$$\theta = \frac{\tau}{2}\sin\kappa \qquad \psi = -\tau\sin^2\frac{\kappa}{2}. \tag{3.22}$$

An example of this oscillating dependence of $\theta(\omega)$ is presented in figure 3.1.

Because both Δn and h depend on the temperature, the rotation angle of the major axis of the polarization has an oscillating temperature dependence too (Valiev *et al* 1987a).

3.4   The magnetooptical Kerr effect (MOKE)

Along with magnetooptical effects that take place during transmission of light through a magnetized substance, there are a number of effects that manifest themselves when the light is reflected from the surface of a magnetized material. These phenomena are conventionally designated magnetooptical Kerr effects. There are three types of Kerr effect, which are differentiated according to the orientation of the magnetization with respect to the direction of the wave propagation and with respect to the normal to the surface.

3.4.1   The three types of Kerr effect

3.4.1.1   The complex polar Kerr effect

This effect consists in both a rotation of the plane of polarization and the appearance of ellipticity, occurring when linearly polarized light is reflected from a sample surface where the sample is magnetized normally to this surface (see figure 1.3(a)).

3.4.1.2   The longitudinal (meridional) Kerr effect

In this effect, a rotation of the plane of polarization occurs and an ellipticity appears when linearly polarized light reflects from a sample surface, provided that the magnetization vector lies both in the sample plane and in the plane of incidence of the light (see figure 1.3(b)). The effect can be used to observe the domain structure of a material whose magnetization lies in the sample plane.

The polar and longitudinal Kerr effects constitute the group of longitudinal magnetooptical effects. Under certain conditions, variations of the intensity of linearly polarized reflected light are observed in the configurations of either the polar or longitudinal Kerr effect geometry (Krinchik and Chepurova 1973).

3.4.1.3   The transverse (equatorial) Kerr effect

Like the above-mentioned effects, this effect is linear in magnetization. The transverse effect can be observed only for absorbing materials. It is manifested as intensity variations and a phase shift of linearly polarized light reflected from a magnetized material, if the magnetization lies in the sample plane but is perpendicular to the plane of incidence of the light (see figure 1.3(c)).

The derivation of the formulae which describe the Kerr effect is straightforward but rather cumbersome. Thus we now confine our considerations to a couple of simple examples in order to illustrate the main ideas and to clarify the meaning of the parameters that we use. The general formulae for the Kerr effects, and the corresponding reflection and transmittance matrices will be presented below.

3.4.2   The polar effect; the case of normal incidence

If the light falls from the vacuum perpendicularly on the sample surface, the expression for the complex polar Kerr effect takes the following form:

$$\tilde{\Phi}_K = \Phi_K + i\Psi_K = \frac{ig}{\sqrt{\varepsilon_1}(\varepsilon_1 - 1)} = \frac{inQ}{n^2 - 1}. \tag{3.23}$$

It is assumed here that the dielectric permeability tensor of the magnetic media is described by (2.17), that the magnetooptical Voigt parameter $Q = g/\varepsilon_1$, and that the magnetic permeability of all of the media is equal to 1. The orientation of the magnetization is assumed to coincide with the propagation direction of the reflected light wave.

Formula (3.23) can be clarified in the following way. Let us decompose the incident, reflected, and refracted waves into the normal modes introduced by formulae (3.4) and (3.5), which are circular polarized:

$$E_i = \frac{1}{\sqrt{2}} E_i e'_+ + \frac{1}{\sqrt{2}} E_i e'_-$$

$$E_r = \frac{1}{\sqrt{2}} E_{r+} e''_+ + \frac{1}{\sqrt{2}} E_{r-} e''_- \tag{3.24}$$

$$E_t = \frac{1}{\sqrt{2}} E_{t+} e'''_+ + \frac{1}{\sqrt{2}} B_{t-} e'''_- .$$

It is assumed here that the incident wave is polarized along the x-axis. $e'_\pm, e''_\pm, e'''_\pm$ are defined in section 2.5. The corresponding refraction vectors are equal to $n'_\pm = (0, 0, -1), n''_\pm = (0, 0, 1), n'''_\pm = (0, 0, n_\pm)$, respectively, where, by equation (2.15), $n_\pm = n(1 \pm \frac{1}{2} Q)$.

The amplitudes $A_\pm$ of the reflected waves can be defined by the well-known Fresnel formulae[1]

$$E_{r\pm} = -E_i \frac{(n_\pm - 1)}{(n_\pm + 1)} \tag{3.25}$$

which follow readily from the boundary conditions (2.5) and (2.6), where, by equation (2.2), $H = [n \times E]$ should be used. Inserting (3.25) into (3.24), one obtains the equation for the reflected wave:

$$E_r = \frac{1}{2} E_i \left(\frac{(n_+ + 1)^{-1}(n_+ - 1) + (n_- + 1)^{-1}(n_- - 1)}{i[(n_+ - 1)/(n_+ + 1) - (n_- - 1)/(n_- + 1)]} \right) e^{-i\omega(t - c^{-1}z)}$$

$$= E_i \left(\frac{(n - 1)/(n + 1)}{ig/(n(n + 1)^2)} \right) \exp[-i\omega(t - c^{-1}z)]. \tag{3.26}$$

With this result and the formulae (2.29)–(2.31), with the proviso that $|g| \ll 1$, one can obtain the orientation angle of the polarization ellipse:

$$\Phi_K = -\mathrm{Im} \frac{g}{n(n^2 - 1)} = -\mathrm{Im} \left(\frac{nQ}{n^2 - 1} \right) \tag{3.27}$$

and the ellipticity angle:

$$\Psi_K = \mathrm{Re} \frac{g}{n(n^2 - 1)} = \mathrm{Re} \left(\frac{nQ}{n^2 - 1} \right). \tag{3.28}$$

The polar Kerr effect is odd with respect to the magnetization, i.e. it changes sign when the sample is remagnetized.

[1] It should be noted that a right-polarized incident wave is transformed into a left-polarized reflected wave and vice versa (in the intrinsic coordinate system of the light beam in which the z-axis is parallel to the wave vector; see section 2.4). The polarizations of the waves (3.24) refer to the laboratory coordinate frame, in which the z-axis is parallel to the magnetization, and these are retained after reflection.

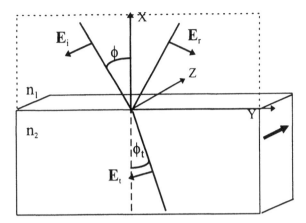

Figure 3.2. A schematic diagram of the transmission of light through an interface between magnetic and non-magnetic media.

3.4.3   The transverse effect

Let us consider transmission of a light wave through an interface between magnetic and non-magnetic media (figure 3.2). We choose the coordinate system shown in figure 3.2, the $x0y$ plane coinciding with the plane of incidence of the light, and the magnetization vector being oriented along the $0z$-axis. Let n_1 be the reflection coefficient of the non-magnetic medium. With the approximation of taking into account only terms linear in the magnetization, the magnetic media are described by the dielectric permeability tensor (2.7); it is assumed here that $\varepsilon_1 = \varepsilon_0 = \varepsilon$.

The magnetic permeability μ of all of the media is assumed to be equal to 1. Let a plane light wave be incident from medium 1 onto medium 2 at the angle ϕ, with the electrical vectors parallel to the plane of incidence (p polarization). The electric field vectors of the incident and reflected waves are written in terms of the projections on the coordinate axes as

$$
\begin{aligned}
E_i &= E_i \begin{pmatrix} -\sin\phi \\ -\cos\phi \\ 0 \end{pmatrix} e^{-i\omega t + i(\omega/c)n_i \cdot r} \\
E_r &= E_r \begin{pmatrix} -\sin\phi \\ \cos\phi \\ 0 \end{pmatrix} e^{-i\omega t + i(\omega/c)n_r \cdot r}
\end{aligned}
\tag{3.29}
$$

where

$$
n_i = n_1(-\cos\phi, \sin\phi, 0) \qquad n_r = n_1(\cos\phi, \sin\phi, 0). \tag{3.30}
$$

When a p-polarized light beam falls on a magnetic medium with the refractive index $n_2 = \varepsilon^{1/2}$, it excites a wave that propagates from the upper interface of the medium. This wave is the transverse normal mode considered in section 2.3, the refraction vector of which is $n_t = n_t(-\cos\phi_t, \sin\phi_t, 0)$, where, according to equation (2.16), $n_t = n_2(1 - Q^2) \approx n_2$, and $\sin\phi_t$ is defined by Snell's law: $\sin\phi_t = (n_1/n_2)\sin\phi$ and $\cos\phi_t = (1 - (n_1/n_2)^2\sin^2\phi)^{1/2}$.

The electric field of this wave can be found from equation (2.13):

$$E_t = E_t \begin{pmatrix} -\sin\phi_t + iQ/\cos\phi_t \\ -\cos\phi_t \\ 0 \end{pmatrix} e^{(-i\omega t + i(\omega/c)(n_t \cdot r))}. \tag{3.31}$$

The wave (3.31) is elliptically polarized in the plane of incidence; in addition to the ordinary transverse component (in the absence of magnetization) it contains a longitudinal component (oriented along the wave-vector direction), which is proportional to Q, i.e. to the magnetization. It is this component in the field vector (3.31) that leads to the magnetooptical effects.

The magnetic fields of the wave can be found from Maxwell's equation (2.1), which in this case takes the form $H = [n \times E]$. From this, and from (3.30) and (3.31) it readily follows that

$$H_{i,r,t} = (0, 0, H_{i,r,t}) \tag{3.32}$$

where $H_i = n_1 E_i$, $H_r = n_1 E_r$ and $H_t = n_2(1 - iQ\tan\phi_t)E_t$.

Now we have to match the tangential components of E and H at the interface:

$$E_{iy} + E_{ry} = E_{ty} \qquad H_{iz} + H_{rz} = H_{tz}. \tag{3.33}$$

Using (3.30)–(3.32) in (3.33), one readily obtains

$$\begin{aligned} (E_i - E_r)\cos\phi &= E_t\cos\phi_t \\ n_1(E_i + E_r) &= n_2 E_t(1 - iQ\tan\phi_t) \end{aligned} \tag{3.34}$$

whence the formulae for the reflection and transmission coefficients in the linear (-in-Q) approximation (Druzhinin *et al* 1983) immediately follow:

$$\tilde{t}^p_{12} = \frac{E_t}{E_i} = t^p_{12} + i\,\delta t^p_{12} = t^p_{12}(1 + i\tau^p_{12}) \tag{3.35}$$

$$\tilde{r}^p_{12} = \frac{E_r}{E_i} = r^p_{12} + i\,\delta r^p_{12} = r^p_{12}(1 + i\rho^p_{12}) \tag{3.36}$$

where the linear (-in-Q) corrections δt, δr, τ^p_{12}, and ρ^p_{12} are equal to

$$\begin{aligned} \delta t^p_{12} &= \frac{2\sin^2\phi\cos\phi_t}{(\eta\cos\phi + \cos\phi_t)^2}Q & \delta r^p_{12} &= -\frac{2\sin\phi\cos\phi}{(\eta\cos\phi + \cos\phi_t)^2}Q \\ \tau^p_{12} &= \frac{(1 + r^p_{12})Q\sin\phi}{2(\eta^2 - \sin^2\phi)^{1/2}} & \rho^p_{12} &= \frac{r^p_{12}Q\sin\phi}{2(\eta^2 - \sin^2\phi)^{1/2}} \end{aligned} \tag{3.37}$$

and t_{12}^p and r_{12}^p are the conventional Fresnel coefficients

$$t_{12}^p = \frac{2\cos\phi}{\eta\cos\phi + \cos\phi_t} \qquad r_{12}^p = \frac{\eta\cos\phi - \cos\phi_t}{\eta\cos\phi + \cos\phi_t} \qquad (3.38)$$

where $\eta = n_2/n_1$.

Thus, the relative change in the reflected light intensity is given by

$$\delta_p = \frac{(I - I_0)}{I_0} = 2\,\mathrm{Im}\,\rho_{12}^p \qquad (3.39)$$

where I and I_0 are the intensities of the reflected light in the magnetized and non-magnetized states, respectively, and ρ_{12}^p is defined by (3.36).

If the incident wave is s polarized, i.e., $E_i = E(0, 0, 1)$, both the reflected wave E_r and the wave in the magnetic medium E_t are also s polarized. All of the electric field vectors are purely transverse, because they do not contain linear (-in-Q) corrections and, therefore, the Kerr effect for s-polarized waves is absent. This conclusion remains valid only if the magnetic permeability of the medium is unity. Otherwise, the transverse Kerr effect for s-polarized waves, which is proportional to the non-diagonal component of the $\widehat{\mu}$-tensor, takes place. Although it is usually much weaker than the same effect for p-polarized light, this effect provides the principal opportunity to separate out the magnetic susceptibility contribution.

Similarly, the formulae for the polar and longitudinal Kerr effects at arbitrary incidence angles, and Jones matrices can be derived.

3.4.4 The Jones matrices for an interface

The Jones matrices for the reflection and transmission of light at the interface of a non-magnetic medium (1) and a magnetic medium (2) are defined in the following way:

$$\begin{pmatrix} E_s^r \\ E_p^r \end{pmatrix} = \begin{pmatrix} r_{ss} & r_{sp} \\ r_{ps} & r_{pp} \end{pmatrix} \begin{pmatrix} E_s^i \\ E_p^i \end{pmatrix} \qquad (3.40)$$

$$\begin{pmatrix} E_s^t \\ E_p^t \end{pmatrix} = \begin{pmatrix} t_{ss} & t_{sp} \\ t_{ps} & t_{pp} \end{pmatrix} \begin{pmatrix} E_s^i \\ E_p^i \end{pmatrix}. \qquad (3.41)$$

In a linear approximation as regards the magnetooptical parameter Q, and when $\mu = 1$, the components of these matrices in terms of the angle of incidence and the optical constants of the media take the following forms (Mayevskii 1985):

$$r_{ss} = r_{ss}^0 \qquad r_{pp} = r_{pp}^0(1 + i\rho_p) \qquad \rho_p = \frac{n_1 r_{12}^p Q \sin\phi}{2g_2} \qquad (3.42a)$$

$$r_{sp} = it_{12}^s t_{21}^p (\cos\alpha + m\cos\beta)\delta_1 \qquad (3.42b)$$

$$r_{ps} = i t_{12}^s t_{21}^p (\cos\alpha - m\cos\beta)\delta_1 \tag{3.42c}$$

$$t_{ss} = t_{ss}^0 \qquad t_{pp} = t_{pp}^0(1 + i\tau_p) \qquad \tau_p = (1 + r_{21}^p)\frac{n_1 Q \sin\phi}{2g_2} \tag{3.42d}$$

$$t_{sp} = i t_{12}^p r_{21}^s(\delta_1 \cos\alpha + m\cos\beta) = \frac{t_{12}^p r_{21}^s}{t_{12}^s t_{21}^p} r_{sp} \tag{3.42e}$$

$$t_{ps} = i t_{12}^p r_{21}^s(\delta_1 \cos\alpha - m\cos\beta) = \frac{t_{12}^p r_{21}^s}{t_{12}^s t_{21}^p} r_{ps} \tag{3.42f}$$

where the corresponding Fresnel coefficients for s- and p-polarized light waves at the interface of the media with complex (in the general case) refractive indexes n_i and n_j are well known to be equal to

$$
\begin{aligned}
r_{j.k}^s &= \frac{g_j - g_k}{g_j + g_k} & r_{j.k}^p &= \frac{g_j n_k^2 - g_k n_j^2}{g_j n_k^2 + g_k n_j^2} \\
t_{j.k}^s &= \frac{2g_j}{g_j + g_k} & t_{j.k}^p &= \frac{2g_j n_j n_k}{g_j n_k^2 + g_k n_j^2}.
\end{aligned}
\tag{3.43}
$$

The parameter g_j in (3.42)–(3.43) is defined by the expressions

$$g_j = \sqrt{n_j^2 - n_1^2 \sin^2\phi} \qquad g_1 = n_1 \cos\phi.$$

The parameter δ_1, which corresponds to the interface magnetooptical effects, is proportional to the magnetization (or to the magnetooptical parameter Q):

$$\delta_1 = \frac{n_2}{4g_2} Q$$

and the parameter m is

$$m = \frac{n_1}{g_2}\sin\phi.$$

Formulae (3.42) are valid when $|\delta_1| \ll 1$, which is usually the case in practice. The direction of the magnetization vector is usually defined by the polar (α-), longitudinal (β-), and transverse (γ-) angles. These angles are defined by

$$\cos\alpha = (b \cdot q) \qquad \cos\beta = (b \cdot \tau) \qquad \cos\gamma = (b \cdot s) \tag{3.44}$$

where b is the unit vector of magnetization, q is the unit vector normal to the sample plane, s is the unit vector normal to the plane of incidence of the light, and the unit vector $\tau = q \times s$ is oriented along the intersection of the above-mentioned planes.

3.4.5 General MOKE formulae

Now it is not difficult to derive the general formulae for the magnetooptical Kerr effects. They take the following forms (Bolotin and Sokolov 1961):

(i) the polar Kerr effect:

$$\Phi_K^{s,p} = \mathrm{Re}\,\frac{r_{sp}}{r_{ss}(r_{pp})} = \mathrm{Im}\,\frac{\eta^2[(\eta^2 - \sin^2\phi)^{1/2} \mp \sin\phi\tan\phi]}{(\eta^2 - 1)(\eta^2 - \tan^2\phi)}Q \qquad (3.45)$$

(ii) the longitudinal Kerr effect:

$$\Phi_K^{s,p} = \mathrm{Re}\,\frac{(\pm r_{sp})}{r_{ss}(r_{pp})} = \mathrm{Im}\,\frac{[\sin\phi\eta^2(\sin\phi\tan\phi \pm \sqrt{\eta^2 - \sin^2\phi})]}{(\eta^2 - 1)(\eta^2 - \tan^2\phi)(\eta^2 - \sin^2\phi)^{1/2}}Q \qquad (3.46)$$

(iii) the transverse Kerr effect:

$$\delta_p = \frac{\Delta I}{I} = -\mathrm{Im}\,\frac{4\tan\phi\eta^2}{(\eta^2 - 1)(\eta^2 - \tan^2\phi)}Q. \qquad (3.47)$$

Here $\eta = n_2/n_1$ is the complex relative refractive index of the adjacent media. The upper sign corresponds to s polarization of light, and the lower sign to p polarization of light. The ellipticities in the polar and longitudinal effects are defined by formulae similar to (3.45) and (3.46), provided that Re is replaced by Im and Im by Re.

In particular, it follows from these formulae that in metals, where usually $|\eta| \gg 1$, the polar effect is the strongest one, and the longitudinal effect is the weakest one for the same magnetooptical parameters.

Formulae (3.45) and (3.47) provide a relationship that binds the polar and transverse Kerr effects (Bolotin and Sokolov 1961):

$$\delta_p = 2\frac{(\Phi_K^p - \Psi_K^p)}{\sin\phi}.$$

This relationship is well confirmed by experiment (Voloshinskaya and Ponomariova 1969).

Krinchik and Chepurova (1973) have discovered and studied magnetooptical intensity effects. These effects take place when the polarization vector of the incident light beam is oriented intermediately between the s- and p-polarization directions. In this case the reflected light intensity is linearly dependent on the magnetooptical parameter Q. These intensity effects extend the opportunities to use the magnetooptical methods in studies of magnetic materials.

3.5 Even magnetooptical reflection effects

It is clear that even magnetooptical effects of the same nature as the complex Cotton–Mouton or Voigt effect must be observed in reflected light.

The anisotropy of the magnetooptical effects of many magnetic materials has been well studied in the transmitted light geometry (section 3.2). The possibility of studying the anisotropy of the reflection magnetooptical effects was predicted when the magnetooptics of Ni was studied (Donovan and Medcalf 1965). It was demonstrated (Hodges *et al* 1967) that the shape of the Fermi surface of Ni depends on the orientation of the magnetization in the crystal. This effect was discovered by Hodges *et al* (1967), and then explained by taking into consideration the spin–orbit interaction (Falicov and Ruvalds 1968). Evidently, this phenomenon can be observed using optical methods—for example, by means of fixing the changes in intensity of the reflected light when the magnetization in the crystal rotates. It was in an experiment that had been especially carried out for this purpose (Krinchik and Gushchin 1969) that this magnetooptical effect was actually discovered, and it was named the orientational magnetooptical effect (OME).

As the magnetization orientation changes from transverse to longitudinal, the reflected light intensity varies quadratically in the magnetization—this is the orientational magnetooptical effect. This effect is defined quantitatively by the formula

$$\delta_{OME} = \frac{I_1 - I_2}{I_0} \tag{3.48}$$

where I_1, I_2, and I_0 are the intensities of the reflected light in the transverse and longitudinal geometries, and in the demagnetized state, respectively. In general the orientational magnetooptical effect is strongly anisotropic, i.e. it depends on the relative orientations of the magnetization, the crystal axes, and the direction and polarization of the light beam (Krinchik and Gan'shina 1973).

The orientational magnetooptical effect proved to be relatively strong, and yet, unlike the transverse Kerr effect, even in magnetization, which makes it possible to distinguish it from the odd transverse Kerr effect. Typical values of the OME for Fe, Ni, and permalloy in the spectral band $\lambda = 0.6$–1.76 μm are of the order of $\delta_{OME} \approx (0.5$–$1) \times 10^{-3}$. The orientational effect has been studied for many magnetic materials: the alloys Fe–Ni, Fe–Ti, Fe–V; the rare-earth orthoferrites, orthochromites, and orthomanganites; and the magnetic semiconductors (Gan'shina 1994).

The phenomenological theory of the orientational magnetooptical effect has been developed by Krinchik and Gan'shina (1973) and Bolotin (1975). According to this theory, the intensity variation of the light reflected from a magnetized crystal is caused by those components of the dielectric permeability tensor $\hat{\varepsilon}$ that are quadratic in magnetization, i.e., the orientational magnetooptical effect is, in fact, the result of the complex linear magnetic birefringence observed in the reflection process. This effect is characterized by the value of (Krinchik and Gan'shina 1973)

$$\delta_{OME} = (\Delta n_\perp - \Delta n_\parallel) f(n, \phi) \tag{3.49}$$

where $\Delta n_\perp$ and $\Delta n_\parallel$ are the changes of the refractive index for the magnetization in the transverse and longitudinal directions, ϕ is the angle of incidence, and $f(n, \phi)$ is a function which is independent of the magnetization.

Figure 3.3 shows the anisotropy of the orientational magnetooptical effect for Ni in comparison with the transverse MOKE. It is seen that the transverse MOKE is almost isotropic, and the orientational magnetooptical effect depends strongly on the orientation of the magnetization. Besides this, the effect is revealed to have a stronger temperature dependence than the transverse effect.

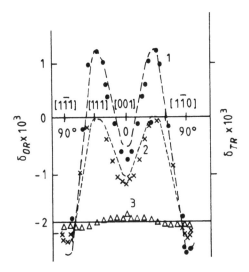

Figure 3.3. The angular dependence of the orientational effect at the photon energies of 0.31 eV (curve 1) and 0.7 eV (curve 2), together with that of the transverse effect at 0.7 eV (curve 3), and the incidence angle 80° for Ni (Gan'shina 1994).

The magnetic linear birefringence is known to be much more sensitive to the microstructural properties of magnetic crystals than odd magnetooptical effects (e.g. the Faraday effect, and the Kerr effect). This is also valid for the quadratic orientational reflection effect[2].

3.6　Thin films and multilayers

Magnetooptical devices and experimental magnetooptical set-ups usually involve multilayered systems. When light passes through these systems, its transformation is governed by the joint influence of several magnetooptical effects. For example, in the simplest case of light transmission through a magnetic plate or a film that is magnetized perpendicularly to its surface, along with the volume magnetic circular birefringence there are effects that are associated with the light reflection at the input and output film surfaces. The interference of many reflecting waves should be taken into account. Surface optical coatings are often used to enhance the magnetooptical effects.

[2] In some antiferromagnetic crystals the orientational magnetooptical effect may reveal *H*-odd behaviour (Krinchik and Kosturin 1982).

Calculations for the multilayered systems require formulae that define the dependence of the magnetooptical effects on the magnetization orientation, on the layer thickness, on the optical parameters of the media, on the polarization properties, and on the wave vector of the incident wave. These formulae can then be solved using the Jones matrices. In the above-mentioned example the resultant electric field vector at the output is presented as a product of the Jones matrices that describe the influence of each layer. In other cases direct solution of (2.1)–(2.6) is a more effective method.

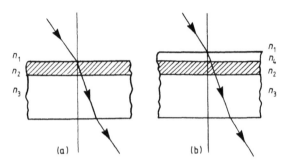

Figure 3.4. Two-layer (a) and multilayer (b) magnetooptical structures.

To emphasize the main physical features of the magnetooptics of multilayers, in the following sections of this chapter we shall consider just the simplest structures, shown in figure 3.4. We shall see that there are essentially new effects in these simple structures, caused by the interaction of light with the interfaces. These effects include, in particular, the transverse effect on the intensity of the transmitted light, and the specific dependence of the rotation of the plane of polarization and ellipticity of the reflected and transmitted waves on the thickness of the magnetic layer.

3.6.1 The thickness dependence

Let us consider the transmission of a light wave through a polar-magnetized film (medium 2) which is adjacent to a non-magnetic medium (figure 3.4). We shall assume that the magnetic film is an optically isotropic medium. Then the $\widehat{\varepsilon}$-tensor is defined by (2.7). In addition, the difference between the ε_0- and ε_1-components can be disregarded, i.e., the value of b in (2.8) is assumed to be zero.

The simplest way of obtaining the formulae for the state of polarization of the reflected and transmitted light is to introduce the circularly polarized modes. Let the incident wave be decomposed into circularly polarized modes. It is important to point out that these modes propagate through the whole multilayered system independently of each other.

The electric field within the magnetic film can according to (2.12) be written as

$$E^\sigma = E^\sigma(0)\frac{\sinh ik^\sigma(h-z)}{\sinh ik^\sigma h} + E^\sigma(h)\frac{\sinh ik^\sigma z}{\sinh ik^\sigma h} \quad (3.50)$$

where $E^\sigma(0)$ and $E^\sigma(h)$ are the fields at the film surfaces, $\sigma = \pm 1$ for right and left circularly polarized waves, $k^\sigma = k_0\tilde{n}_2$, $k_0 = \omega/c$, and $\tilde{n}_2 \equiv n_2(1+\sigma Q)^{1/2}$. The corresponding magnetic field is according to Maxwell's equation

$$H^\sigma = -\frac{\sigma c}{\omega}\frac{d}{dz}E^\sigma. \quad (3.51)$$

Substituting (3.50) into (3.51) yields

$$H^\sigma(z) = i\tilde{n}_2\sigma\left[E^\sigma(0)\frac{\cosh ik^\sigma(h-z)}{\sinh ik^\sigma h} - E^\sigma(h)\frac{\cosh ik^\sigma z}{\sinh ik^\sigma h}\right]. \quad (3.52)$$

The boundary conditions (2.5) and (2.6) connect the fields $E^\sigma(0)$ and $E^\sigma(h)$:

$$
\begin{aligned}
&E_i^\sigma + E_r^\sigma = E^\sigma(0) \\
&n_1(E_i^\sigma - E_r^\sigma) = \tilde{n}_2(E^\sigma(0)\operatorname{cotanh} ik^\sigma h - E^\sigma(h)\sinh^{-1} ik^\sigma h) \\
&E_3^\sigma = E^\sigma(h) \\
&n_3 E_3^\sigma = \tilde{n}_2(E^\sigma(0)\sinh^{-1} ik^\sigma h - E^\sigma(h)\operatorname{cotanh} ik^\sigma h)
\end{aligned}
\quad (3.53)
$$

where E_i^σ, E_r^σ, and E_3^σ are the amplitudes of the incident, reflected, and transmitted circularly polarized components. The formulae for the reflection and transmission coefficients follow from (3.53) (for $n_1 = 1$):

$$r^\sigma = \left(\frac{E_r^\sigma}{E_i^\sigma}\right) = \frac{1 - n_3 - (n_3\tilde{n}^{-1} - \tilde{n})\tanh ik_0\tilde{n}h}{1 + n_3 - (n_3\tilde{n}^{-1} + \tilde{n})\tanh ik_0\tilde{n}h} \quad (3.54)$$

$$t^\sigma = \left(\frac{E_3}{E_i^\sigma}\right) = 2[(1+n_3)\cosh ik_0\tilde{n}h - (n_3\tilde{n}^{-1} + \tilde{n})\sinh ik_0\tilde{n}h]^{-1}. \quad (3.55)$$

Thus, general formulae for the polarization states θ and ψ of both the reflected and the transmitted waves can be obtained directly from the formulae (2.32)–(2.34). The asymptotic behaviour of Φ_K and θ_F is given by

$$\Phi_K - i\Psi_K = \begin{cases} (in_2 Q)/(n_2^2 - 1) & h \gg c\omega^{-1} \\ 2n_3 Q\omega h/c(n_3^2 - 1) & h \ll c\omega^{-1} \end{cases} \quad (3.56)$$

$$\theta_F - i\psi_F = \begin{cases} (-Q/2)\left(\dfrac{n_2\omega h}{c} + \dfrac{i(n_2^2 - n_3)}{(1+n_2)(n_2+n_3)}\right) & h \gg c\omega^{-1} \\ -(n_2^2\omega h Q)/(c(1+n_3)) & h \ll c\omega^{-1} \end{cases} \quad (3.57)$$

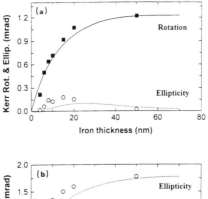

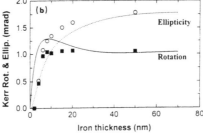

Figure 3.5. The measured saturated values of the Kerr rotation (■) and the ellipticity (○) as functions of the iron layer thickness for s polarization (a) and p polarization (b). The curves show the calculated Kerr rotation (full lines), and Kerr ellipticity (dotted lines). After Postava *et al* (1996).

Formulae (3.54)–(3.57) illustrate the typical thickness dependences of the magnetooptical reflection and transmission effects. More general formulae for an arbitrary orientation of the magnetization and angle of incidence of light can be obtained similarly, but the calculations are more tedious.

The most striking feature of the MOKE in the thin-film layer is that it is proportional to the film thickness h. In the limit of small thickness, this is also true for the longitudinal Kerr effect (Moog *et al* 1990):

$$\pm\Phi_K + i\Psi_K = (2\omega/c)[n_3/(1 - n_3^2)]Q\phi h$$

where the $\pm$ signs are for the s and p polarizations, respectively, and the angle of incidence ϕ is assumed to be small. It is interesting to note that in the thin-film approximation the Kerr signal is independent of the refractive index of the film.

The thickness dependence of the Kerr rotation and ellipticity for the longitudinal geometry of iron layers on a silicon substrate is shown in figures 3.5(a) and 3.5(b) (Postava *et al* 1996). The thicknesses of the iron films vary from 2 nm to 50 nm. These measurements were compared by Postava *et al* with a magnetooptical calculation of the longitudinal Kerr rotation and ellipticity. They used the following optical constants: $n_{Si} = 3.88 - i\,0.02$, $n_{Fe} = 2.36 - i\,3.48$, and $Q_{Fe} = 0.034 - i\,0.003$. The thickness-dependent MOKE measurements for the Fe/Au system have been discussed by Moog *et al* (1990) in the terms of the figure-of-merit parameter $\theta_m R^{1/2}$, where $\theta_m = (\Phi_K^2 + \Psi_K^2)^{1/2}$ is the effective MOKE response, and R is the reflectivity of the system. The Fe/GaAs(100) system has been studied by Daboo *et al* (1993).

3.6.2 Odd transverse effects in transmitted light

As the second example, we now describe magnetooptical effects that are odd in magnetization, which are manifested as variations in the intensity and polarization of light when it passes through a transversely magnetized ferromagnetic film or plate. These effects were theoretically predicted by Mayevskii and Bolotin (1973) and experimentally studied in ferromagnetic metal films by Druzhinin *et al* (1981, 1985). The transverse effects consist in relative intensity variations of the transmitted light (for both p and s polarizations):

$$\delta_{p,s}^{123} = \left\{ \frac{I(M) - I_0}{I_0} \right\}_{p,s}$$

where $I(M)$ and I_0 are the intensities of the transmitted light for the magnetized and non-magnetized samples. Figure 3.6 shows the measured values of these effects in nickel and iron films which were deposited on glass substrates.

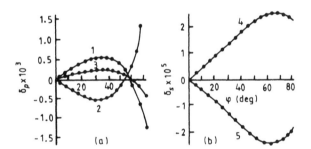

Figure 3.6. Angular dependences of the transverse effects δ_p (a) and δ_s (b) in transmitted light ($\lambda = 0.63 \ \mu$m) for the following cases: (curve 1) iron film of thickness $h = 25$ nm, with incidence from the film side; (curve 2) the same case as for curve 1, but with incidence from the substrate side; (curve 3) nickel film of thickness $h = 60$ nm; (curve 4) iron film of thickness $h = 1.1 \ \mu$m, with incidence from the film side; and (curve 5) the same case as for curve 4, but with incidence from the substrate side (after Druzhinin *et al* 1981, 1985).

It is obvious that, in the geometry considered, no birefringence that is odd in magnetization is present. Therefore, the origins of these effects are associated with the boundary conditions at both of the interfaces of the magnetic layer.

The formulae that define the transverse effects of the rotation of the plane of polarization in transmitted light can be derived using formulae (3.42) for the coefficient of transmission through the interfaces of the magnetic and non-magnetic media.

Let $t^p_{123} = E_3/E_1$ be the coefficient of transmission through the three-layer structure (see figure 3.4). In the main-ray approximation

$$t^p_{123} = \tilde{t}^p_{12}\tilde{t}^p_{23}F_2$$

where $F_2 = \exp(-i\omega c^{-1}g_2d_2)$ is the phase multiplier which determines the phase build-up of the wave and its decay in the magnetic layer with thickness d_2, $\tilde{t}^p_{12}$ is defined by (3.35), (3.37), and (3.42), and $\tilde{t}_{23}$ can be obtained by a method quite similar to that used to obtain $\tilde{t}^p_{12}$:

$$\tilde{t}^p_{23} = t^p_{23}(1 + i\tau^p_{23})$$

where

$$\tau^p_{23} = -\frac{n_1(1 + r^p_{23})Q\sin\phi}{2g_2}$$

and the coefficients r^p_{23} and t^p_{23} are defined by (3.38) and (3.43).

As a result, from the above equations, the relative change in intensity found after the wave has emerged from the film is given by (Druzhinin *et al* 1983)

$$\delta^{123}_p = -\,\mathrm{Im}\left[(r^p_{21} - r^p_{23})\frac{n_1Q\sin\phi}{g_2}\right] \tag{3.58}$$

where r^p_{21} and r^p_{23} are the reflection (Fresnel) coefficients of the p-polarized wave at the 1–2 and 2–3 interfaces, and $g_2 = (n_2^2 - n_1^2\sin^2\phi)^{1/2}$.

The change in intensity is therefore proportional to the difference between the reflection coefficients of the lower and upper faces of the film, and vanishes if the media on either side of the film are the same. Formula (3.58), derived using the main-beam approximation, describes the main features of the transverse δ_p-effect in transmitted light. The effect exists only for oblique incidence ($\phi \neq 0$). It vanishes if the refractive indexes n_1 and n_3 of the adjacent media are equal. It is also not difficult to verify the sign symmetry of the effect with respect to reversed beams, i.e., the validity of the condition $\delta^{321}_p = -\delta^{123}_p$.

The formula for the intensity variation δ^{123}_p of the s-polarized wave looks the same as (3.58), with the slight differences that the index p is changed to s and that the Voigt parameter of the magnetic permeability tensor Q_M is substituted for the magnetooptical parameter Q (see section 2.2). This means that for p polarization the intensity transverse-transmission effect is purely gyroelectric, while for s polarization it is gyromagnetic. This is the second known (the first is the transverse Kerr effect in reflected light) magnetooptical effect which allows us to decide definitely on whether or not a ferromagnetic medium exhibits gyromagnetic properties.

Figure 3.6 shows the angular dependence of the transverse effect in transmitted light, which was measured for ferromagnetic metal films deposited on glass substrates. The effect is antisymmetric (non-reciprocal) with respect to the reversal of the light beam: $\delta_p^{321} = -\delta_p^{123}$ (medium 1: air; medium 2: the magnetic film; medium 3: the substrate). It is also seen that the effect vanishes at the incidence angle $\phi = 56°$, which is equal to the Brewster angle for glass ($n_3 = 1.48$). This agrees well with the formulae (3.42) and (3.44). The order of magnitude of the δ_p-effect is 10^{-3} at most, which can be reliably detected by modern experimental set-ups for optical frequencies. The value of the δ_s-effect is two orders of magnitude lower, $\sim 10^{-5}$.

Note that the dependences of the δ_p- and δ_s-effects on the incidence angle for transmitted light significantly differ from those for reflected light (the transverse Kerr effect).

3.6.3 The Jones matrix of a multilayer

Here the general formulae are given for the odd-magnetization MOKE for the most important case among multilayer systems—a three-layer system (figure 3.4) consisting of a ferromagnetic film (medium 3) and non-magnetic layers on either side of it (media 2 and 4), which can perform the protective or amplifying roles of the film and substrate (Mayevskii 1985).

First, we need to make some definitions. The reflection and transmitted matrices of the system, which connect the s and p components of the amplitudes of the reflected and transmitted waves with the components of the incident wave, are defined by formulae (3.40) and (3.41). All of the media are assumed to be optically isotropic and, except for the first, can be absorbing.

The following recurrence formulae are satisfied by the coefficients of reflection and transmission of single-layer, two-layer, ... systems:

$$r_{jkl}^{s(p)} = \frac{r_{jk}^{s(p)} + F_k^2 r_{kl}^{s(p)}}{1 + F_k^2 r_{jk}^{s(p)} r_{kl}^{s(p)}} \qquad t_{jkl}^{s(p)} = \frac{t_{jk}^{s(p)} t_{kl}^{s(p)} F_k}{1 + F_k^2 r_{jk}^{s(p)} r_{kl}^{s(p)}} \tag{3.59}$$

$$r_{jklm}^{s(p)} = \frac{r_{jk}^{s(p)} + F_k^2 r_{klm}^{s(p)}}{1 + F_k^2 r_{jk}^{s(p)} r_{klm}^{s(p)}} \qquad t_{jklm}^{s(p)} = \frac{t_{jk}^{s(p)} t_{klm}^{s(p)} F_k}{1 + F_k^2 r_{jk}^{s(p)} r_{klm}^{s(p)}} \tag{3.60}$$

and so on, where j, k, l, m are the numbers labelling the media and

$$F_k = \exp(-i\omega c^{-1} g_k d_k) \tag{3.61}$$

are phase multipliers which determine the phase build-up of the wave and its decay in the kth layer with thickness d_k. The Fresnel coefficients $r_{jk}^{s(p)}$, $t_{jk}^{s(p)}$, and g_k are defined by (3.43). There are certain parameters

$$\delta = \frac{\omega}{2c} d_3 n_3 Q \qquad \delta_1 = (4g_3)^{-1} Q \tag{3.62}$$

which are proportional to the magnetization. The condition $|\delta| \ll 1$ is normally called the 'thin'-film approximation. The direction of the magnetization is given by the polar (α-), longitudinal (β-) and transverse (γ-) angles, which are introduced by the formulae (3.44).

The diagonal elements of the Jones matrices (3.40), (3.41) for multilayer systems are

$$r_{ss} = r_{ss}^0 \qquad r_{pp} = r_{pp}^0(1 + \rho_p)$$
$$t_{ss} = t_{ss}^0 \qquad t_{pp} = t_{pp}^0(1 + \tau_p) \tag{3.63}$$

where $r_{ss(pp)}^0 = r_{12345}^{s(p)}$ and $t_{ss(pp)}^0 = t_{12345}^{s(p)}$, and the corrections to them associated with the transverse magnetization component ($\cos\gamma \neq 0$) are given by

$$\rho_p = i(1 - F_3^2)\left(\frac{r_{345}^p - \bar{r}_{321}^p}{1 - F_3^2\bar{r}_{321}^p r_{345}^p} - \frac{r_{345}^p - r_{321}^p}{1 - F_3^2 r_{321}^p r_{345}^p}\right)\frac{n_1 Q \sin\phi \cos\gamma}{2g_3} \tag{3.64}$$

$$\tau_p = \frac{i(1 - F_3^2)(r_{321}^p - r_{345}^p)}{1 - F_3^2 r_{321}^p r_{345}^p}\frac{n_1 Q \sin\phi \cos\gamma}{2g_3} \tag{3.65}$$

$$\bar{r}_{321}^p = (r_{32}^p r_{21}^p + F_2^2)(r_{21}^p + F_2^2 r_{32}^p)^{-1}. \tag{3.66}$$

The non-diagonal elements are

$$r_{sp} = it_{123}^s t_{321}^p N^{-1}[(A^- + B^-)\cos\alpha + (A^+ + B^+)m\cos\beta]$$
$$r_{ps} = it_{123}^s t_{321}^p N^{-1}[(A^- + B^-)\cos\alpha - (A^+ - B^+)m\cos\beta]$$
$$t_{sp} = iF^3 t_{123}^p t_{345}^s N^{-1}[(C_s^- + D_s^-)\cos\alpha + (C_s^+ + D_s^+)m\cos\beta]$$
$$t_{ps} = iF^3 t_{123}^s t_{345}^p N^{-1}[(C_p^- - D_p^-)\cos\alpha - (C_p^+ + D_p^+)m\cos\beta] \tag{3.67}$$

where

$$N = (1 - F_3^2 r_{321}^s r_{345}^s)(1 - F_3^2 r_{321}^p r_{345}^p) \qquad m = g_3^{-1} n_1 \sin\phi$$
$$A^\pm = (1 - F_3^2)(1 \pm F_3^2 r_{345}^s r_{345}^p)\delta_1 \qquad B^\pm = iF_3^2(r_{345}^s \pm r_{345}^p)\delta$$
$$C_{s(p)}^\pm = (1 - F_3^2)(r_{321}^{s(p)} \pm r_{345}^{p(s)})\delta_1 \qquad D_{s(p)}^\pm = i(1 \pm F_3^2 r_{321}^{s(p)} r_{345}^{p(s)})\delta.$$

Using the formulae (3.62)–(3.66) and (2.28)–(2.30), the states of polarization and intensities of the reflected and transmitted light can be obtained.

In certain special cases, formulae (3.62)–(3.66) can be simplified. Suppose that the layer 2 is omitted; then $r_{321}^{s(p)} = r_{31}^{s(p)}$, and so on. In the case of very absorbent films, interference effects can be neglected by putting $F_3^2 = 0$. In that case we can put $n_2 = n_1$ and $F_2 = 1$. Then, according to formulae (3.58) and (3.59) we will have

$$r_{321}^{s(p)} = r_{31}^{s(p)} \qquad t_{321}^{s(p)} = t_{31}^{s(p)} \qquad r_{12345}^{s(p)} = r_{1345}^{s(p)}$$

and so on. If a certain layer, layer 4 for instance, is transparent, and its thickness is much greater than the wavelength of light, then as a result of the random variation of the thickness of the layer from point to point along its surface, and the finite width of the group of waves of the incident light beam, the value $\operatorname{Re} F_4 = \operatorname{Re} \exp(-2\pi\lambda^{-1}g_4d_4)$ changes from -1 to 1. In that case, averaging of the reflection and transmission coefficients with respect to the phase of F_4 gives the result that formally one can put $\langle F_4 \rangle = 0$ in formulae (3.59) and (3.59).

It is not difficult to generalize formulae (3.62)–(3.66) to the case of a system with more than two non-magnetic layers using the recursion formulae (3.58) and (3.59).

There are formalisms for the analysis of electromagnetic wave propagation in multilayers based on (4×4)-matrix algebra (Smith 1965, Yeh 1980, Višňovský 1991a, b, Zak *et al* 1990, Atkinson and Lissberger 1992, 1993, Heim and Scheinfein 1996), and (2×2)-matrix algebra (Mansuripur 1990). These formalisms are convenient for computer simulation of the magnetooptical properties of multilayers.

It should be noted that the magnetooptical macroscopic approach described above has not yet been verified for describing the various artificial magnetic structures with atomically thin layers down to a monolayer. There are well-known problems with the sharpness of interfaces and Maxwell's boundary conditions (Agranovich and Ginzburg 1984)[3]. The microscopical magnetooptical approach may be an adequate alternative (Kosobukin 1996).

A rather sophisticated magnetization distribution along the z-direction may be produced by exchange coupling across non-magnetic interlayers. Consequently, Q may not remain spatially invariable. It is therefore important to consider the effect of a spatially dependent parameter Q on the magnetooptical behaviour of magnetic multilayers. This problem was explored in depth by Traeger *et al* (1992, 1993), Hubert and Traeger (1993), Wenzel *et al* (1995), Atkinson and Kubrakov (1995), and Atkinson *et al* (1994b).

To describe the magnetooptical properties of metal multilayers it is often assumed that for each layer a dielectric tensor is known in advance. Then the solution of the routine boundary-value problem allows the state of the polarization to be easily obtained. However, if the electron mean free path exceeds the layer thickness, this method is inappropriate. The problem *can* be solved within the framework of the effective-medium approach.

Atkinson *et al* (1996) tried to solve this problem for materials which exhibit giant magnetoresistance (GMR). The conductivity tensor was found from the kinetic Boltzmann equation and the boundary conditions for the electron distribution functions in each layer, taking into account the spin-dependent scattering of the electrons at the rough interfaces. As a result, an analytical representation for the conductivity, in the framework of the effective medium, suitable for the study of materials that exhibit GMR, was found.

---

[3] Besides this, there have been reports of thickness-dependent optical constants, and the electronic and optical properties are expected to be different at surfaces and interfaces (see, e.g., Freeman and Fu 1987).

3.7 Non-linear magnetooptical effects

Non-linear magnetooptical effects—and, in particular, the second-harmonic-generation (SHG) effect—are becoming increasingly important, because they are non-destructive, and can be remotely sensed *in situ* with high spatial and temporal resolution at any interface accessible to light.

The field of non-linear optics was first broached experimentally in the 1960s, but it was not until the 1980s that physicists became really interested in non-linear magnetooptical phenomena[4]. Non-linear optical susceptibilities of magnetic origin possess quite different transformation properties under space and time symmetry operations to non-linear susceptibilities of electric origin (Pershan 1963, Shen 1984, Kielich and Zawodny 1973, Akhmediev *et al* 1985, Graham and Raab 1992). Akhmediev and Zvezdin (1983), Akhmediev *et al* (1985), and Girgel and Demidova (1987) suggested a magnetism-induced mechanism for non-linear magnetooptical effects. In this case the non-linear electrical polarization vector contains terms depending on the magnetization, and vanishing if the magnetization (or staggered magnetization in antiferromagnets) vanishes. The most interesting materials as regards observing magnetism-induced non-linear magnetooptical effects are the so-called magnetoelectrical materials (Cr_2O_3, $BiFeO_3$, etc) whose magnetic structures are odd in space. Agal'tsov *et al* (1989) observed a strong magnetism-induced increase of the SHG in $BiFeO_3$ below the point of transition into the antiferromagnetic state. Fiebig *et al* (1994) explored in depth the non-linear magnetooptical effects in the magnetoelectrics Cr_2O_3. Aktsipetrov *et al* (1990), Pisarev *et al* (1993), and Petrocelli *et al* (1993) investigated the optical SHG in thin films of magnetic garnets using reflected and transmitted light.

3.7.1 The non-linear Kerr effect

In centrosymmetric media, an electric field contribution to the second-order optical polarizability is forbidden by symmetry. At a surface or at the boundaries between centrosymmetric media, the inversion symmetry is broken, resulting in the high surface sensitivity of SHG (Shen 1984). The magnetization of a material does not usually break[5] the inversion symmetry, but it can modify the form of the non-linear susceptibility for surface SHG.

Pan *et al* (1989), and Hübner and Bennemann (1989) have shown theoretically that a magnetic effect should be detectable via surface SHG. The surface non-linear optical polarization of second order can be written in the form (Pan *et al* 1989)

$$P_x^{2\omega} = \chi_{ijk}^{(2)}(M) E_j E_k \tag{3.68}$$

---

[4] Veselago *et al* (1984) observed for the magnetic semiconductor $CdCr_2Se_4$ Faraday rotation that was dependent on the intensity of light. Nersisyan *et al* (1983) suggested a mechanism for non-linear Faraday rotation based on the temperature dependence of the value of the gyration vector.

[5] This is so except for the above-mentioned magnetoelectrical materials.

where the surface non-linear susceptibility tensor $\chi_{ijk}^{(2)}$ is a function of the magnetization M, and E is the electric field of the light wave. The symmetry of $\chi^{(2)}$ is defined by the time-reversal symmetry and by the symmetry of the particular surface under consideration. The time-reversal properties, neglecting dissipation, require that the real part of $\chi^{(2)}$ is an even function of M, while the imaginary part is an odd function of M. This latter feature can be particularly useful for probing the surfaces of magnets.

Linear and non-linear magnetooptical responses are strongly connected to each other. To illustrate the common features and the differences between them, let us consider the surface of an optically isotropic medium, whose linear permittivity tensor $\widehat{\varepsilon}$ is given by formula (1.6). From a symmetry viewpoint the second-order surface polarization (3.68) in the linear (-in-M) approach can be written as

$$P_s^{2\omega} = P_{s0}^{2\omega} + P_{sm}^{2\omega} \tag{3.69}$$

where

$$P_{s0}^{2\omega} = \chi_1 E(E \cdot N) + \chi_2 E^2 N \tag{3.70}$$

$$P_{sm}^{2\omega} = \chi_3 E(E \cdot [m \times N]) + \chi_4 E^2[m \times N] + \chi_5[E \times m](E \cdot N) \\ + \chi_6[E \times N](E \cdot m) \tag{3.71}$$

are contributions that are independent of the magnetization ($P_{s0}^{2\omega}$) and linear in the magnetization ($P_{sm}^{2\omega}$), where χ_i, $i = 1, 2$, are non-linear optical parameters, and χ_i, $i = 3, 4, 5, 6$, are nonlinear magnetooptical parameters; $m = M/M$, where M is the magnetization; and N is the vector normal to the surface. It is clear that only these two independent combinations that are second order in E and have the symmetry of the polar vector can be composed from the polar vectors N and E to get $P_{s0}^{2\omega}$; also only four independent combinations can be composed from the polar vectors E and N and the axial vector m to get $P_{sm}^{2\omega}$.

The signal reflected from a centrosymmetric medium actually consists of not only a surface contribution but also a bulk contribution. The latter, however, tends to be weaker than the former in the case of metals (Pan *et al* 1989).

Let us consider the surface SHG effect for the surface of a ferromagnet in transverse geometry (figure 3.2). The incident beam (of p polarization, for definiteness) with frequency ω produces reflected and transmitted beams with frequency 2ω. The non-linear reflected signal contains the field $E_{r0}^{2\omega}$, which is generated by $P_{s0}^{2\omega}$, and the odd-in-magnetization field $E_{rm}^{2\omega}$, which is induced by $P_{sm}^{2\omega}$.

The non-linear transverse Kerr effect is defined by

$$\delta_{p(s)}^{2\omega} = \frac{I_r^{2\omega}(M) - I_r^{2\omega}(0)}{I_r^{2\omega}(0)} \tag{3.72}$$

where $I_r^{2\omega}(0) = |E_{r0}^{2\omega}|^2$, and $I_r^{2\omega}(M) = |E_{r0}^{2\omega} + E_{rm}^{2\omega}|^2$. The reflected and transmitted fields are defined by (3.29)–(3.31) for $\omega \to 2\omega$. The amplitudes $E_{r0}^{2\omega}$,

$E_{rm}^{2\omega}$, $E_{tm}^{2\omega}$, and $E_{t0}^{2\omega}$ can be found from the boundary conditions (2.5) and (2.6), which should be modified to take into account the surface polarization $P_s^{2\omega}$. To understand the effect under consideration clearly, studying just a simplified case for which the angles of incidence and refraction $\phi \ll 1$, $\psi \ll 1$ will be sufficient. The modified boundary conditions, readily following from Maxwell's equations (2.1) and (2.2) (see also (3.33)) are (Shen 1989, Jérôme and Shen 1993)

$$
\begin{aligned}
E_{iy} + E_{ry} &= E_{ty} - 4\pi\varepsilon_2^{-1}\frac{\partial P_{sx}}{\partial x} \\
H_{iz} + H_{rz} &= H_{tz} + (i4\pi\omega c^{-1})(P_{sy} - \varepsilon_{2yx}\varepsilon_2^{-1}P_{sx}).
\end{aligned}
\tag{3.73}
$$

Substituting (3.70), (3.71), and $E_i(2\omega) = 0$ into (3.73) yields

$$
\begin{aligned}
E_r + E_t &= -4\pi i\omega c^{-1}\psi n_2^{-1}P_{sx} \\
n_1 E_r - n_2(1 - iQ\psi)E_t &= i4\pi\omega c^{-1}(P_{sy} - iQP_{sx}).
\end{aligned}
\tag{3.74}
$$

From (3.74) it readily follows that

$$
\begin{aligned}
E_{r0}^{2\omega} &= i8\pi\omega c^{-1}\frac{\chi_1 - \chi_2}{n_1 + n_2}\psi E_t^2 \\
E_{rm}^{2\omega} &= i8\pi\omega c^{-1}\frac{\chi_3 + \chi_4 - i(Q^\omega\chi_1 + Q^{2\omega}\chi_2)}{n_1 + n_2}E_t^2.
\end{aligned}
\tag{3.75}
$$

Substituting (3.75) into (3.72) yields

$$
\delta_p^{2\omega} = 2\operatorname{Re}\left[\frac{\chi_3 + \chi_4 - i(Q^\omega\chi_1 + Q^{2\omega}\chi_2)}{(\chi_1 - \chi_2)\psi}\right]
\tag{3.76}
$$

where Q^ω and $Q^{2\omega}$ are the Voigt parameters for the frequencies ω and 2ω, respectively. One can get the corresponding formulae for s polarization of the light beam similarly.

Formulae (3.36)–(3.39), (3.72), and (3.76) give the total linear and non-linear magnetooptical responses of the surface of the magnetic medium for transverse geometry. In a like manner (see also section 3.6), one can derive the formulae for multilayers and buried layers, to take into account the contributions of the non-linear electrical polarization from many interfaces.

The first experimental evidence for the non-linear surface Kerr effect was given by Reif *et al* (1991) and Spierings *et al* (1993). Vollmer *et al* (1995) directly compared linear magnetooptical Kerr effect measurements with non-linear Kerr effect measurements on Co films grown epitaxially on Cu(001). The magnetization-induced SHG measurements were carried out in the transverse geometry. The angle of incidence was $\approx 35°$. It was verified that the second-harmonic outgoing light is purely p polarized in accordance with the above theory. Figure 3.7 shows the relative magnetic SHG signal, defined by formulae (3.72), and the linear Kerr effect as a function of Co film thickness. While

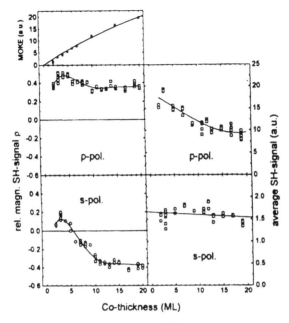

Figure 3.7. The dependence of the linear MOKE (left-hand topmost panel) and the non-linear Kerr effect (bottom four panels) for Co films on Cu(001) as functions of the layer thickness. The linear measurements were taken in the longitudinal geometry, while for the measurements of the magnetization-induced SHG, the transverse geometry was chosen (Vollmer *et al* 1995).

the linear MOKE signal increases nearly linearly with the film thickness (see also subsection 3.6.1 and figure 3.6) the non-linear Kerr effect does not, directly providing the interface sensitivity of the non-linear Kerr effect. The source of the SHG is bound to the interface within a region of less than two monolayers.

It should be noted that the transverse non-linear Kerr effect is considerably stronger than the linear effect. In line with this, the non-linear Kerr rotation can be enhanced by up to one order of magnitude compared to the linear Kerr angle in the longitudinal geometry (Pustogowa *et al* 1994, Vollmer *et al* 1995).

Chapter 4

Light waveguiding in thin magnetic films

4.1 Normal modes, and waveguiding conditions

A uniform dielectric film surrounded by dielectric media is an optical waveguide if the refractive indexes of the surrounding media are lower than that of the film. Dielectric waveguides are usually formed by diffusion, sputtering, or epitaxy on the substrate surface. In this case the surrounding media are the substrate and the air. The propagation of a light wave in a thin-film waveguide can be represented as a zigzag propagation of a plane wave (figure 4.1) occurring as a result of successive reflections at the interfaces of the media.

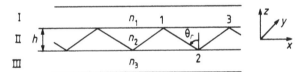

Figure 4.1. The thin-film dielectric waveguide: I: the overlayer; II: the waveguiding film; and III: the substrate.

Let us consider the normal modes in such a three-layer medium. We shall assume that all three media are absolutely transparent and optically isotropic, and for the sake of simplicity we will take the magnetic permeability to be equal to 1. Let the z-axis be perpendicular to the layers (see figure 4.1), and the x-axis be parallel to the wave propagation direction. At the interfaces of the film and the substrate, and of the film and the overlayer there are discontinuities of the dielectric constant. In an isotropic waveguide two systems of modes (TE and TM modes) that do not interact can exist: the TE mode has the components E_y, H_x, and H_z, while TM mode has the components E_x, E_z, and H_y. Given that all of the media are isotropic, the solution of the Maxwell equations for the waveguide propagation of light in the film takes the following form:

$$E_i = E_{0i} \exp(-i\omega t + i\beta x \pm ibz) \qquad (4.1)$$

where

$$\beta^2 + b^2 = k_2^2 \qquad \beta = k_2 \sin \theta_r \qquad b = k_2 \cos \theta_r \qquad k_2 = kn_2.$$

Here n_2 is the refractive index of the film, and θ_r is the angle of incidence of the light on the film–substrate and film–overlayer interfaces. In the surrounding media, evanescent waves of the type

$$E_i = E_{0i} \exp(-i\omega t + i\beta x \pm pz) \tag{4.2}$$

propagate, where $p = (k^2 - \beta^2)^{1/2}$ is the penetration constant. The guided waves must have certain phase relationships: the total phase accumulated on the zigzag path must be a multiple of π; that is,

$$\begin{aligned} \kappa_{TE} &= \kappa_{TE}^{12} + \kappa_{TE}^{23} + 2kn_2 h \cos \theta_{rq} = 2\pi q \\ \kappa_{TM} &= \kappa_{TM}^{12} + \kappa_{TM}^{23} + 2kn_2 h \cos \theta_{rq'} = 2\pi q'. \end{aligned} \tag{4.3}$$

Here h is the film thickness, q and q' are the integers that determine the mode order (the mode indexes), and $\kappa_{TE(TM)}^{12}$ and $\kappa_{TE(TM)}^{23}$ are the phase shifts that occur during the total internal reflection from the film–overlayer and film–substrate interfaces; κ has the value

$$\kappa = -2 \tan^{-1}\{[(k_2^2 - \beta^2)^{-1}(\beta^2 - (kn')^2)]^{1/2}(n_2/n')^u\} \tag{4.4}$$

where $u = 0$ for TE modes and $u = 2$ for TM modes, and $n' = n_1$ or n_3. Formulae (4.3) and (4.4) are consequences of the boundary conditions (2.5) and (2.6).

Equations (4.3) define either the θ_r-angles for TE and TM modes or the corresponding propagation constants $\beta = n_2(\omega/c) \sin \theta_r$. Figure 4.2 shows the dependence of the propagation constants β on the waveguide thickness h (so-called dispersion curves). As seen from the figure, in the general case the propagation constants of TE and TM modes are different. Of particular importance is the difference between the values of β_{TE} and β_{TM} for modes of the same order (deviation):

$$\Delta\beta = \beta_{TE} - \beta_{TM}. \tag{4.5}$$

If $h < h_{min}$ the film supports no localized modes. The symmetric waveguide, where the zero-order TE mode propagates at any waveguide thickness, is particularly noticeable. In asymmetric waveguides based on epitaxial films of iron garnet, $h_{min} \approx 0.2 \ \mu$m.

For the isotropic waveguiding structure, the mode propagation condition is reduced to the following:

$$n_1, n_3 < \beta < n_2.$$

Anisotropy in at least one of the layers substantially complicates the propagation condition. Let us consider a waveguiding structure with all three of the layers being double-axis crystalline media, which have their major optical axes coinciding with the coordinate system x, y, z (figure 4.1). Let ε_{1i}, ε_{2i}, and ε_{3i}, where $i = x, y, z$, be the diagonal elements of the dielectric permittivity tensor for the ith layer. Then the waveguiding conditions will take the following forms:

$$(\varepsilon_{3y})^{1/2} \quad \text{and} \quad (\varepsilon_{1y})^{1/2} < (\varepsilon_{2y})^{1/2} \qquad \text{for TE modes}$$

and

$$(\varepsilon_{3x})^{1/2} \quad \text{and} \quad (\varepsilon_{1x})^{1/2} < (\varepsilon_{2x})^{1/2} \qquad \text{for TM modes.}$$

Therefore, there are the waveguides that support waves of only one polarization. These waveguides have been named 'semileaky'-type waveguides (Yamamoto and Makimoto 1974). This means that when in such waveguides the TM ↔ TE mode transition takes place, the propagating wave (for instance, TE) will in the course of the transition to the TM mode radiate ('leak') from the waveguide. In this waveguiding structure the following conditions should be satisfied:

$$\varepsilon_{2x} < \varepsilon_{3x}, \varepsilon_{2x} \qquad \text{but} \qquad \varepsilon_{2y} > \varepsilon_{3y}, \varepsilon_{1y}$$

or

$$\varepsilon_{2y} < \varepsilon_{3y}, \varepsilon_{1y} \qquad \text{but} \qquad \varepsilon_{2x} > \varepsilon_{3x}, \varepsilon_{1x}.$$

In the first case the TE mode is the waveguiding one and the TM mode is the 'leaky' one; in the second case it is the other way round. The first type of structure can be used as a TE rectifier and the second one as a TM rectifier.

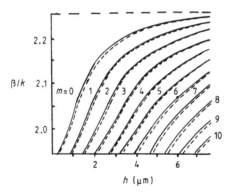

Figure 4.2. The dependence of the effective refractive indexes β/k of the modes in an isotropic waveguide on the film thickness. Solid lines: TE modes; dashed lines: TM modes. Overlayer: air. Refractive index of the film: $n_2 = 2.25$; refractive index of the substrate: $n_3 = 1.945$. $\lambda = 1.15$ μm.

If there is a perturbation in the waveguide (non-symmetric components of the ε_{ik}-tensor, inhomogeneities, etc), the modes interact. For instance, the mode that was excited in the waveguide scatters in the course of propagation, i.e., excites other modes. Modes that are closer (in the β-parameter) interact more strongly.

As seen from figure 4.2, the propagation constants of different modes differ substantially. Thus the problem of light propagation in the waveguide with different perturbations can be solved in the two-mode approximation, i.e., taking into account interaction of only adjacent TE and TM modes and disregarding all other modes (the coupled-mode method).

Note the important and fruitful similarity of a dielectric waveguide and a birefringent medium, which we considered in section 3.3. Let us represent the electromagnetic field in a waveguide Ψ_w as a superposition of two adjacent TE and TM modes:

$$\Psi_w = A\Psi_{TE} + B\Psi_{TM}. \tag{4.6}$$

Generally speaking, the Ψ_w-field vector comprises six components (E, H), but practically it is sufficient to confine consideration to two: $E_y(z, x, t)$ and $H_y(z, x, t)$, because the other components can be expressed via this pair using the Maxwell equations. The phase velocities of the TE and TM waves $v = (\sin\theta)^{-1}c$ differ, like the corresponding velocities of the normal modes of a birefringent medium. If we form the column vector with coordinates A and B, then the propagation matrix of the waveguide (the Jones matrix) looks like the propagation matrix of a birefringent plate where $\Delta\beta$ is substituted for the value $(\omega/c)\Delta n$.

4.2 Gyrotropic waveguides

Light propagation in a gyrotropic waveguide is a peculiar phenomenon, because off-diagonal elements in the dielectric tensor result in interaction between the TE and TM modes. Consider a waveguide made of a gyrotropic material. The substrate and the overlayer are assumed to be isotropic, so the dielectric tensors of these media contain only diagonal elements which are equal, that is, $\varepsilon_{11} = \varepsilon_{22} = \varepsilon_{33}$.

The magnetization direction (and the direction of the gyration vector) with respect to the film's plane can differ, and can be controlled by external fields. The most interesting and practically very important case arises if the film is magnetized in the plane along the light propagation direction, i.e., when $M \parallel x$ (the Faraday geometry). The rotation of the plane of polarization in this case is manifested in transformation (or conversion) of the modes. If a TE mode is excited at the input of a transparent waveguide, then due to gyrotropy it gradually (in the course of propagation) transforms into a TM mode (usually not completely), then back into a TE mode, etc. This is the way in which the Faraday effect is manifested in a thin-film waveguide. The mode transformation can be defined in terms of the mode conversion coefficient R (see below). The transformation efficiency is naturally restricted by the absorption coefficient and, as we will see below, by the mode mismatch $\Delta\beta$.

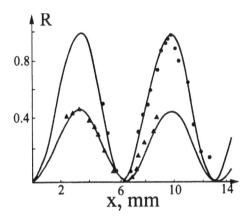

Figure 4.3. The oscillating dependence of the transformation coefficient R for the zero-order TE and TM modes on the distance x between the input and output along the waveguide in the Faraday effect geometry. Solid lines: theory; dots and triangles: experiment (Prokhorov *et al* 1984).

In a gyrotropic waveguide the eigenmodes are, like in the above-discussed (section 3.3) birefringent medium, elliptically polarized:

$$\Psi_W = \left[\frac{1}{2}A(x)\begin{pmatrix} e_y(z) \\ 0 \end{pmatrix} + \frac{1}{2}B(x)\begin{pmatrix} 0 \\ h_y(z) \end{pmatrix}\right] \exp i(\tilde{\beta}x - \omega t) + \text{CC} \qquad (4.7)$$

where $\tilde{\beta} = \beta_{TE} - \Delta\beta/2$, which is a combination of the TE and TM modes. The amplitudes of the TE and TM modes $A(x)$ and $B(x)$ change periodically with the x-coordinate, and are defined by the propagation matrix $\widehat{T}$, which, as we noted above, looks the same as for a birefringent medium (section 3.3):

$$\begin{pmatrix} A(x) \\ B(x) \end{pmatrix} = \begin{pmatrix} T_{11} & T_{12} \\ T_{21} & T_{22} \end{pmatrix} \begin{pmatrix} A(x_0) \\ B(x_0) \end{pmatrix} \qquad (4.8)$$

where

$$\begin{aligned}
T_{11} &= \cos v(x - x_0) - i\cos\theta \sin v(x - x_0) \\
T_{22} &= \cos v(x - x_0) + i\cos\theta \sin v(x - x_0) \\
T_{12} &= -T_{21} = -\sin\theta \sin v(x - x_0)
\end{aligned} \qquad (4.9)$$

$$\begin{aligned}
v &= [(\Delta\beta/2)^2 + \kappa^2]^{1/2} & \cos\theta &= (1/2)v^{-1}\Delta\beta \\
\sin\theta &= \kappa/v & \kappa &= (2nc)^{-1}g\omega = \Phi_F.
\end{aligned} \qquad (4.10)$$

Φ_F is the specific Faraday rotation in the gyrotropic film. Assuming that $A(0) = 1$ and $B(0) = 0$, the mode conversion coefficient may be defined as

$$R = |B(x)|^2.$$

From equation (4.8) we get R:

$$R = \frac{|T_{12}|^2}{|T_{11}|^2 + |T_{12}|^2} = \sin^2 \theta \sin^2 vx. \tag{4.11}$$

We can see that the maximum R-coefficient is

$$R = \frac{\kappa^2}{(\Delta\beta/2)^2 + \kappa^2}. \tag{4.12}$$

The maximum transformation ($R = 1$) may be reached if $|\Delta\beta/2| \ll |\kappa|$ and is achieved, as follows from (4.9) and (4.10), at the distance

$$L = \pi/2\Phi_F.$$

In this case, when $\Phi_F \approx (10^3)° \text{ cm}^{-1}$, $L \approx 1$ mm.

Figure 4.3 shows the oscillating dependence of the mode conversion coefficient in a gyrotropic epitaxial iron garnet film waveguide (Prokhorov *et al* 1984). In order to increase the conversion coefficient one should match the phase velocities of the TE and TM modes, i.e., reduce the mismatch $\Delta\beta/2$. The other interesting possibility arises from the availability of waveguiding structures with the gyration vector (or magnetization) of the magnetic film varying periodically along the structure. The increase of R is achieved due to periodic matching of the phases of the interacting modes (see below).

The important feature of the gyrotropic waveguide is its non-reciprocity. This means that the propagation constants of the mixed (elliptically polarized) modes (4.7), and the TE–TM transformation coefficients for the light propagation directions x and $-x$ are different. Especially high non-reciprocity can be reached in combined (or cascade) waveguides, in which different parts have different directions of the gyration vector. This follows from the fact that the transformation matrices for different parts do not commute. The 'semileaky' waveguides, which were considered above and in which the waveguiding conditions at $\kappa = 0$ are satisfied only for one of the modes—say, for the TE mode—are also interesting from this point of view. In this case, during the propagation the TE mode transforms into a TM mode, which radiates ('leaks') from the waveguide. This structure is of interest for possible application in integrated optical devices of rectifier type.

4.3 Dielectric waveguides with graded $\varepsilon(z)$ profiles; the quantum mechanics analogy

Let us consider another approach to the problem of the mathematical description of a multilayered waveguide with gradual variation of the dielectric constant.

Let us represent the Maxwell equation (2.1), (2.2) in the matrix form

$$
\begin{pmatrix}
k\varepsilon_0 & i\,\partial_x & -i\,\partial_z & 0 & k\varepsilon_{23} & 0 \\
i\,\partial_x & k & 0 & 0 & 0 & 0 \\
-i\,\partial_z & 0 & k & 0 & 0 & 0 \\
0 & 0 & 0 & k & -i\,\partial_x & i\,\partial_z \\
k\varepsilon_{32} & 0 & 0 & -i\,\partial_x & k\varepsilon_0 & 0 \\
0 & 0 & 0 & i\,\partial_z & 0 & k\varepsilon_0
\end{pmatrix}
\begin{pmatrix}
E_y \\ H_z \\ H_x \\ H_y \\ E_z \\ E_x
\end{pmatrix}
= 0.
\tag{4.13}
$$

When $\varepsilon_{23} = 0$, equation (4.13) can be separated into two independent systems that define the TE and TM waves. Therefore the ε_{23}-term provides the interaction between them. Let us express the components $E_{x,z}$ and $H_{x,z}$ via E_y and H_y:

$$
H_{x,z} = \pm k^{-1} i\,\partial_{z,x} E_y
$$
$$
E_x = -(k\varepsilon_0)^{-1} i\,\partial_z H_y \qquad E_z = (k\varepsilon_0)^{-1}(i\,\partial_x H_y - k\varepsilon_{32} E_y).
$$

Substituting those values into (4.13) and introducing a new term $\tilde{H} = H_y \varepsilon^{-1/2}$, and disregarding the second-order terms in ε_{23}, we have

$$
\begin{pmatrix}
k^2 \varepsilon_0 + \partial_x^2 + \partial_z^2 & k\varepsilon_{23}\varepsilon_0^{-1/2} i\,\partial_x \\
k\varepsilon_{32}\varepsilon^{-1/2} i\,\partial_x & k^2 \varepsilon + (1/2)\varepsilon^{1/2}\partial_z(\varepsilon^{-3/2}\partial_z\varepsilon) + \partial_x^2 + \partial_z^2
\end{pmatrix}
\begin{pmatrix}
E \\ \tilde{H}
\end{pmatrix}
= 0.
$$

Let us solve these equations using the perturbation theory. If $\varepsilon_{23} = \varepsilon_{32} = 0$ we have waves of two types:

$$
\Psi_{TE}^n = \begin{pmatrix} E_n \\ 0 \end{pmatrix} \exp i(\beta_{TE}^n x - \omega t)
\tag{4.14}
$$
$$
\partial_z^2 E_n + [-(\beta_{TE}^n)^2 + k^2 \varepsilon_0] E_n = 0
$$

$$
\Psi_{TM}^n = \begin{pmatrix} 0 \\ \tilde{H}_n \end{pmatrix} \exp i(\beta_{TM}^n x - \omega t)
\tag{4.15}
$$
$$
\partial_z^2 \tilde{H}_n + [-(\beta_{TM}^n)^2 + k^2 \varepsilon_0 + (1/2)\varepsilon^{1/2}\partial_z(\varepsilon^{-3/2}\partial_z\varepsilon)]\tilde{H}_n = 0.
$$

Equations (4.14) and (4.15) coincide with the Schrödinger equations for particles that move in a potential well (figure 4.4):

$$
V_{TE} = -k^2 \varepsilon_0(z)
$$
$$
V_{TM} = -k^2 \varepsilon_0(z) + \delta V(z)
$$
$$
\delta V(z) = -\frac{1}{2}(\varepsilon^{-1}\partial_z^2\varepsilon) + \frac{3}{4}(\varepsilon^{-1}\partial_z\varepsilon)^2.
$$

The propagation constants $(\beta_{TE}^n)^2$ and $-(\beta_{TM}^n)^2$ correspond to different energy levels. Usually $|\delta V| \ll |k^2 \varepsilon_0|$; therefore, the value of $-(\beta_{TM}^n)^2$ can be defined using perturbation theory:

$$
-(\beta_{TM}^n)^2 = -(\beta_{TE}^n)^2 - (\langle n|n\rangle)^{-1}(\langle n|(1/2)(\varepsilon_0^{1/2}\partial_z^2(\varepsilon_0^{-3/2}\partial_z\varepsilon_0))|n\rangle)
$$

where the matrix element

$$\langle n|A|n \rangle = \int_{-\infty}^{\infty} dz \, E_n^*(z) A E_m(z).$$

The eigenfunctions $E_y(z)$ and $H_y(z)$ are presumed to be normalized: $\langle E_y^* E_y \rangle = 1$ and $\langle \varepsilon_0^{-1} H_y^* H_y \rangle = 1$.

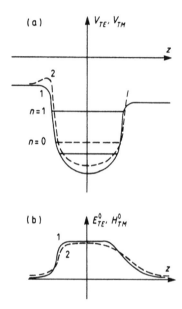

Figure 4.4. Potential wells (a) and the distribution of electric and magnetic fields (b) in the TE (curves 1) and TM (curves 2) modes.

In this approximation we have a system of quasi-doublets that are rather widely separated in energy (see figure 4.2). The perturbation ($\varepsilon_{23} \neq 0$)—that is, the gyration—results in the mixing of states inside each quasi-doublet—that is, Ψ_w^n is defined by (4.6) and (4.7). In the first-order perturbation theory let us derive the so-called coupling mode equations (Yariv 1975) for the coefficients A and B. The coupling equations can be presented in matrix form:

$$i \partial_x \begin{pmatrix} A \\ B \end{pmatrix} = \mathcal{H} \begin{pmatrix} A \\ B \end{pmatrix} \tag{4.16}$$

where

$$\mathcal{H} = (\Delta\beta/2)\sigma_z - \kappa\sigma_y = (s \cdot \nu) \tag{4.17}$$

and where

$$\sigma_z = \begin{pmatrix} 1 & 0 \\ 0 & -1 \end{pmatrix} \qquad \sigma_y = \begin{pmatrix} 0 & i \\ -i & 0 \end{pmatrix}$$

are the Pauli matrices, and

$$\kappa = -\mathrm{i}(1/2)k\langle E_y^* \varepsilon_{23}\varepsilon_0^{-1} H_y\rangle$$

is the coupling coefficient. The vector

$$\nu = (0, -\kappa, \Delta\beta/2) = (0, -\nu\sin\theta, \nu\cos\theta)$$

where

$$\sin\theta = \kappa/\nu \qquad \cos\theta = \Delta\beta/2\nu \qquad \nu = ((\delta\beta/2)^2 + \kappa^2)^2$$

may be considered as the 'effective field'. The operator $s = (\sigma_x, \sigma_y, \sigma_z)$ is the quasi-spin.

The solution of (4.16) for $\nu = $ constant takes the following form:

$$\begin{pmatrix} A(x) \\ B(x) \end{pmatrix} = \mathbf{T} \begin{pmatrix} A(x_0) \\ B(x_0) \end{pmatrix} \tag{4.18}$$

where

$$\mathbf{T} = \exp[-\mathrm{i}\nu(s \cdot n)(x - x_0)] = \cos\nu(x - x_0)\mathbf{I} - \mathrm{i}\nu(s \cdot n)\sin\nu(x - x_0)$$

in which $\widehat{\mathbf{I}}$ is the unit matrix, and $n = (0, -\sin\theta, \cos\theta)$. The matrix $\mathbf{T}$ obviously coincides with (4.8). The conversion coefficient R is here

$$R = \frac{\kappa^2}{\nu^2} \sin^2 \nu x. \tag{4.19}$$

Equation (4.16) coincides with the Schrödinger equation that describes a two-level system with the Hamiltonian $\mathcal{H}$ (4.17), only t should be substituted for x. Therefore, for the study of a particular waveguiding structure, one can use known results from two-level-system theory. In particular, formula (4.19) for the mode conversion coefficient is equivalent from this point of view to the well known Rabi formula for the probability of transition under the influence of a perturbation $V = -\kappa\sigma_y$, and the value κ plays the role of the Rabi frequency.

As we mentioned in section 2.5, it is convenient to use the Bloch vector S_B or the Stokes vector to describe polarization of the electromagnetic waves, in particular A- and B-waves. In the course of the evolution described by (4.16) the vector S_B will trace a trajectory on the Poincaré sphere (or more precisely on the Bloch sphere).[1]

The matrix $\mathbf{T}$ in (4.18) can be considered as the matrix of rotation of the vector S_B around the direction of the effective field ν. Figure 4.5 depicts the rotation of the vector S_B around a uniform direction of the effective field ν when the x-coordinate changes. The mode conversion coefficient oscillates with

[1] In fact, the distinction between the Poincaré and the Bloch spheres is only that of the choice of coordinate system. This difference can be disregarded here.

changing x, and reaches its maximum κ/ν when the end-point of the S_B-vector is at the maximum distance from the north pole of the Poincaré sphere.

If the effective field $n\nu$ depends on the x-coordinate, the trajectory of the evolution of $S_B(x)$ can be found from the Heisenberg equation for the quasi-spin operator s and the commutation rules for the Pauli matrices. From equation (4.17) we can obtain the equation for the quasi-spin s:

$$\partial_x s = s \times \nu. \tag{4.20}$$

The equation for the Bloch vector S_B follows directly from (4.20):

$$\partial_x S_B = S_B \times \nu. \tag{4.21}$$

According to equation (4.21), the Bloch vector describing the polarization of the electromagnetic wave in the waveguide precesses around the 'instant' effective field $\nu(x)$.

Equations (4.16) and (4.21) can be used for the description of waveguides made of anisotropic optical materials when there is gyration in both the $\hat{\varepsilon}$- and $\hat{\mu}$-tensors. Only particular values of $\Delta\beta$ and κ must be changed in them (the corresponding formulae are presented, for example, by Prokhorov *et al* 1984). The coupling coefficient κ increases gradually with decreasing film thickness, which is associated with growth of the mode overlap integral.

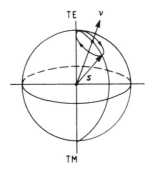

Figure 4.5. The precession of the quasi-spin s around the direction of the effective field ν, which characterizes the mode conversion in a homogeneous gyrotropic waveguide.

4.4   The efficiency of mode conversion; modulated waveguides

Let us return to the question of the efficiency of the mode conversion. Formula (4.12) shows that in order to get a higher coefficient of the mode transformation one should reduce the mismatch $\Delta\beta$. Another approach is based on the matching of the interacting mode's phases using the spatial periodicity of the magnetization vector in the film. In this case, in order to calculate the propagation matrix and the mode conversion coefficient, one must solve (4.16) or (4.18). Intuition hints that if the periodicity of $\kappa(x)$ is close to $(\Delta\beta)^{-1}\pi$, there is a resonance, and one can expect a significant increase in the conversion coefficient R.

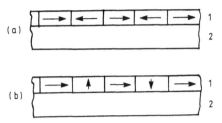

Figure 4.6. Magnetooptical waveguides with periodic changes of magnetization in the waveguiding films along the x-axis (a) and along the x- and z-axes (b). 1: the film; 2: the substrate. Arrows show the magnetization directions.

In practice, waveguides with periodically changing magnetization directions are often used (figure 4.6). When the sign of the projection of the magnetization onto the propagation direction of the electromagnetic wave changes, the direction of the effective field ν—that is, the sign or the value of the θ-angle—changes too.

Let us follow the sequence of rotations of S_B on the Poincaré sphere (figure 4.7(a)) when the light propagates through the waveguide with periodically changing magnetization along the x-axis (figure 4.6(a)).

Let the ratio of the gyrotropy and the mismatch $\Delta\beta$ be such that $|\theta| = |\theta_0| = \pi/6$, and a TE wave is excited at the input. The north pole of the Poincaré sphere—point A_0—corresponds to this wave. After rotation around the axis

$$n_1 = (0, \sin\theta_0, \cos\theta_0)$$

by half of the period, which corresponds to the condition $\nu\,\Delta x = \pi$, the state of the wave will correspond to the point A_1 with coordinates ($\alpha = \pi/3, \beta = \pi/2$); further rotation by a half-period around the axis

$$n_2 = (0, -\sin\theta_0, \cos\theta_0)$$

transforms the point A_1 into the point A_2 with coordinates ($\alpha = 2\pi/3, \beta = 3\pi/2$), and finally rotation by a half-period around the axis

$$n_1 = (0, \sin\theta_0, \cos\theta_0)$$

transforms the point A_2 into the south pole of the Poincaré sphere, A_3, which corresponds to the TM mode.

This example shows that the process can provide a 100% transformation of a TE mode into a TM mode (and back) if $\theta_0 = \pi/N$ where N is an integer.

Figure 4.7(b) shows the rotation scheme for S_B for the waveguide that is pictured in figure 4.6(b) (when $\theta_0 = \pi/8$), which results in 100%

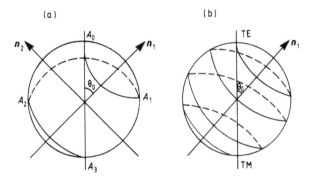

Figure 4.7. An illustration of the mode conversion in the waveguides with periodic variations of the magnetization, with the aid of Poincaré spheres. The variations of the magnetization correspond to those depicted in figure 4.6.

mode conversion. Periodical structures of this type are fabricated using ion implantation.

Let us consider harmonic modulation of the coupling coefficient. Let $\kappa = \kappa_0 \cos kx$ and $\Delta\beta \gg \kappa_0$. Such a modulation of the coupling coefficient can be caused, for example, by exciting a magnetostatic spin wave with the wavevector k in the film. For the resonance condition $\Delta\beta = k$—i.e., when the modes' phases are synchronized—S_B reorients from the north pole of the Poincaré sphere to the south pole practically along the meridian. The light propagates through the distance $l = \kappa^{-1}\pi$. This process corresponds to the dynamic process in quantum mechanics that converts the system from one level to another using the so-called 180° pulse (that is, the one that rotates the spin by 180°).

In order to take into account losses in the waveguide, one should add the term

$$i\alpha \begin{pmatrix} A \\ B \end{pmatrix}$$

where α is the loss coefficient, into the left-hand part of equation (4.16). For the practical use of gyrotropic waveguides it is desirable for the magnetooptic conversion length $l = \kappa^{-1}\pi$ to be less than the loss length α^{-1}—that is, $\alpha < \kappa^{-1}\pi$.

Chapter 5

Microscopical mechanisms of magnetooptical activity

The understanding of the microscopical mechanisms that define magnetooptical effects is based on the knowledge of the electronic structure and electron wave functions in the crystals. Of particular importance are the selection rules for light-induced quantum transitions. It is known that right or left circularly polarized photons excite electron transitions, which, in accordance with the law of angular momentum conservation, change the appropriate projection of the system's angular momentum by $\Delta m = \pm 1$. As we shall show below, off-diagonal elements of the polarizability tensor in the electric dipole approximation and, consequently, the Faraday and Kerr effects are defined by the difference between the contributions of the 'right' and 'left' transitions (i.e., the transitions with $\Delta m = +1$ and with $\Delta m = -1$). In non-magnetic materials at $H = 0$ these processes have become equalized; therefore, no odd magnetooptical effects take place. In a magnetic field and in magnetically ordered materials the balance between the total contributions of the 'right' and 'left' processes to the permittivity is broken; this results in a rotation of the plane of polarization of light or in a difference between the absorption coefficients of the light with different polarizations.

The implementation of this simple idea in particular cases strongly depends on the character of the wave functions and the electron energy spectrum in the magnetic medium. In order to illustrate the main mechanisms, in this chapter we shall consider three classes of system.

- Materials in which the magnetooptical properties are mainly governed by ions with unfilled f or d shells. In this case the wave functions of the f and d electrons are localized and can be approximated by the atomic wave functions.
- Materials in which a substantial role is played by the charge transfer from a non-magnetic ion (O, Cl, F, etc) to a magnetic one (usually of d type), which usually takes place in the transition metal oxides (ferrites and others). Such

complexes as octahedral FeO_6^{9-} or tetrahedral FeO_4^{5-} are the main structural elements in these materials. The electron wave functions in this case are partially delocalized and are described in the ligand theory approximation.

- Metal systems with delocalized electrons that can be described by the Bloch wave functions.

Before we look into these questions, we have to consider the simplest models of the magnetooptical activity in a material. By this, we mean models based on the so-called paramagnetic and diamagnetic contributions to the Faraday and Kerr effects, and some general questions that are associated with the microscopical mechanisms of the magnetooptical activity.

5.1   The permittivity tensor

The magnetooptical phenomena are governed by the permittivity and magnetic permeability tensors or by the corresponding polarizabilities of the medium. The polarizability is for most magnetooptical materials assumed to be the sum of the polarizabilities of the d and f ions, which govern the magnetic properties of the materials. The components of the polarizability tensor α_{ij} of the ion, in the electronic dipole approximation, are determined by the Kramers–Heisenberg formula, which can be derived by means of time-dependent perturbation theory (see, e.g., Schiff 1955):

$$\alpha_{ij} = -\hbar^{-1} \sum_{ab} \rho_a \left\{ \frac{d_{ab}^i d_{ba}^j}{\omega_{ab} + \omega - i\Gamma_{ab}} + \frac{d_{ab}^j d_{ba}^i}{\omega_{ab} - \omega + i\Gamma_{ab}} \right\} \qquad (5.1)$$

where the sum is taken over all of the ground (a-) and excited (b-) states of the ion, ρ_a is the probability of the electron being located at the level of energy E_a, d_{ab}^i is the matrix element of the ith component of the dipole moment connecting the a- and b-states ($i = x, y, z$), $\hbar\omega_{ab} = E_b - E_a$, and Γ_{ab} is the half-width of the spectral band of the $a \rightarrow b$ transition.

The contribution of magnetic ions of concentration N to the material's permittivity is

$$\delta\varepsilon_{ij} = 4\pi N \left(\frac{n^2 + 2}{3} \right)^2 \alpha_{ij} \qquad (5.2)$$

where n is the average refractive index of the material (here $\delta\varepsilon_{ij} \ll n^2$). As an approximation, the Lorentz–Lorenz factor

$$L = \left[\frac{n^2 + 2}{3} \right]^2$$

is used to take into account the renormalization of the electric field of the wave. This field affects the ion, due to the polarization of the medium. Expression

(5.2) can easily be derived using the Lorentz–Lorenz relation, which is well known in the theory of dielectrics:

$$\varepsilon = \frac{3 + 8\pi\kappa}{3 - 4\pi\kappa}.$$ (5.3)

Assuming that

$$\varepsilon = \varepsilon_0 + \delta\varepsilon \qquad \alpha = \alpha_0 + \delta\alpha$$

and that $|\delta\varepsilon| \ll \varepsilon_0$, $|\delta\alpha| \ll \alpha_0$, and varying (5.3) with respect to $\delta\widehat{\alpha}$, we get the relationship (5.2).

According to (3.6), the off-diagonal components of the $\delta\varepsilon_{ij}$-tensor govern the Faraday effect, the circular dichroism, and other magnetooptical effects. Substituting (5.1) and (5.2) into (2.4), we get, for example, for the Faraday effect

$$\Phi_F = \frac{\pi N e^2 L}{mcn} \sum_{ab} \rho_a \frac{f_{ab}^+ - f_{ab}^-}{\omega_{ab}} \varphi(\omega, \omega_{ab})$$ (5.4)

where

$$f^\pm = (\hbar e^2)^{-1} m\omega_{ab} |(d_{ab}^x \pm id_{ab}^y)|^2 = (\hbar e^2)^{-1} m\omega_{ab} |d_{ab}^\pm|^2$$ (5.5)

are the oscillator strengths for the right and left circularly polarized radiation. These dimensionless terms are the probabilities of electric dipole transitions. The function

$$\varphi(\omega, \omega_{ab}) = \frac{\omega_{ab}\omega(\omega_{ab}^2 - \omega^2 - \Gamma_{ab}^2)}{(\omega_{ab}^2 - \omega^2 + \Gamma_{ab}^2)^2 + 4\omega^2\Gamma_{ab}^2}$$ (5.6)

characterizes the shape of the dispersion curve of the Faraday effect for the $a \to b$ transition. The formulae for the ellipticity Ψ, which characterizes the magnetic circular dichroism, differ from (5.4)–(5.6) in that the numerator in (5.6) is equal to $2\omega^3\Gamma_{ab}^2$ for the case of the ellipticity.

As is seen from formula (5.4), the Faraday effect and magnetic circular dichroism result from the difference between the polarizabilities of the quantum system in the fields of right and left circularly polarized waves. This difference, in turn, results from the differences between the frequencies of the right and left transitions (i.e., the Zeeman effect), between their oscillator strengths $f^\pm$, and between the populations of their initial levels ρ_a. This asymmetry in the polarizabilities appears in a magnetic field and in magnetically ordered materials—in contrast to the natural optical activity effect in chiral objects (molecules, crystals, etc), in which the spatial parity is broken. Although formula (5.4) corresponds to the Faraday effect, relationships (5.1) and (5.2) define ε_{xy}, and, therefore, apply to all odd magnetooptical effects.

5.2 The main interactions

Let us briefly consider the interactions that govern the magnitudes and features of the magnetooptical effects. For the sake of simplicity we shall consider only those particular examples of the interactions in which one can isolate a separate ion in the crystal and use the concepts and methods of the free-ion theory to describe it in terms of quantum mechanics. Of course, we shall note only the main points of the theory, referring our readers to well-known textbooks (see, for detail, Condon and Shortley 1935, Sobel'man 1972).

The Hamiltonian of a free ion can be presented in the form

$$\widehat{H} = \widehat{H}_0 + \widehat{H}'_{ee} + \widehat{H}_{so} \tag{5.7}$$

where H_0 is the sum of single-particle Hamiltonians with potential energies that correspond to the self-consistent central-field approximation, H'_{ee} is the so-called correlational interaction—i.e. the 'non-central' part of the electrostatic interaction between electrons, which cannot be reduced to a self-consistent field, and H_{so} is the Hamiltonian of the spin–orbit interaction.

In the zero-order approximation to H'_{ee} and H_{so}, the state of the ion is determined by its electron configuration $n_1 l_1^{N1} n_2 l_2^{N2}, \ldots$, i.e., by the distributions of the electrons among the single-electron states, which are in turn determined by the Pauli–Fermi principle and by the minimum energy of the ion.

In the practically important cases (for 3d and 4f ions), the spin–orbit interaction energy H_{sl} does not exceed the value of H'_{ee}, and, therefore, the Russell–Saunders approximation ($L–S$ coupling) may be used for the classification of the multielectron states in the ion. In this scheme the ground configuration, defined by the Hamiltonian H_0, splits into terms with particular values of the total (spin) S- and (orbital) L-momenta, and the state is described by the wave functions $|\kappa; S, L, M_S, M_L\rangle$, where κ is the configuration index. Note that in this basis of wave functions the matrix of the H'_{ee}-operator is diagonal and determines the energies of the ^{2S+1}L-terms, the degree of degeneracy of the latter being $(2L + 1)(2S + 1)$. According to Hund's rule, the ground term has maximal S, and, at the same time, maximal L.

The spin–orbit interaction, which in terms of the basis of functions of the given ^{2S+1}L-term takes the form

$$H_{so} = \lambda S \cdot L \tag{5.8}$$

where λ is the spin–orbit coupling constant, splits this term into $^{2S+1}L_J$-multiplets, each having the energy

$$\Delta E_{SLJ} = E_{SL} + \frac{\lambda}{2}[J(J + 1) - S(S + 1) - L(L + 1)]. \tag{5.9}$$

Apart from splitting terms into multiplets, H_{so} also causes SL-mixing of terms with different values of S and L. The role of the SL-mixing expands as the element number increases. For the iron-atom group, $H'_{ee} \sim 1$ eV, and $H_{so} \sim 10^{-2}$–10^{-1} eV; for the f elements, $H'_{ee} \sim 1$ eV, $H_{so} \sim 10^{-1}$–1 eV.

The interaction between an ion and its surroundings is analysed within the approximation of the crystalline-field theory. For the Hamiltonian of the crystalline field, the following parametrization is used:

$$H_{CF} = \sum_{kq} B_{kq} C_q^k. \tag{5.10}$$

The coefficients

$$C_q^k = \left(\frac{4\pi}{2k+1}\right)^{1/2} Y_{kq}$$

are the so-called tensor spherical harmonics, which are defined in the functional space of a given term or multiplet, and the B_{kq} are the parameters of the crystalline field.

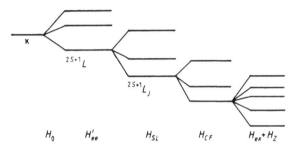

Figure 5.1. The hierarchy of interactions splitting the configuration κ. H_0 is the self-consistent central-field Hamiltonian, H'_{ee} is the correlation interaction, H_{SL} is the spin–orbit interaction, H_{CF} is the crystal field, H_{ex} is the exchange interaction, and H_Z is the Zeeman energy.

For rare-earth ions with unfilled 4f shells, $H_{CF} \ll H_{SL} \ll H'_{ee}$ (the case of a weak crystalline field); therefore, diagonalization of the Hamiltonian (5.8) in the functional space of a given J is a suitable approximation for determining the energy spectrum of the f ion. For the iron-group ions $H_{CF} \sim H'_{ee}$; therefore, in order to find the energy levels and the wave functions one should use the basis of all SL-functions of a given definite configuration. This approximation is an intermediate-crystalline-field approximation.

Finally, if the hybridization of the wave function of a particular ion (cation) with the surrounding ions (anions) is strong, one should use the ligand theory (see below).

The interaction of an ion with the external field (the Zeeman interaction) takes the following form:

$$H_Z = \mu_B \sum_i (l_i + 2s_i) \cdot H \tag{5.11}$$

where l_i and s_i are the angular and spin momenta of the ith electron respectively, and the summation is taken over all of the electrons of a particular configuration. Projection of the Zeeman Hamiltonian onto the functional space of the particular term ^{2S+1}L and the multiplet $^{2S+1}L_J$ provides other approximate forms of the Zeeman Hamiltonian:

$$H_Z = \mu_B(L + 2S) \cdot H \tag{5.12}$$

for the term and

$$H_Z = g_L \mu_B J \cdot H \tag{5.13}$$

for the multiplet, where g_L is the Landé factor of the ion, which depends on the quantum numbers S, L, J. The Hamiltonian of the Heisenberg exchange between the ions:

$$H_{ex} = \sum_{ij} I_{ij} S_i \cdot S_j \tag{5.14}$$

where I_{ij} the exchange integral of the ith and jth ions, is the spin Hamiltonian, which parametrizes the low-energy part of the energy spectrum of the electrostatic interaction in the system of ions. In materials having substantial concentrations of magnetic ions, this interaction accounts for the development of the magnetic order and, in particular, determines the Curie temperature T_C and the Néel temperature T_N.

In frequent use is the concept of an exchange field, which affects a particular ion selected from all of the other ions of the sample, and which in the simplest case is defined as

$$H_{ex} = -\frac{1}{2\mu_B} \sum_j I_{ij} \langle S_j \rangle \approx -\frac{z I \langle S_j \rangle}{2\mu_B} \tag{5.15}$$

where z is the number of the nearest neighbours, and $\langle S_j \rangle$ is the average spin of the surrounding ions.

It is important to note that the exchange field affects only the spin of the ion, and in this sense is not equivalent to a magnetic field, which also affects the orbital momentum. This difference is substantial for magnetooptics.

We have considered the main interactions which determine the energy spectrum and wave functions of an ion in a crystal (figure 5.1). In some cases it is necessary to take into consideration more subtle interactions of a relativistic nature: the dependence of the exchange integrals in (5.14) on the orbital momentum operator L, the Dzyaloshinskii–Moriya antisymmetrical exchange, and the interaction of the spin and orbital momenta that belong to different ions and configurations.

Let us consider briefly the question of the oscillator strengths of the transitions between the localized states that are described by atomic orbitals. The selection rule allows only those electric dipole transitions in which the quantum number S is unchanged whereas the quantum numbers L and J change by $0, \pm 1$, the transition from $J = 0$ to $J = 0$ being excluded. In addition, the transitions must satisfy the Lamport rule, which allows only transitions between terms of different parity. The oscillator strength of an allowed electric dipole transition for—say—the f ions is $|f| \approx 10^{-1}$–10^{-2}. Electric quadrupole and magnetic dipole transitions obey the other selection rules and take place between terms of the same parity. The oscillator strengths of those transitions are 6–7 orders of magnitude lower than the oscillator strength of an allowed electric dipole transition.

5.3 The simplest models; diamagnetic and paramagnetic mechanisms

Let us first consider the magnetooptical activity mechanisms in simple schemes for electric dipole optical transitions. We shall characterize the state of an ion by the quantum numbers J and M, which are the eigenvalues of the angular momentum operator of the ion and of its z-projection. A magnetic field removes the degeneracy of the multiplet in M, splitting each J-multiplet into $2J + 1$ levels. The selection rule for electric dipole transitions can be easily obtained by taking into account that the angular momentum and its z-projection for photons of left (right) angular polarization are $j = 1$ and $m = +1$ (-1), respectively. Therefore, the left- (right-) polarized photon will induce a transition of the ion to the state with $\Delta J = \pm 1, 0$ and $\Delta M = +1$ (-1).

Let us consider the transition from $^1S_0$ $(J = 0)$ to $^1P_1$ $(J = 1)$ (figure 5.2(a)). If we measure the spectra separately for the left- and right-polarized radiation, we obtain two identical lines of the type shown in figure 5.2(a), which would be displaced with respect to each other by the Zeeman splitting of the excited multiplet. According to (5.4), a difference between these dispersion curves causes the presence of the Faraday effect and circular dichroism (figure 5.2). The dispersion of magnetooptical effects of this type is called diamagnetic, because it is often met in studies of diamagnetic substances.

Another situation arises in the $^1P \rightarrow {}^1S$ transition (figure 5.3). The elementary transitions here are the same as those described above, but the result of the superimposition of the transitions is substantially different, because the population difference of the ground multiplet levels plays an important role. When $T = 0$ only the f_+ transition is possible. At a specific temperature the overall effect is determined by the difference in population of the ground multiplet levels—in other words, by the magnetization of the ion (or paramagnetic susceptibility in the case of a weak magnetic field). The total dispersion curves of the magnetooptical effects are shown in figures 5.3(b) and 5.3(c). They are referred to as paramagnetic. Of course, in this case there is also a diamagnetic contribution to the magnetooptical effects due to the difference in transition frequencies, but it is usually far smaller than the paramagnetic one.

We should note that the terms *paramagnetic* and *diamagnetic* are merely conventional, because—depending on the particular transition—the corresponding values of the rotation of the plane of polarization, which are determined by these mechanisms, may be either positive or negative (in contrast to the well-known properties of diamagnetic and paramagnetic susceptibilities).

Let us estimate the rotation angles of the plane of polarization in these cases. For paramagnetic dispersion, setting $\omega_+ \sim \omega_- = \omega_0$, $\Gamma_+ = \Gamma_- = \Gamma_0$,

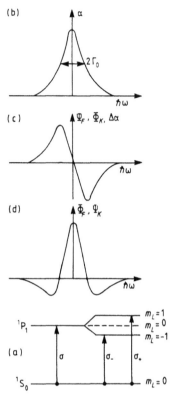

Figure 5.2. The energy level scheme (a) which gives rise to spectral dependences of the absorption (b), of the Kerr rotation and Faraday ellipticity (c), and of the Faraday rotation and Kerr ellipticity (d) of diamagnetic type.

and $f^{\pm} = f_0$ in formula (5.4), we have

$$\Phi_F = \frac{\pi N e^2}{mcn\omega_0} f_0 L\varphi(\omega, \omega_0)m(T)$$

where

$$m(T) = M(T)/M(0) = \rho_1 - \rho_{-1}$$

is the reduced average magnetic moment of the ion. Φ takes its extreme values at $\omega = \omega_0 \pm \Gamma_0$:

$$(\Phi_F)_{max} = \frac{\pi N e^2}{4mcn\Gamma_0} f_0 Lm(T).$$

When $N = 10^{22}$ ions cm^{-3}, $n = 2$, $\hbar\Gamma_0 = 1$ eV, $f_0 = 1$, $m(T) = 1$, we find $\Phi_F = 6 \times 10^6$ deg cm^{-1}. Note that, for example, in EuS Φ_F reaches 2×10^6 deg cm^{-1} (Suits 1972).

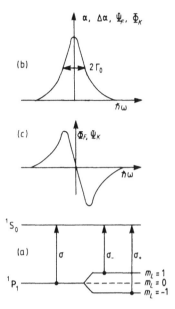

Figure 5.3. The energy level scheme (a) which gives rise to dispersions of the absorption (b), of the Kerr rotation and Faraday ellipticity (c), and of the Faraday rotation and Kerr ellipticity (d) of paramagnetic type.

For diamagnetic dispersion, substituting the following values:

$$\omega_{11} = \omega_0 - \Delta \qquad \omega_{12} = \omega_0 + \Delta \qquad f^+ = f^- = f_0 \qquad \Gamma_+ = \Gamma_- = \Gamma_0$$

into the formula (5.4), and assuming that $\Delta \ll \Gamma_0 \ll \omega_0$, we find that at $\omega \sim \omega_0$

$$\Phi_F = \frac{\pi N e^2}{4(mcn\omega_0)[(\omega_0 - \omega)^2 + \Gamma^2]}\omega \Delta f_0 L.$$

The maximum rotation, which is reached at $\omega = \omega_0$ and $m(T) = 1$, is

$$(\Phi_F)_{max} = \frac{\pi N e^2}{4mcn\Gamma_0^2}\Delta f_0 L.$$

The maximum rotation in the case of diamagnetic dispersion differs from the value for paramagnetic dispersion by a factor of $4\Delta/\Gamma_0$, which is always less than 1 for intense optical transitions. In the wings of the line, that is, where $|\omega - \omega_0| \gg (\Delta, \Gamma)$, the difference between the rotations in these two cases increases even more. In the bismuth-substituted iron garnet $Bi_{2.2}Tm_{2.5}Fe_{3.9}Ga_{1.1}O_{12}$, in the 0.37 μm spectral band at $T = 295$ K, $\Phi_F = 1.2 \times 10^5$ deg cm^{-1}; for the $Bi_{2.2}Gd_{0.8}Fe_5O_{12}$ composition one should expect $\Phi_F = 7 \times 10^5$ deg cm^{-1}.

5.4   Transitions between the groups of levels (terms, multiplets, etc)

We have considered extremely simplified schemes. In practice, the energy
spectrum and the wave functions of an ion are substantially more complicated.
However, the actual characteristic mechanisms of magnetooptical activity are
similar to those discussed above. This is because the allowed ionic electric
dipole transitions often lie in the ultra-violet spectral region, although the visible
and infra-red spectral bands are usually of practical importance. Therefore,
we actually deal with the 'long-wavelength wings' of the allowed transitions
of the type that we considered above. The condition $|\Delta\omega_{ab}| \ll |\omega_0 - \omega|$,
where ω_0 is the characteristic frequency of the allowed transition group, ω is the
radiation frequency, and $\Delta\omega_{ab}$ is the difference between the frequencies of the
elementary transitions, usually holds true. This condition is used very effectively
in calculations of the polarizability (5.1).

The following formula is often used to describe the rotation of the plane of
polarization caused by the transition between the ground and excited multiplets
(Buckingham and Stephens 1966):

$$\Phi_F = -\frac{4\pi L N \omega^2}{\hbar c n} \left\{ \frac{2\omega_0[(\omega_0^2 - \omega^2)^2 - \omega_0^2\Gamma_0^2]}{\hbar[(\omega_0^2 - \omega^2)^2 + \omega_0^2\Gamma_0^2]^2} A \right.$$
$$\left. + (B + C/kT)\frac{\omega_0^2 - \omega^2}{(\omega_0^2 - \omega^2)^2 + \omega_0^2\Gamma_0^2} \right\} H \tag{5.16}$$

where ω_0 and Γ_0 are the frequency and the half-width of the absorption band for
the $L_0 S_0 J \rightarrow LSJ$ transition, ω is the radiation frequency, H is the magnetic
field, and L is the Lorentz–Lorenz factor (see (5.2)). The constants A, B, C
are defined by the matrix elements of the electric dipole moment; they do not
depend on the temperature, magnetic field, or frequency.

The analogous expression for the magnetic circular dichroism (Buckingham
and Stephens 1966) is

$$\Psi_F = -\frac{4\pi L N \omega^3 \Gamma_0}{\hbar c n[(\omega_0^2 - \omega^2)^2 + \omega_0^2\Gamma_0^2]}$$
$$\times \left\{ \frac{4\omega_0(\omega_0^2 - \omega^2)A}{\hbar[(\omega_0^2 - \omega^2)^2 + \omega_0^2\Gamma_0^2]^2} + (B + C/kT) \right\} H. \tag{5.17}$$

Expressions (5.16) and (5.17) are derived on the assumption that the Zeeman
energy is much less than the average thermal energy and than the distance
between the centres of gravity of the $L_0 S_0 J$- and LSJ-multiplets involved in
the electric dipole transition.

Now we are going to explain how formulae (5.16) and (5.17) can be derived.
The basic idea is used for even more general situations than the above-discussed
transitions between the multiplets. So, we shall explain it without mentioning
any particular type of transition.

Let us consider the allowed electric dipole transitions from a group of levels that is adjacent to the ground state ($E_a = E_0 \pm \Delta E_a$) to a group of levels having an average energy E_1 ($E_b = E_1 \pm \Delta E_b$) (figure 5.4). Here E_0 and E_1 are the energies of the ground and exited multiplets (or terms), and ΔE_a and ΔE_b are their splittings resulting from crystalline, magnetic, and exchange fields and from spin–orbital interaction (in the case of the terms). For example, the levels E_0 and E_1 may belong to different configurations ($3d^n$ and $3d^{n-1}4p^1$ for d ions and $4f^n$, $4f^{n-1}5d^1$ for rare-earth ions). If $|\Delta\omega_{ab}| \ll |\omega - \omega_0|$ and $\Gamma_{ab} \ll |\omega - \omega_0|$, where $\Delta\omega_{ab} = \Delta(E_b - E_a)/\hbar$, the function $\varphi(\omega, \omega_0)$ can be expressed as

$$\varphi(\omega, \omega_0) \approx \varphi_1(\omega, \omega_0) + \varphi_2(\omega, \omega_0)\,\Delta\omega_{ab}\,\omega_0^{-1} + O((\omega_0^{-1}\,\Delta\omega_{ab})^2)$$

$$(5.18)$$

where

$$\varphi_1(\omega, \omega_0) = (\omega_0^2 - \omega^2)^{-1}\omega^2$$
$$\varphi_2(\omega, \omega_0) = -(\omega_0^2 - \omega^2)^{-2}2\omega^2\omega_0^2.$$

Substituting (5.18) into (5.4) yields

$$\Phi_F = A\varphi_2(\omega, \omega_0) + [B(h) + Cm(H, T)]\varphi_1(\omega, \omega_0) \qquad (5.19)$$

where the first term, $A = A_1(H) + A_2m(H, T)$, is the diamagnetic contribution to the Faraday effect, the third term, Cm (where m is the reduced magnetic moment of the ion), is the paramagnetic contribution, and the second term, B (the mixing term), takes into account the admixing the other multiplets with the ground and excited multiplets under the influence of the magnetic and exchange fields. Admixture is depicted in figure 5.4 as a wavy line.

The coefficients A, B, and C were first introduced by Serber (1932) and now are widely used in the literature on magnetooptics. For the ions with $L \neq 0$ in the ground state, the paramagnetic term C is the largest in (5.19). For example, for the rare-earth ions the A- and B-coefficients are estimated as follows:

$$\left|\frac{A_1}{C}\right| \sim \frac{\mu_B H_{eff}}{(\hbar\omega_0)} \qquad \left|\frac{A_2}{C}\right| \sim \frac{\Delta_{so}}{(\hbar\omega_0)} \qquad \left|\frac{B}{C}\right| \sim \frac{\mu_B H_{eff}}{\Delta_{so}} \qquad (5.20)$$

where Δ_{so} is the spin–orbit interaction energy, and H_{eff} is the external or exchange field. For the rare-earth ions $\hbar\omega_0 \approx (1\text{–}10) \times 10^4$ cm^{-1}, $\Delta_{so} \approx 10^3\text{–}10^4$ cm^{-1}, and $\mu_B H_{eff} \approx 1\text{–}10$ cm^{-1} at $H_{eff} \approx 10^3\text{–}10^4$ Oe; thus

$$|A_1/C| \sim 10^{-4} \qquad |A_2/C| \sim 10^{-2} \qquad |B/C| \sim 10^{-2}\text{–}10^{-3}. \qquad (5.21)$$

The main feature of the term B is that it is independent of temperature. In its origin and physical sense, this contribution to the Faraday effect is fully

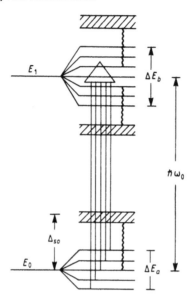

Figure 5.4. A schematic drawing of d- or f-ion structure in a crystalline field. The solid arrows show the allowed electric dipole transitions.

analogous to the well-known Van Vleck temperature-independent contribution to the magnetization of paramagnetics (the Van Vleck paramagnetism). The contribution to the Faraday effect, which is described by the term B, is important in studying the magnetooptical phenomena in strong magnetic fields.

It is easy to see that the structures and physical contents of formulae (5.16) and (5.19) are quite close. Of course, the terms A, B, and C are somewhat different in these formulae, but, if necessary, their mutual correspondence can be easily established. Note also the slight difference between the frequency dependencies of $\Phi_F(\omega)$ in the two cases. They coincide when $\Gamma_0 \to 0$.

It is convenient to determine the magnetooptical parameters A, B, and C using the moment method (Stephens 1970, Henry *et al* 1965). This method is based on the utilization of magnetic circular dichroism spectra.

5.5 S ions; the effective 'spin–orbit field'

Let us consider in more detail ions with orbital singlet ground states. This class of ions embraces those with the orbital momentum $L = 0$ (for example Fe^{3+}, Mn^{2+}, Gd^{3+}, and Eu^{3+}) as well as those with 'frozen' orbital momentum (Cr^{3+}, Ni^{2+} in octahedral complexes, and others). Let us consider the main magnetooptical features of the ions within a simplified model of the transitions between the ^2S and ^2P terms (figure 5.5). These terms split under the influence of spin–orbit interaction and a magnetic field. The crystalline field in the first

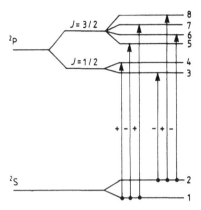

Figure 5.5. A schematic diagram of the electric dipole transitions between the ^{2}S and ^{2}P terms, which illustrates the formation mechanism of the Faraday effect for the ions with zero orbital momentum in the ground state.

approximation in $|\Delta\omega_{ab}|/\omega_0$ does not contribute to the Faraday effect; therefore it can be disregarded here. The energy levels E_{1-8} are

$$E_{1,2} = E_0 + 2\mu_B H m \qquad\qquad m = -1/2, +1/2 \qquad\qquad (5.22)$$

$$E_{3,4} = E_1 - \lambda_{SL} + (2/3)\mu_B H m \qquad m = -1/2, +1/2 \qquad\qquad (5.23)$$

$$E_{5-8} = E_1 + \lambda_{SL}/2 + (4/3)\mu_B H m \qquad m = -3/2, -1/2, +1/2, +3/2$$
$$\qquad\qquad\qquad\qquad\qquad\qquad\qquad\qquad\qquad\qquad\qquad (5.24)$$

where λ_{SL} is the spin–orbit coupling constant of the excited term. The non-zero matrix elements of the dipole moments $d_{ab}^{\pm}$ are

$$|d_{14}^+|^2 = |d_{23}^-|^2 = (2/9)d^2$$
$$|d_{15}^-|^2 = |d_{28}^+|^2 = (1/3)d^2 \qquad\qquad (5.25)$$
$$|d_{17}^+|^2 = |d_{26}^-|^2 = (1/9)d^2$$

where $d = $ constant.

First we assume that the temperature $T = 0$. In this case only the low level E_1 is populated, and only three transitions—1 to 4, 1 to 5, and 1 to 7—should be taken into account. If we disregard the differences between the frequencies of these transitions, the Faraday effect (and circular dichroism) are nullified, because according to (5.25) $|d_{14}^+|^2 - |d_{15}^-|^2 + |d_{17}^+|^2 = 0$. Thus, Φ_F is non-zero only because of the difference between the frequencies of the elementary transitions (as in the above-discussed case of diamagnetic dispersion). At a finite temperature ($T \neq 0$) the Faraday effect is contributed by the transitions 2–3, 2–8, and 2–6. Their total contribution is equal in value (but opposite in sign) to that of the first-level transitions (to an accuracy that is the same as that of the population of the ρ_2-level). This is the same situation as in the case of paramagnetic dispersion—that is, here the angle Φ_F, which is determined by the difference between the total transitions from the E_1- and E_2-levels, is also proportional to the difference $\rho_1 - \rho_2$, i.e., to the average magnetic moment of the ion. Substituting (5.24) and (5.25) into (5.4) yields

$$\Phi_F = -\frac{4\pi N e^2 L f_0}{3mcn\omega}\varphi_2(\omega)(\hbar\omega_0)^{-1}(-\mu_B H_Z + \lambda_{SL}\langle S_z\rangle) \qquad (5.26)$$

where $\langle S_z\rangle = (1/2)(\rho_1 - \rho_2)$ is the averaged spin in the ground state of the ion, and f_0 is the oscillator strength of the $^2S \rightarrow {}^2P$ transition.

Returning to the S ions Fe^{3+}, Mn^{2+}, Gd^{3+}, and Eu^{2+}, we note that although their spin numbers in the ground state are 5/2 and 7/2, and the number of elementary transitions that contribute to the Faraday effect is greater than in the case of the $^2S \rightarrow {}^2P$ transition, formula (5.26) is valid for these ions too, but with substitution of the appropriate magnetic moment of the ion, or the value $\langle S_z\rangle$. Formula (5.26) can also be used for d ions with 'frozen' orbital momentum. It follows from (5.26) that the effect of spin–orbit interaction can be considered here in the terms of the effective axial field

$$H_{SL} = -\mu_B^{-1}\lambda_{LS}\langle S\rangle \qquad (5.27)$$

where λ_{SL} is the spin–orbit coupling constant of the excited term ($H_{SL} \approx 10^6$ Oe for d ions). This axial field is acting only on the orbital momentum of the ion. Since

$$\langle S\rangle = (2\mu_B)^{-1}M_S = (2\mu_B)^{-1}\chi H$$

where χ is magnetic susceptibility (per ion), the effect of the spin–orbit interaction may be considered here as an amplification of the external field by the factor

$$\eta_{SL} = \frac{\lambda_{SL}\chi}{2\mu_B^2} \approx \frac{\lambda_{SL}S(S+1)}{3kT}. \qquad (5.28)$$

The paramagnetic Faraday effect (commonly, it is greater than the diamagnetic one) appears only when the strength of the spin–orbit interaction exceeds that of the interaction with the crystalline environment, because only in this case is the average orbital momentum non-zero in the ground state. Thus, in order to construct materials with larger Faraday effects, one should use ions with strong spin–orbit coupling (for example, Bi^{3+}, Pb^{2+}, and U^{3+}).

Once more it should be noted that (5.19) and (5.26) are derived for the allowed electric dipole transition between the ground and excited terms of d or f ions. But in general there may be many allowed transitions like this. If that is the case, one should sum these expressions over all allowed transitions. The summation effects will modify only the frequency dispersion, with the structure of the formulae obtained—i.e., the dependence on m, H, and T—remaining the same as that of (5.19) and (5.26).

5.6   Magnetic dipole transitions, and the gyromagnetic Faraday effect

In a number of cases one should take into account the magnetic dipole transitions, i.e., the gyromagnetic contribution to the Faraday effect (Krinchik and Chetkin 1960). It can be especially important in the infra-red spectral region—for instance, in the transparency band of iron garnets.

A characteristic feature of the gyromagnetic Faraday effect in the IR and near-visible spectral bands in garnets is that the rotation of the plane of polarization does not depend in practice on the radiation frequency. As has been shown by Krinchik and Chetkin (1961), the reason for this is that the gyromagnetic effect is determined by the off-diagonal components of the magnetic permeability tensor $\chi_m^{\pm}$, i.e., it is caused by the precession of the magnetization vector under the influence of the magnetic field of the electromagnetic wave:

$$\Phi_{Fm} = c^{-1}\pi\omega n(\chi_m^- - \chi_m^+).$$ (5.29)

Substituting the values of $\chi_m^{\pm}$, which are well known from ferromagnetic resonance theory, into (5.29) one obtains

$$\Phi_{Fm} = \frac{2\pi n\gamma M}{c} = \frac{g\pi neM}{mc^2}$$ (5.30)

where M is magnetization and g is the g-factor of the magnetic ions.

The Faraday effect that is observed in the IR region takes place in the short-wavelength wing of the magnetic resonance band, where $\chi_m^{\pm} \sim \omega^{-1}$; this is why Φ_{Fm} is independent of the frequency. In some cases the Faraday effect is contributed by the exchange resonance.

The size of the gyromagnetic Faraday effect varies within the 40–70 deg cm^{-1} region for different iron garnets. In the region where $\lambda \leq 1$ μm, the gyroelectric contribution to the Faraday effect in garnets overlaps the frequency-independent part of the effect.

5.7   The polarizability of f ions

The magnetooptical properties of rare–earth materials in the visible and UV spectral bands are mainly governed by the allowed electric dipole transitions[1] $4f^N \rightarrow 4f^{N-1}5d$. The energy of these transitions in the free rare-earth ions R^{3+} is $\approx 10^5$ cm^{-1}. The electron structure of the $4f^{N-1}5d$ configuration in a crystal is determined by several interactions, which are comparable to each other (see section 5.2). In particular, for the $4f^N$ and $4f^{N-1}5d$ configurations the correlation interaction, i.e., the part of the Coulomb interaction that is not

---

[1] We should note here that a number of authors (see, e.g., Judd and Pooler 1982) have shown that in certain cases the transitions in configurations of the $4f^{N-1}5g$ and $nd^9 4f^{n+1}$ type (where $n = 3, 4$) should be taken into account.

reduced to a self-consistent field, and the spin–orbit interaction are comparable, which leads to some difficulties in the calculations of the spectrum and wave functions of the ion.

Energy level schemes for the $4f^N$ and $4f^{N-1}5d$ configurations were calculated and compared with observed absorption and fluorescence spectra for rare-earth compounds, and are presented by Dieke (1968). Moreover, the centre points of the energy spectrum for the $4f^n$ and $4f^{n-1}5d$ configurations of trivalent ions were calculated and plotted versus the atomic number of the element through the rare-earth series, and these are presented by Dieke and Crosswhite (1963). The dependency shows that when the nuclear charge grows by one, the difference between the binding energies of the 4f and 5d electrons, i.e., the difference between the energies of the ground and excited configurations, increases[2] by 4000–5000 cm^{-1}. The characteristic values of the $4f^N \rightarrow 4f^{N-1}5d$ transition wavelengths vary from ≈ 300 nm for Ce^{3+} to ≈ 140 nm for Yb^{3+} for CaF_2, and from ≈ 450 nm to ≈ 200 nm for YAG. Note that Ce^{3+}, Tb^{3+}, Pr^{3+}, and Nd^{3+} ions exhibit relatively small energy differences between the ground $4f^N$ and excited $4f^{N-1}5d$ configurations[3].

The wave functions of the ground configuration are usually described in terms of the Russell–Saunders approximation. As a rule, the wave functions of the ground multiplet are 'pure'; this is not true for the excited multiplets (Dieke 1968). The wave functions of the term for the $4f^{N-1}5d$ configuration can be built up in the framework of a genealogical scheme (Sobel'man 1972) as a superposition of the products of the wave functions of the $4f^{N-1}$ shell and the wave functions of a single 5d electron, taking into account the antisymmetry principle. In order to calculate the matrix elements of the electric dipole moments, one also often uses the genealogical scheme for the ground-state wave function (Kolmakova *et al* 1990).

Consider the contribution to the polarizability tensor α_{ij} of an f ion arising from transitions from the states of the $S_0 L_0$-term of the ground $4f^N$ configuration to the excited $S_0 L\nu$-terms of the $4f^{N-1}5d$ configuration (ν is an additional index which characterizes the term with a particular value of L in the $4f^{N-1}5d$ configuration). The difference between the ground energy level (E_g) and excited energy level (E_e) can be written as

$$E_e - E_g = \hbar\omega_{L\nu} + (\Delta E_b - \Delta E_a)$$

where $\hbar\omega_{L\nu}$ is the distance between the 'centres of gravity' of the b- and a-terms, and ΔE_b and ΔE_a are the values of the term splitting under the effect

[2] This conclusion has been confirmed in many experiments, in particular via absorption spectra of the rare-earth-doped CaF_2 and YAG crystals in the 130–300 nm region (Loh 1966, Bagdasarov *et al* 1982, Meil'man *et al* 1984, Gorban' *et al* 1985).

[3] The main feature of 4f–5d spectra of many rare-earth ions is that one can distinguish two absorption bands. The existence of two bands is probably what causes the strong splitting of the 5d state ($\sim 10^4$ cm^{-1}) into the t_{2g} and e_g states in an ambient crystalline field. For example, the spectrum of the Tb^{3+} ion in YAG includes two intense absorption bands with $\lambda = 229$ nm and $\lambda = 275$ nm and oscillator strengths $f = 0.002$ and $f = 0.0053$, respectively (Valiev *et al* 1985a, b).

of spin–orbit and exchange interactions, and crystalline and magnetic fields.
The main small parameter of this theory is

$$\frac{|\Delta E_b - \Delta E_a|}{\hbar|\omega - \omega_{Lv}|} \ll 1 \qquad (5.31)$$

where ω is the frequency of the electromagnetic field. Hence expression (5.1) can be expanded in terms of this small parameter. As a result, in the lowest approximation in terms of this small parameter one obtains (Zvezdin *et al* 1986)

$$\alpha_{ij} = a_0\delta_{ij} + ia_1 e_{ijk}\langle L_k\rangle + a_2\langle Q_{ij}\rangle \qquad (5.32)$$

where $\langle L_k\rangle$ and $\langle Q_{ik}\rangle$ are the components of the orbital momentum L and the quadrupolar moment $\hat{Q}$ averaged over the ground state of the ion; summation over repeated indices is assumed. The coefficients a_0, a_1, a_2 are defined in terms of the matrix elements of the optical transition considered:

$$a_n = -2\hbar^{-1}\sum \frac{\langle L_0\|n_i\|L_0v\rangle^2}{(2L_0 + 1)(\omega^2 - \omega_{Lv}^2)} P_{nl}. \qquad (5.33)$$

Here

$$P_{0L} = (1/3)\omega_{Lv} \qquad P_{1L} = -2\omega C_1(L) \qquad P_{2L} = \omega_{Lv}C_2(L)$$

$$C_1(L_0 + 1) = (1/2)(L_0 + 1)^{-1}$$
$$C_1(L_0) = -[2L_0(L_0 + 1)]^{-1}$$
$$C_1(L_0 - 1) = -(1/2)L_0^{-1}$$
$$C_2(L_0 + 1) = [(L_0 + 1)(2L_0 + 3)]^{-1}$$
$$C_2(L_0) = [L_0(L_0 + 1)]^{-1}$$
$$C_2(L_0 - 1) = -[L_0(2L_0 - 1)]^{-1}.$$

The first term in (5.32) is the isotropic contribution to the polarizability of the ion, the second term is the gyrotropic contribution, and the third term represents the even magnetooptical effects.

Projecting the orbital momentum $\hat{L}$ onto the ground multiplet of the ion, and taking into account the admixing of the excited multiplet wave functions with the ground-multiplet wave functions due to the Zeeman interaction (jj-mixing) one can obtain

$$\langle L\rangle = (2 - g_L)\langle J\rangle + \langle\Delta L\rangle_{mix} = \mu_B^{-1}\left(\frac{2 - g_L}{g_L}M_0 + \chi_{VV}H\right) \qquad (5.34)$$

where M_0 is the average magnetic moment of the f ion, which is calculated or measured without taking into account the jj-mixing, g_L is the Landé factor of the ground multiplet, and χ_{VV} is the contribution arising from the jj-mixing to the permeability of the ion, which is of the same nature as the van Vleck permeability (related to one ion). This latter contribution can be estimated as $\chi_{VV} \sim \mu_B^2/\Delta E$, where ΔE is the characteristic value of the energy that separates the ground multiplet from the excited ones. For example, for the Tb^{3+} ion, $g_L = 3/2$, and $\Delta E \approx 2000$ cm^{-1}.

It can be seen that the structure and physical sense of the second term in (5.32), with $\langle L \rangle$ defined by (5.34), are similar to those of the formula (5.16), with the exception of the diamagnetic contribution in (5.19), which was omitted in deriving the formula (5.32). For rare-earth ions, except Gd^{3+} and Eu^{3+}, the diamagnetic contribution is usually small.

It should be noted that the gyrotropic contribution to the polarizability tensor of an ion is determined by the average value of the orbital momentum, $\langle L \rangle$, of the ion, and that the last term in (5.32), which represents the even magnetooptical effects, is proportional to the average quadrupole moment, $\langle \widehat{Q} \rangle$, of the ion. This leads to the interesting conclusion that the odd effects measure the average orbital momentum of an ion (if it is non-zero, and allowing a correction for small diamagnetic and magnetic dipole contributions), whereas using the even effects one can measure the quadrupole moments of ions (Vedernikov *et al* 1987, Kolmakova *et al* 1990). This result is applicable not only to the f ions, but also to the d ions, to clusters of MeO_n type where Me is a transition metal, to molecular objects, etc. In general, this approach can be used if both the ground and the excited electronic states belong to localized atomic (ionic) orbitals. It is very important that in the case of d ions the orbital momentum is usually 'frozen', i.e., in the first approximation $\langle L \rangle = 0$. It is partially 'unfrozen' under the influence of spin–orbital interaction, but in this case the diamagnetic contribution can be of the same order as or exceed the 'unfrozen' contribution from $\langle L \rangle$.

5.8   The Faraday effect in rare-earth garnets

In order to illustrate the above-discussed theoretical statements, let us take a look at the experimental data on the Faraday effect in rare-earth garnets.

The rare-earth garnets—aluminates and gallates—are excellent objects for use in studying rare-earth-induced magnetooptical activity, because the diamagnetic basis (matrix) of the materials (for example, of $Y_3Al_5O_{12}$) is transparent over a wide spectral range, including the visible and ultra-violet spectral bands. In addition, they exhibit well-determined and stable magnetic properties, which provide the basis for a reliable comparison of the magnetic and magnetooptical measurements.

Valiev *et al* (1983, 1985a, b, 1987b) have studied the Faraday effect and the magnetic susceptibility of the rare-earth garnets over wide wavelength and temperature ranges. The results of their measurements were processed using

(5.19) to obtain the Verdet constant, which can be conveniently written as

$$V = -\frac{\omega^2}{\omega_0^2 - \omega^2}\left[\frac{2\omega_0 A}{\hbar(\omega_0^2 - \omega^2)} + B + C\chi\right] + V_{gm} \qquad (5.35)$$

where χ is the paramagnetic susceptibility of a rare-earth subsystem, ω_0 is the effective transition frequency, and V_{gm} is the gyromagnetic contribution to the Verdet constant. The diamagnetic contribution (the term involving A) is small for all of the garnets studied (except the gadolinium garnets). This agrees well with the estimates considered above (see (5.20)).

Throughout the wavelength interval studied, 0.4–1 μm, the spectral dependence $V(\omega)$ of the garnets that contain Nd^{3+}, Tb^{3+}, Dy^{3+}, Ho^{3+}, Er^{3+}, Tm^{3+}, and Yb^{3+} is well described by (5.35) with $A = 0$. The interdependence between V and χ also complies with (5.35).

Figure 5.6 illustrates the interdependence of the Verdet constant and the susceptibility for the terbium aluminium garnet.

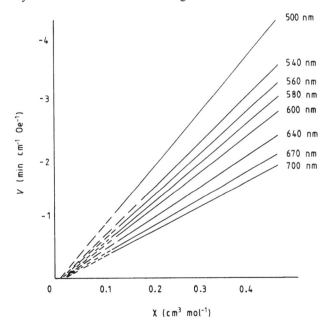

Figure 5.6. The dependence $V(\chi)$ for TbAlG (Valiev *et al* 1983).

Dependencies like this provide the possibility of explicitly defining the mixing term B using the coordinates of the crossing point on the x-axis of figure 5.6 (Valiev *et al* 1983).

The main contribution to the Verdet constant of garnets arises from the temperature-dependent 'paramagnetic' mechanism. For the garnets that contain Nd^{3+}, Tb^{3+}, and Dy^{3+} ions, the jj-mixing contribution is ~ 5–10% of the paramagnetic contribution at room temperature. For other rare-earth ions, except Eu^{3+} and, possibly, Sm^{3+}, the mixing mechanism does not contribute substantially to the Faraday effect.

The Gd^{3+}, Eu^{3+}, and Sm^{3+} ions stand out. The Gd^{3+} ion has a ground state of S type ($L = 0, S = 7/2$). As has been mentioned earlier, both the paramagnetic contribution and the mixing term for this state are zero. Therefore, the Faraday effect for the ion is relatively small, and exhibits diamagnetic dispersion. However, according to (5.26), its temperature and field dependences, which are mainly determined by the magnetization, have the same character as for the ions with $L \neq 0$. It should be noted that there is a strong field dependence of the Faraday effect in gadolinium garnets (figure 5.7). First, Φ_F increases with the field, and then, in the field region where the magnetization tends to saturate (curve 2), it begins to decrease. In addition, the Faraday effect at this wavelength changes its sign at higher temperature: it is negative at 4.2 K and positive at room temperature. Consequently, the Faraday effect cannot under these conditions be considered as being proportional to the magnetization. These peculiarities are naturally described by (5.35), which takes into account two terms contributing to the Faraday effect having opposite signs. One of these terms is proportional to the magnetization, while the other is proportional to the magnetic field.

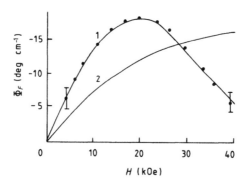

Figure 5.7. The field dependence of the Faraday effect (curve 1) and the magnetization (curve 2) for $Gd_3Ga_5O_{12}$ ($\lambda = 0.47$ μm, $T = 4.2$ K) (Zvezdin *et al* 1983).

The ground state of the Eu^{3+} ion is a singlet: $^7F_0$ with $J = 0$. The first excited state (the $^7F_1$ triplet) is relatively close to the ground state, being only $\Delta_1 \approx 350$ cm^{-1} apart from it. The following multiplets, $^7F_2$ and $^7F_3$, are separated from the ground state by $\Delta_2 = 925$ cm^{-1} and $\Delta_3 = 2 \times 10^3$ cm^{-1}, respectively.

The singlet character of the ground state of the Eu^{3+} ion substantially affects the magnetic properties of the europium-substituted compounds. Using (5.32), it is easy to show that the expression for the Verdet constant of the Eu^{3+} ion, which is determined by all of the allowed electric dipole transitions from the ground state, $^7F_0$, and both thermally populated, $^7F_1$ and $^7F_2$, multiplets to the terms ^7G and ^7F, of the $4f^2 5d$ configuration, may be written as

$$V \approx -\chi' \sum_{L_0 S_0 \to LS} C(LS) \frac{\omega^2}{\omega_{0LS}^2 - \omega^2} \qquad (5.36)$$

where the magnetic-field-induced jj-mixing of the $^7F_0$, $^7F_1$, and $^7F_2$ multiplets in the magnetic field is taken into account, and $\chi' = \chi - \frac{4}{3}(\chi_1 - \chi_2)$, where χ is the magnetic susceptibility of Eu^{3+}, and χ_1 and χ_2 are the contributions from the thermal population of the $^7F_1$ and $^7F_2$ levels to the magnetic susceptibility (Valiev and Popov 1985, Valiev *et al* 1987b). Small frequency-independent additions from the magnetic dipole transitions are omitted here.

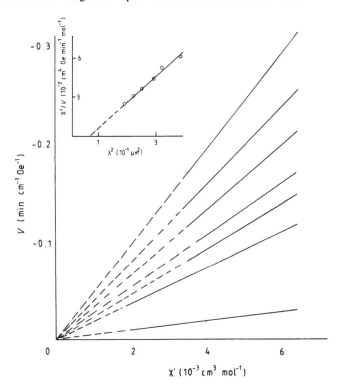

Figure 5.8. The dependences $V(\chi')$ for EuGaG at different wavelengths. In the upper inset, the dependence $(\chi'/V)(\lambda^2)$ is shown (Valiev *et al* 1987b).

Figure 5.8 (Valiev *et al* 1987b) illustrates the validity of the relation $V \sim \chi'$ for europium gallium garnet. The dependence of the ratio χ'/V on the square of the wavelength, λ^2, which is explained well with (5.36), is also presented in an inset.

Thus, while for most of the rare earths the main contribution to the gyrotropy arises from the paramagnetic mechanism, and the mixing term provides only a small addition, for Eu^{3+} the mixing mechanism is dominant. It should be noted that both the magnetic susceptibility and the magnetooptical activity are determined here by the jj-mixing of the multiplets. However, the magnetic susceptibility is determined by the matrix elements of the operator $L + 2S$. The Verdet constant is determined by the matrix elements of the L-operator; this is the reason for the different behaviours of the values.

5.9   The nature of the gyrotropic properties of the ferrites and other oxide compounds

The theory of the magnetooptical properties of ferrites and other transition-metal-based oxide compounds is far from being complete. There is no clear understanding of the nature of the electron transitions that would explain the main magnetooptical effects, and could be used to identify particular transitions in magnetooptical spectra.

Clogston (1959, 1960) explained the large magnetooptical effects and high absorption in the visible and ultra-violet spectral bands for iron garnets in terms of allowed electric dipole charge-transfer transitions from the oxygen ions to the d orbitals of the Fe^{3+} ion in octahedral FeO_6^{9-} or in tetrahedral FeO_4^{5-}, and in terms of transitions of the $3d^5 \rightarrow 3d^4 4p^6$ type in the Fe^{3+} ions. Later these ideas were widely used and developed (Kahn *et al* 1969, Suits 1972, Wittekoek *et al* 1975, Zenkov 1990, Moskvin *et al* 1993).

The octahedral FeO_6^{9-} and tetrahedral FeO_4^{5-} clusters are frequently used to describe the charge-transfer transitions in the ferrites.

The single-electron wave functions in the complex are chosen as molecular orbitals—linear combinations of atomic orbitals (the MO-LCAO method) of various symmetries. The configuration of the ground state consists of: all of the occupied molecular orbitals, which mainly exhibit ionic character (O_{1S-2P}); filled 'cation' molecular orbitals Fe_{1S-3P}; and partially occupied 3d-type orbitals (t_{2g} and e_g). All of them, except the 3d t_{2g} and 3d e_g orbitals, are mainly the ligand's ones. The t_{2g} and e_g orbitals consist mainly of the d atomic orbitals of the d ion:

$$e_g^1 = \langle d, 0| \qquad e_g^2 = \frac{(\langle d, +2| + \langle d, -2|)}{\sqrt{2}}$$

$$t_{2g}^1 = -\langle d, +1| \qquad t_{2g}^2 = -\langle d, -1|$$

$$t_{2g}^3 = \frac{(\langle d, +2| - \langle d, -2|)}{\sqrt{2}}.$$

The distribution of electrons among various molecular orbitals defines the multielectron configuration of the complex. The filling of the single-electron states is chosen in accordance with Hund's rule, i.e., according to the requirement of maximal spin. The ground term—$^6A_{1g}$—has spin $S = 5/2$, and is a combination of the $\kappa_0 3d^5$ orbitals of the filled ligands and the central ion, where κ_0 is an index that corresponds mainly to ligands.

The role of the orbital momentum of the electrons in the complex is played by the quasi-momentum, which is defined by the indices of the irreducible representation of the corresponding symmetry group. The crystalline term $^6A_{1g}$ can be attributed to the effective orbital momentum with $L = 0$.

The excited states are constructed in the same manner. The excited configuration is conventionally designated as $\tilde{\gamma}_{2P} 3d_0^6$, where $\tilde{\gamma}_{2P}$ represents a 'hole' in the basis of the anion molecular orbital. The transition between the ground and excited configurations can in this case be written as the charge-transfer transition $\tilde{\gamma}_{2P} 3d_0^6 - \kappa_0 3d^5$. The excited-configuration terms are designated as $^{2S'+1}\Gamma'$, where S' is the spin of the excited configuration. Among the excited terms of the $\tilde{\gamma}_{2P} 3d_0^6$ configuration, the $^6T_{1u}$ term and the spin $S = 5/2$ are of considerable importance for magnetooptics.

The strength of the $^6A_{1g} - ^6T_{1u}$ transition is given by the matrix elements $\langle \tilde{\gamma}_{2P} | l r | e_g \rangle$, and $\langle \tilde{\gamma}_{2P} | l r | t_{2g} \rangle$. The oscillator strengths of these transitions are $f \sim 0.01$–0.1, which is several orders higher than the oscillator strengths of d–d transitions.

The theoretical analysis of the polarizability of the FeO_6^{9-} complex (and similar complexes), which is based on the Kramers–Heisenberg formula (5.1), is quite similar to the one presented above for the f ions, because in both cases we deal with the localized states in crystals. Like in section 5.4, here one can separate out the contributions to the gyration vectors from several mechanisms:

(i)   the paramagnetic effect;
(ii) the diamagnetic effect, which results from the splitting of the ground and excited terms under the influence of various interactions; and
(iii) the mixing term, including the terms having different quasi-momenta, and, consequently, contributions from 'forbidden transitions'.

The paramagnetic contribution to the gyration vector, like that for the f ions, is proportional to the average value of the orbital momentum $\langle L \rangle$; therefore, for the FeO_6^{9-} complex it is equal to zero, because the ground state here is an orbital singlet. The same is true for the complexes with Mn^{2+}, Cr^{3+}, and Ni^{2+} ions. The main mechanism here is the diamagnetic one. In the first-order perturbation theory only the interactions that include the odd-power terms in the orbital momentum and that provide orbital splitting or mixing of the excited terms of T_{1u} type are substantial. All of the terms of this type can be described as an interaction with an axial field that affects the electron orbital momenta.

We have now encountered a situation that we have already considered in section 5.5, with the one difference that here we deal with the crystalline rather than the atomic terms. Both the g-factor and the effective spin–orbit interaction constants for the orbital singlet $^6A_{1g}$ and the orbital triplet $^6T_{1u}$ are defined by their particular wave functions. The spin–orbit interaction (5.8) splits the $^6T_{1u}$ triplet into three multiplets with effective momenta $J = 3/2, 5/2, 7/2$. The constant of the spin–orbit coupling λ is a linear combination of the λ_{2p} and λ_{3d} constants for the 2p and 3d electrons, respectively. Since λ_{2p} is much lower than λ_{3d}, the 3d shell is the main contributor to λ.

In this case the Faraday effect can be determined using (5.26), in which λ is the value of the $^6T_{1u}$ term and the effective orbital momentum l_{eff} is the quantum-mechanical average $\langle T_{1u}| \sum_i l_i \cdot h|T_{1u}\rangle$, where the summation is performed over all of the electrons of the excited configuration, and h is the unit vector along the direction of the magnetic field. According to Zenkov (1990), $l_{eff} = -3/4$, and $\lambda \approx 0.1\lambda_{3d}$ for the $^6T_{1u}$ term for the FeO_6^{9-} complexes. The isotropic exchange interaction of the central Fe^{3+} ion with the ambient magnetic ions in Heisenberg form contributes only to the scalar part of the polarizability of the FeO_6^{9-} complex:

$$\Delta\alpha_0 \sim D \sum J_k \langle S \cdot S_k \rangle$$

where $\langle S \cdot S_k \rangle$ is the thermodynamic average which characterizes the near magnetic order. The presence of this contribution to the isotropic polarizability in magnetic materials can cause abnormal behaviour of optical parameters near the magnetic ordering temperature. Note also the substantial effects due to light scattering in antiferromagnets that are caused by this mechanism (Borovik-Romanov *et al* 1977).

The exchange relativistic interaction of Dzyaloshinskii–Moriya type can contribute directly to the gyrotropic part of the polarizability, because it is directly associated with the orbital momentum. The exchange relativistic interactions of the 'spin–same orbit' and 'spin–other orbit' type, which contribute to the effective orbital field of the type described by (5.27), were studied by Zenkov (1990) and Moskvin *et al* (1993). Interactions of this type cause peculiarities in the anisotropic coupling of the gyration vector of a given ion with the average magnetic moments of other ions. These latter may lead to an anisotropic contribution to the gyration vector for antiferromagnetics and ferrites (Moskvin *et al* 1993).

5.10 Gyrotropic properties of metals

Many experimental investigations have been devoted to the magnetooptical properties of transition metals and alloys, and rare-earth and transition metal compounds. First-principles theoretical studies of the magnetooptical effects have usually lagged behind the experiments. *Ab initio* calculations of the magnetooptical Kerr rotation of the elemental ferromagnetic metals Fe and Ni, and some d compounds, which demonstrate good agreement with the experiments, will be considered in section 5.12. Here we shall confine our discussion to a semiquantitative theory, which provides qualitative insight into the microscopical mechanisms of the magnetooptics of metals.

The traditional approach to this problem is based on consideration of the intra-band and inter-band electric dipole transitions. The former are assumed to be responsible for the long-wavelength spectral band ($h\nu \leq 2$ eV), whereas the latter account for the short-wavelength region (Voloshinskaya and Fedorov 1973, Erskine and Stern 1973, Voloshinskaya and Bolotin 1974, Buschow 1988).

Let us consider the intra-band effects first. As we have seen above, for the case of localized electrons the source of magnetooptical effects is associated with the influence of a magnetic field or magnetic ordering on the orbital motion of electrons. Of course, this statement is valid for conduction electrons as well. The direct effect of a magnetic field on the electron motion is given by the interaction

$$V = (e/c)\boldsymbol{v} \cdot \boldsymbol{A}_0$$

where $\boldsymbol{v}$ is the electron velocity operator, and $\boldsymbol{A}_0$ is the vector potential of the external magnetic field. This interaction manifests itself as the Lorenz force

$$\boldsymbol{F} = (e/c)[\boldsymbol{v} \times \boldsymbol{H}]$$

which 'swirls' the electron trajectories.

In magnetically ordered metals the spin–orbit interaction yields a substantially stronger 'swirling' effect. This effect is called the antisymmetric or the skew-scattering effect. It can be explained qualitatively as follows. Like in the case of localized electrons in a metal, the effect of the spin–orbital interaction can be described by using the effective axial field

$$H_{SL} = (2\mu_B)^{-1}\lambda\langle S\rangle$$

where $\langle S \rangle$ is the average value of the spin momentum, and λ is the spin–orbit interaction constant ($H_{SL} \sim 10^6$ Oe in d metals).

One can try to calculate the conductivity tensor σ_{ik} in the framework of the classic Drude–Lorenz theory, substituting in the frequency governed by the effective field H_{SL} as the Larmor frequency in the well-known formulae for σ_{ik}. This frequency, $\Omega_S = \gamma H_{SL}$, is called the antisymmetric scattering or skew-scattering frequency.

Rigorous analysis confirms that the electron collision frequency must be substantially higher than the skew-scattering frequency. This is true for the d metals (see below). The conduction band structure and (in particular) the density of states in the d metals and alloys are very complicated, and differ from the simplest structure of an isotropic parabolic spectrum. Some authors (Voloshinskaya and Bolotin 1974) have successfully described the Kerr effect in the metals Ni and Fe, and related alloys, phenomenologically. They considered the electrons located near the Fermi surface as a set of two or more subsystems of electrons and holes, which are characterized by their effective masses, densities, and relaxation times.

Following this approach it is easy to derive the following formulae for the off-diagonal components of the permittivity tensor $\varepsilon_{xy} = \varepsilon_1' + i\varepsilon_2'$:

$$\varepsilon_1' = \sum_n \frac{2C_n\gamma_n}{(\omega^2 + \gamma_n^2)^2} \tag{5.37}$$

$$\varepsilon_2' = \sum_n \frac{C_n(\omega^2 - \gamma_n^2)}{\omega(\omega^2 + \gamma_n^2)^2} \tag{5.38}$$

where the index n corresponds to the contributions of various groups of electrons, γ_n is the relaxation frequency for the nth carrier type, and the constants C_n are proportional to their skew-scattering frequency Ω_{sn}.

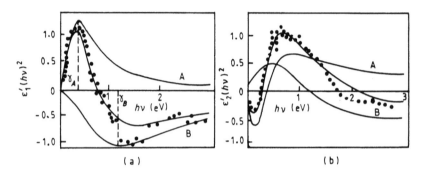

Figure 5.9. (a) The frequency dependence of $\varepsilon_1'(\hbar\omega)^2$. (b) The frequency dependence of $\varepsilon_2'(\hbar\omega)^2$. Solid circles represent experimental values (Voloshinskaya and Bolotin 1974).

The dependences $\varepsilon_1'(\omega)$ and $\varepsilon_2'(\omega)$ which were measured by Voloshinskaya and Bolotin are shown in figure 5.9. As follows from the figure, in order to explain the dispersion of the measured value of $\varepsilon'(h\nu)$ one must suggest the existence of two groups of carriers with their respective contributions, which are represented by the curves A and B. The maxima and minima of the curves correspond to the relaxation frequencies γ_A and γ_B. The values C_A and C_B must have opposite signs. Figure 5.9 shows the frequency dependence $\varepsilon_2'(\nu)(h\nu)^2$. The theoretical curves which are also represented in this figure were calculated using the values of γ_A and γ_B, and the C_A- and C_B-parameters determined by the theoretical processing of the experimental dependence shown in figures 5.9. The values of the parameters for Ni and Fe are summarized in Buschow's review (Buschow 1988).

One's attention is attracted by the very short relaxation times of the electrons: $\gamma \sim 10^{14}$–10^{15} s^{-1}. This agrees well with the suggestion made above that $\Omega_S \ll \gamma$. In order to verify this mechanism, Voloshinskaya and Fedorov (1973) and Voloshinskaya and Bolotin (1974) studied the correlation between the magnetooptical properties and the abnormal Hall effect. The conductivity tensor and the permittivity tensor are related by the well-known expression

$$\sigma_{ij}(\omega) = -\frac{i\omega}{4\pi}\varepsilon_{ij}(\omega). \tag{5.39}$$

Using (5.37) and (5.38), one can easily derive the following formula for the off-diagonal component:

$$\sigma_{xy}(0) = \frac{1}{4\pi}\left(\frac{C_1}{\gamma_1^2} + \frac{C_2}{\gamma_2^2}\right). \tag{5.40}$$

On the other hand, this value can be determined from the known relation via the saturation magnetization M_S, the electric resistivity ρ, and the coefficient of the abnormal Hall effect R_s:

$$\sigma_{xy}(0) = 4\pi R_s M/\rho^2. \tag{5.41}$$

Substituting the values of γ_n and C_n for Ni into formula (5.40) yields $\sigma_{xy}(0) = -6.8 \times 10^{14}$ s^{-1}. Using the values $4\pi R_s = -3.8 \times 10^{-20}$ s and $\rho = 7.5 \times 10^{-18}$ s for Ni (Stoll 1972), one can get $\sigma_{xy}(0) = -6.8 \times 10^{14}$ s^{-1}. This amazingly good agreement, noted by Voloshinskaya *et al* (1974) and Voloshinskaya and Bolotin (1974), provides additional confirmation that at relatively low frequencies ($h\nu < 2$ eV) magnetooptical effects are defined by the intra-band mechanism, in which the main role is played by the skew scattering of electrons under the influence of spin–orbit interaction.

5.11 Plasma resonance and the polar Kerr effect

Feil and Haas (1987) drew attention to one interesting effect, which was associated with free electrons. From the expression for the complex polar Kerr effect

$$\Phi_K + i\Psi_K = \frac{\varepsilon_{xy}}{\sqrt{\varepsilon_{xx}}(1 - \varepsilon_{xx})} \tag{5.42}$$

it readily follows that a substantial increase in Φ_K and Ψ_K can be expected in the frequency band where the denominator of (5.42), $\sqrt{\varepsilon_{xx}}(1 - \varepsilon_{xx})$, is small.

Using a simple model, Feil and Haas demonstrated that resonant growth of the Kerr effect is possible near the plasma resonance, where $\mathrm{Re}\,\varepsilon_{xx} = 0$. The diagonal part of the dielectric constant for a metal can be approximated by the formula

$$\varepsilon'_{xx}(\omega) = \varepsilon_0(\omega) + \varepsilon_{intra}(\omega)$$

where $\varepsilon_0(\omega)$ is the contribution to the dielectric constant from the inter-band transitions, and $\varepsilon_{intra}(\omega)$ is the contribution from the intra-band transitions. In the Drude–Lorenz model the latter takes the form

$$\varepsilon_{intra}(\omega) = 1 - \frac{\omega_{p0}^2}{(\omega^2 - i\omega/\tau)} \tag{5.43}$$

where τ is the carrier relaxation time, and ω_{p0} is the unscreened plasma frequency of the conduction electrons, which in the simplest approximation is equal to

$$\omega_{p0}^2 = \frac{4\pi n e^2}{m^*} \tag{5.44}$$

where n is the free-carrier concentration, and m^* is the effective mass. The relaxation time τ can be related to the conductance σ via the well-known formula $\sigma = ne^2\tau/m^*$. The coupled plasma frequency is determined by the condition $\varepsilon_1(\omega) = 0$.

The off-diagonal component ε_{xy} is related to the appropriate component of the conductivity tensor as follows: $\varepsilon_{xy}(\omega) = 4\pi i \sigma_{xy}/\omega$. Feil and Haas in their calculations assumed for the sake of simplicity that σ_{xy} was independent of the frequency. They assumed also that the spectral dependence of $\varepsilon_0(\omega)$ has a shape that corresponds to the simplest oscillator model:

$$\varepsilon_0(\omega) = 1 + \frac{A}{\omega_1^2 - \omega^2 + i\omega\tau_1}. \tag{5.45}$$

Formulae (5.42), (5.43), and (5.45) display the strong resonance-like behaviour of the Kerr effect and the ellipticity. In addition, there is a possibility of achieving here a substantial increase in the angle of rotation of the reflected light at insignificant ellipticities. Thus, the dispersion dependence of the diagonal part of the permittivity tensor can strongly affect the value and the dispersion character of the Kerr effect and the ellipticity.

Feil and Haas suggest that the resonance-like peculiarities of the Kerr effect in PtMnSb (Van Engen *et al* 1983a, b, Van der Heide *et al* 1985) and in TmS (Reim *et al* 1984) are caused by the dispersion of $\varepsilon_1(\omega)$ in the frequency region near the plasma resonance rather than by the inter-band transitions.

However, the detailed interpretation of the effect of bulk plasmons on magnetooptical properties still remains controversial (see section 5.12). Surface plasmons—electronic resonances of another type—are very strong in noble metals (Raether 1988). These charge oscillations propagating along a metal surface can be excited by the p component of light when the sample is illuminated in total reflection conditions. Fergusson *et al* (1977) studied the influence of surface plasmons on the Kerr effect in ferromagnetic metals, where surface plasmon resonances are not well defined due to the large degree of damping.

Safarov *et al* (1994) reported a precise analysis of the magnetooptical properties, in the total reflection geometry, of the Au/Co/Au multilayer structure where the well-defined noble-metal surface resonance can be excited. The experiment shows that, in this case, a resonance-like characteristic feature is observed in the magnetooptical response. This feature actually corresponds to a strong enhancement of the magnetooptical figure of merit of the multilayer system due to surface plasmon resonances.

5.12 *Ab initio* studies of magnetooptical spectra in 3d-metal-based materials

Ab initio magnetooptical calculations are of importance in the detailed interpretation of the magnetooptical spectra of Fe, Co, and Ni, and 3d-metal-based compounds. These calculations are based on the standard random-phase-approximation expression for the inter-band part of the conductivity tensor (see, for details, Wang and Callaway 1974):

$$\sigma_{\alpha\beta}^{inter}(\omega) = -\frac{i\pi e^2\hbar^2}{\Omega} \sum_{i,f} \int dk \; \frac{f(E_f(k)) - f(E_i(k))}{\hbar\omega - E_{fi}(k) + i\delta} \frac{j_\alpha^{if}(k) j_\beta^{fi}(k)}{E_{fi}(k)}$$

(5.46)

where the subscripts i and f refer to the initial and final relativistic band states, $E_{fi}(k) = E_f(k) - E_i(k)$, $f(E_i(k))$ is an occupation number for the state $|ik\rangle$, $j_\alpha^{if}(k) = \langle ik|\hat{j}_\alpha|fk\rangle$ is a matrix element of the current operator, and Ω is the unit-cell volume. The exchange–correlation processes are taken into account in (5.46) through their respective contributions to the self-consistent potential acting on the states $|ik\rangle$ and $|fk\rangle$, while the local field contribution is neglected in this expression.

In view of the important role that intra-band electronic transitions play for metals, they are usually added to the conductivity tensor in the form of the diagonal Drude term (5.43) with the intra-band plasma frequency and relaxation time defined on the basis of optical measurements or calculated *ab initio*.

The first calculations of the conductivity tensor of Ni and Fe (Wang and Callaway 1974, Singh *et al* 1975) were made in the 1970s. The subsequent calculations for the elemental 3d ferromagnets (Uspenskii and Halilov 1989, Ebert 1990) reinforced the notion that (5.46), with band wave functions and band energies obtained from density functional theory, gives the correct descriptions of $\sigma_{\alpha\beta}(\omega)$ for Fe, Co, and Ni.

The next step towards direct *ab initio* calculations of magnetooptical properties was made by Oppeneer *et al* (1992); in their work the polar Kerr rotation in Fe and Ni was calculated from (5.42) taking into account (5.39), and the conductivity tensor found from (5.46). The results obtained are shown in figure 5.10 together with the available experimental data. They clearly demonstrate the high degree of accuracy of this type of calculation for 3d-metal-based materials, and show that the origin of the Kerr rotation in these materials appears to be inter-band transitions in combination with spin–orbit coupling and exchange splitting of bands.

Numerical calculations performed by Oppeneer *et al* (1992) confirm that the Kerr effect depends primarily on the spin–orbit interaction, and the value of the effect scales in proportion to the spin–orbit coupling strength. Nonetheless, a spin–orbit interaction by itself is not sufficient to produce a large magnetooptical Kerr effect. For example, the strengths of the spin–orbit couplings in Fe and Ni are similar in size, but the Kerr rotation of Fe is about three to six times larger than that in Ni. Intra-band effects are important at energies smaller than 1–2 eV (see section 5.10). The existence of a plasma resonance effect could not be confirmed on the basis of this work, but neither can it be completely ruled out.

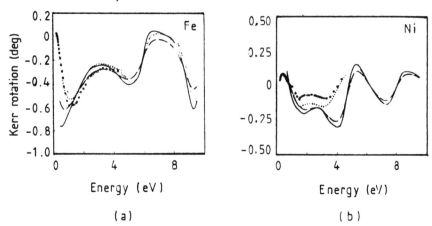

Figure 5.10. (a) Experimental and calculated results for the Kerr rotation of Fe. The experimental results shown are those of (○) Krinchik and Artem'ev (1968) and (dashed line) van Engen (1983). Calculated results are given for two inverse lifetimes: $\delta = 0.03$ Ryd for the full curve and $\delta = 0.05$ Ryd for the chain curve. The effect of an empirical Drude conductivity on the calculated Kerr rotation is illustrated by the dotted curve, which is the result (for $\delta = 0.03$ Ryd) with the Drude term. (b) As (a), but for Ni (Oppeneer *et al* 1992).

Ab initio magnetooptical calculations were also carried out for several groups of 3d-metal-based compounds (Uspenskii *et al* 1995, Oppeneer *et al* 1995). Below we consider the results for compounds with the $C1b$ crystal structure (PtMnSb, PdMnSb, NiMnSb, and PtMnSn) and MPt$_3$ (M = Cr, Fe and Co) compounds, being the most characteristic and well studied. Figure 5.11 shows calculated and experimental spectra of the polar Kerr rotation in PtMnSb. The theory (Uspenskii *et al* 1995) reproduces the large resonance-like extremum $\Phi_K(\omega)$ at $\hbar\omega = 1.7$ eV adequately.

The noted trend and roles which different atoms play in the magnetooptical spectrum formation were investigated in detail by Uspenskii *et al* (1995). Following this work, we will consider the dependence of $\sigma_{xy}(\omega)$ on the main factors, such as the spin polarization, spin–orbit interaction, and atomic composition, using an expression for $\sigma_{xy}(\omega)$ which is correct to first order in the spin–orbit interaction (Argyres 1955):

$$
\begin{aligned}
\sigma_{xy}^{inter}(\omega) \approx &-\frac{\mathrm{i}\pi e^2\hbar^2}{\Omega}\sum_{i,f\neq l}\int \mathrm{d}k\ \frac{f(E_f(k))-f(E_i(k))}{\hbar\omega-E_{fi}(k)+\mathrm{i}\delta}\frac{1}{E_{fi}(k)}\\
&\times\ (H_{SO.z}^{il}(k)j_x^{lf}(k)j_y^{fi}(k)/E_{il}(k)\\
&+\ H_{SO.z}^{fl}(k)j_x^{lf}(k)j_y^{fi}(k)/E_{fl}(k))\\
&-\ [x \rightleftharpoons y]
\end{aligned}
\tag{5.47}
$$

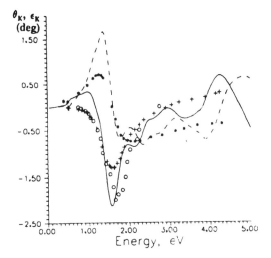

Figure 5.11. The calculated Kerr rotation (solid line) and ellipticity (broken line) for PtMnSb. Also shown are the experimental values of the Kerr rotation (+) and the ellipticity (∗) given by van Engen *et al* (1983b), and the Kerr rotation (○) given by Takanashi *et al* (1987). After Kulatov *et al* (1995).

where

$$H^{il}_{SO.z}(\mathbf{k}) = \frac{\hbar^2}{4m^2c^2} \langle i\mathbf{k}|\hat{\sigma}_z(\nabla_x V(\mathbf{r}) \nabla_y - \nabla_y V(\mathbf{r}) \nabla_x)|l\mathbf{k}\rangle$$

is the matrix element of the spin–orbit interaction operator, and $V(\mathbf{r})$ is the crystal potential. Since the Pauli matrix σ_z is diagonal in the spin indices, expression (5.47) describes the difference between spin-up and spin-down transitions which exactly cancel each other in non-magnetic solids (see section 5.1).

It was shown by Uspenskii *et al* (1995) that the main extremum of the Kerr rotation at 1.7 eV in PtMnSb is due to the corresponding extremum of $\sigma_{2.xy}(\omega)$. The value of $\sigma_{2.xy}(\omega)$ is little changed if spin–orbit interaction at Mn and Sb sites is completely neglected, i.e. spin–orbit interaction at Pt sites is dominant. Replacement of the Pt atoms by Pd or Ni atoms strongly reduces the effective spin–orbit interaction in a compound; because of this the magnetooptical activity of PdMnSb and NiMnSb is much smaller. It was also shown that Sb gives the largest contribution to the matrix elements of the current operator for PtMnSb, while the contributions of Pt and Mn atoms are somewhat smaller. This is not accidental, and is closely connected to the electronic structure of PtMnSb.

One of the earlier explanations (de Groot *et al* 1983) of the high magnetooptical activity of PtMnSb was related to the fact that this compound is a half-metallic ferromagnet, i.e. to the existence of a gap for spin-minority electrons at E_F. This feature of PtMnSb is readily distinguished in figure 5.12, which presents the densities of states for spin-up and spin-down electrons. Note

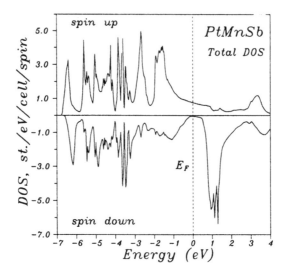

Figure 5.12. The total spin-up and spin-down densities of states for PtMnSb (Uspenskii *et al* 1995).

that the Pt 5d↑, Mn 3d↑ and Sb 5p↑ bands have close energies, giving rise to strong band mixing, and metallic conductivity for spin-up electrons.

The strong exchange splitting of the Mn 3d bands results in the isolation of the Mn 3d↓ band lying at 0.7 eV above E_F, while the Pt 5d↓ and Sb 5p↓ bands are strongly mixed. Uspenskii *et al* (1995) have shown that the most important contributions to $\sigma_{1xx}(\omega)$ and $\sigma_{2xy}(\omega)$ for PtMnSb are connected with spin-down electronic excitations across the gap from the mixed Pt 5d↓–Sb 5p↓ band to the narrow Mn 3d↓ band. That is, these excitations give rise to the large magnetooptical resonance in PtMnSb at $\hbar\omega = 1.7$ eV.

According to calculations of Uspenskii *et al* (1995), a similar resonance of $\sigma_{2xy}(\omega)$ with an even larger amplitude exists also in PtMnSn at $\hbar\omega = 1.2$ eV. This compound has an electronic structure which is similar to that shown in figure 5.12, but the magnetooptical resonance is strongly reduced in PtMnSn.

This difference can be explained by different behaviours of the factor

$$A(\omega) + iB(\omega) = 1/[\sqrt{\varepsilon_{xx}(\omega)}(1 - \varepsilon_{xx}(\omega))]$$

for PtMnSb and PtMnSn (see (5.42)). The factor $A(\omega)$ is much larger (at $\hbar\omega = 1$–2 eV) for PtMnSb than for PtMnSn, while the factor $B(\omega)$ can be neglected for both compounds. That is, this behaviour of $A(\omega)$ in the vicinity of the magnetooptical resonance energy strongly enhances the value of $\Phi_K(\omega)$ for PtMnSb and reduces it for PtMnSn (see also section 5.11).

Among the MPt₃ (M = Cr, Mn, Fe and Co) compounds, the highest magnetooptical activity is observed for MnPt₃ (Kato *et al* 1995a, b). The calculations (Kulatov *et al* 1996) correctly reproduced the large values of the

Kerr rotation and ellipticity for this compound, and predicted a rather large Kerr rotation in $CrPt_3$ and a much smaller one in $FePt_3$ and $CoPt_3$. Analogous results for some of the MPt_3 compounds were obtained by Kubler (1995). The numerical experiments for $MnPt_3$, which were analogous to the ones for PtMnSb, showed that atoms of Pt are dominant in both the effective spin–orbit interaction and the matrix elements of the current. In contrast to the case for PtMnSb, the excitations of spin-majority electrons play a leading part in the formation of $\sigma_{1xx}(\omega)$ and $\sigma_{2xy}(\omega)$.

Chapter 6

Magnetooptical effects in the x-ray region

6.1 Introduction

Attempts to register very weak magnetooptical interactions in the x-ray region using common x-ray sources (e.g. x-ray tubes) encountered significant experimental difficulties. This is because the usual (non-resonant) polarizability of materials in the x-ray region is very weak (see, for example, Landau *et al* 1984). This polarizability is governed by the Thompson scattering of x-rays by atomic electrons, with the amplitude $f_0 \sim -Z r_0$, where Z is the atomic nuclear charge, and r_0 is the classical electron radius. The amplitude of the non-resonant magnetic scattering for x-rays is several orders of magnitude less than that for charges: $f^{mag} \sim (\lambda_c/\lambda) Z_M r_0$, where λ_c is the Compton wavelength and Z_M is the number of electrons in the magnetic shell. An estimate for the incident radiation wavelength of $\lambda = 1$ Å gives $f^{mag} \simeq 10^{-2} f_0$. Hence the contribution of non-resonant magnetic scattering to the magnetization-sensitive polarizability of a magnetic medium is very small.

On the other hand, the x-ray resonant magnetic scattering amplitude may be much larger (up to $100 r_0$, according with Hannon *et al* 1988). The general expression for the polarization dependence of the x-ray scattering amplitude has been derived by Blume and Gibbs (1988). A number of authors (Gibbs *et al* 1985, 1988) have pointed out that the extension of the x-ray scattering to the resonance regime where the photon energy is near to the excitation energy of a solid will produce novel effects and open up a variety of directions for new kinds of synchrotron experiment. Hannon *et al* (1988) predicted that electric multipole transitions, with their sensitivity to the magnetization, arising from the exchange, will give rise to very strong magnetooptical effects in magnetic samples. Therefore, the basis of x-ray magnetooptics is provided by resonances caused by magnetization-sensitive electric multipole transitions. The spectral region in which these effects can be observed is, therefore, limited to narrow bands near the x-ray absorption edges.

6.2   Near-edge magnetization-sensitive transitions

The energy of x-ray photons connected with a sudden increase in x-ray absorption for a given sample is referred to as absorption edge energy. This occurs due to the energy becoming sufficient to excite a core-level atomic electron into an unoccupied state. The latter may be localized (e.g. a state in the open atomic shell), or a band-like state (e.g. a state in the conduction band of a metal), or it may be a state of the continuous spectrum (i.e. reached via a photoionization process). Absorption edges are classified in accordance with the quantum numbers nl_j of the core hole ($n = 0, 1, 2, \ldots, l = s, p, d, \ldots, j = l \pm 1/2$), as shown in table 6.1.

Table 6.1. Absorption edges classified in accordance with the quantum numbers nl_j of the core hole.

Shell	$1s_{1/2}$	$2s_{1/2}$	$2p_{1/2}$	$2p_{3/2}$	$3s_{1/2}$	$3p_{1/2}$	$3p_{3/2}$	$3d_{3/2}$	$3d_{5/2}$
Edge	K	L_1	L_2	L_3	M_1	M_2	M_3	M_4	M_5

Thus the initial state is a well-known spin–orbit-split state denoted as nl_j. But the final state of an electron may be magnetization sensitive. This occurs when this state is a polarized atomic shell or a band split due to interatomic exchange or spin–orbit interaction. These magnetization-sensitive transitions give rise to near-edge resonant enhancement of the x-ray magnetic scattering intensity (Gibbs *et al* 1988), and lead to another phenomenon—magnetic x-ray dichroism (MXD). So we may try to use MXD signals to obtain information about the structure and microscopic properties of magnetization-sensitive states in a magnetic medium. Among these, the 3d and 4f electronic states in transition metals (TM) and rare-earth (RE) elements respectively are of special interest. To probe these states with the aid of MXD, the dipole-allowed transitions $2p^6 3d^n \longrightarrow 2p^5 3d^{n+1}$ and $3d^{10} 4f^n \longrightarrow 3d^9 4f^{n+1}$ should be used. This corresponds to collecting MXD data near $L_{2,3}$ edges of TM and $M_{4,5}$ edges of RE. We should note here that other shells (or bands) may be magnetized in a solid through exchange. This makes it useful to measure MXD near the K edge of TM and the $L_{2,3}$ edges of RE (the 1s $\rightarrow$ 4p and 2p $\rightarrow$ 5d transitions respectively).

6.3   Magnetic x-ray dichroism

MXD near the K, L edges of TM (for example, see Tobin *et al* 1992, Wu *et al* 1992, van der Laan *et al* 1992, Stähler *et al* 1993, Kuiper *et al* 1993, Idzerda *et al* 1993) and L, M edges of RE (for example, see Kappert *et al* 1993, Chaboy *et al* 1994) is widely under investigation as a local, element-specific probe of magnetic structure. It allows one to distinguish between localized and itinerant magnetic moments by collecting MXD signals near appropriate edges of the ion in the magnetic compound (Chaboy *et al* 1994).

Spin-polarization measurements in the x-ray region require the use of SR as a powerful and tunable source. Circularly polarized SR can be obtained from a helical or crossed undulator, a quarter-wave plate, a Goedkoop filter (Goedkoop *et al* 1988a, b, c), or, most commonly, from a bent magnet via an inclined-angle view of the tangent point of the stored electron beam. When observed in the plane of the electron orbit, the SR from a bent magnet is linearly polarized, with the E-vector in the orbit plane. Out of the plane, the radiation intensity decreases, but is elliptically polarized: the E-vector has horizontal and vertical components that are 90° out of phase. The inclination angle above or below the orbit plane determines the sense and degree of the circular polarization, with large angles producing essentially 100% circular polarization but with low intensity. Typical curves for the normalized intensity and degree of circular polarization are shown in figure 6.1. Besides the source of polarized SR, the complete experimental set-up should include collimation and monochromatization systems, a sample unit, and detectors. For example, let us consider in outline the schemes for MCXD measurements that are most widely used.

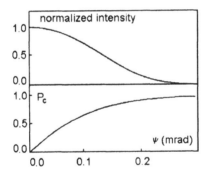

Figure 6.1. The dependence on the inclination angle ψ of the normalized intensity and degree of circular polarization P_c of the SR from the bent magnet at the 3.5 GeV, 100 mA storage ring Doris II. The photon energy = 8 keV. After Stähler *et al* (1993).

6.3.1 The two-beam transmission mode

The following method is that used at the Hamburger Synchrotron Strahlungslabor HASYLAB at the electron storage rings Doris II and Doris III[1]. The principle of the set-up is shown in figure 6.2. The incident photons pass a vertically adjustable double slit and a double-ionization chamber I_w. Since the slits are adjusted symmetrically with respect to the electron orbit plane, two beams with

---

[1] In this description we follow Stähler *et al* (1993).

opposite senses of circular polarization are obtained. In the correct position the intensities of the two beams have to be equal, yielding equal counting rates in the upper and lower ionization chambers of I_w. A typical degree of circular polarization P_c of the white beam is $P_c = \pm 0.8$. The photons are monochromatized using a double-crystal monochromator. The samples are placed inside a solenoid, producing a magnetic field. The absorption measurements are carried out by measuring the incident and transmitted intensities I_0 and I_1 using the double-ionization chambers. A third double-ionization chamber I_2 is placed behind the second target, enabling measurements to be made with a reference sample. The measured value is the MCXD signal μ_c/μ_0:

$$\frac{\mu_c}{\mu_0} = \frac{\ln(I_0/I_1)^+ - \ln(I_0/I_1)^-}{\ln(I_0/I_1)^+ + \ln(I_0/I_1)^-} \frac{1}{|P_c|}$$

where μ_c and μ_0 denote the difference and sum of the absorption coefficients for the circularly polarized radiation of the beams with the opposite senses.

This method requires the high penetration power of hard x-rays, and allows bulk-sensitive highly precise direct measurements of the MCXD signal. It does not require ultra-high-vacuum (UHV) conditions. At HASYLAB, circularly polarized hard x-rays are available over an energy range of 5–30 keV, which allows measurements near the K edges of 3d, 4d TM and the L edges of 5d TM and RE to be performed.

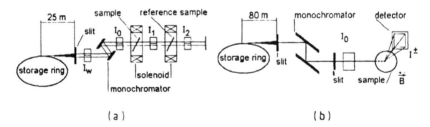

Figure 6.2. (a) The two-beam transmission mode experimental set-up. (b) The fluorescence detection experimental set-up.

6.3.2 Fluorescence detection

This method is that used at the Daresbury storage ring[2]. A schematic diagram of the experimental set-up is shown in figure 6.2. The polarized x-rays are obtained by selecting a beam with the aid of the 'inclined-view' method described above, and monochromatized by a double-crystal monochromator. The incident flux is controlled by the ionization chamber I_0. The fluorescence yield is counted by a solid-state Ge detector at 90° to the incident beam. The magnetic field in the

---

[2] In this description we follow Collins *et al* (1989).

sample is 'flipped'. So the measured value is the so-called 'flipping ratio':

$$\frac{I^+ - I^-}{I^+ + I^-} = \frac{\mu_c}{\mu_0}.$$

I^+ and I^- are the fluorescence intensities for parallel ($+$) and antiparallel ($-$) orientation of the photon spin and the spin of the magnetic electrons in the sample. The fluorescence method allows both hard- and soft-x-ray measurements. It is best suited to work on diluted magnetic systems and foils. The signal measured arises entirely from processes in the atomic shell under consideration.

When the detection of Auger electrons is required, the experimental set-up is similar to that described above, but instead of fluorescence photons, Auger electrons are counted. This method *does* require UHV conditions.

6.4 Theoretical background

6.4.1 The one-particle approach

The theoretical basis of the description of MXD spectra depends on the character of the quantum state to which the core electron is excited. The spectral shape of the MXD signal can be calculated either by using a one-particle (a band-like final state) or a many-particle (a multiplet or satellite fingerprint in the final state) approach. The Korringa–Kohn–Rostoker Green's function method (Ebert *et al* 1988a, b, 1989, Collins *et al* 1989, Ebert and Zeller 1989, 1990, Baudelet *et al* 1991) and multiple-scattering theory (Brouder and Nikam 1991) give good agreement in the former case for L edges of RE and K edges of TM. The origin of the spin-dependent absorption in this case can be understood on the basis of the ejection of a spin-polarized photoelectron after absorption of a circularly polarized photon in an unpolarized atomic core state. For a free atom, the photoelectron polarization

$$P_e = (n^\uparrow - n^\downarrow)/(n^\uparrow + n^\downarrow)$$

where $n^\uparrow$ ($n^\downarrow$) denotes the number of photons ejected with the spin parallel (antiparallel) to the photon, is (Stähler *et al* 1993) given in table 6.2.

According to Fermi's golden rule, the absorption coefficient $\mu(E) \sim |M|^2 \rho(E)$ is related to the transition matrix element M and the density ρ of final states which can be populated according to the dipole selection rules. Application of this rule to the spin-resolved absorption channels permits one to write

$$\frac{\mu_c}{\mu_0} = P_e \frac{\Delta\rho}{\rho}$$

where $\Delta\rho = \rho^\uparrow - \rho^\downarrow$ and $\rho = \rho^\uparrow + \rho^\downarrow$. We assume here that M does not depend on the spin direction of the photon related to the electron spin. In

Table 6.2. The photoelectron polarization.

Edge	Transition	P_e
K	$1s_{1/2} \rightarrow p$	< 0.01
L_1	$2s_{1/2} \rightarrow p$	-0.1 to -0.2
L_2	$2p_{1/2} \rightarrow d$	-0.5
L_3	$2p_{3/2} \rightarrow d$	$+0.25$

many cases the ratio of the measured MCXD signal μ_c/μ_0 to the calculated spin-polarization density profile $\Delta\rho/\rho$ gives a value for P_e very similar to the free-atom one. This shows that the simplified approach described above correctly reflects the general physical phenomena behind the MCXD spectra when transitions to the band-like states are considered.

6.4.2 The many-particle approach

To explain the properties of the MXD signal related to transitions into atomic-like states, a many-electron approach is required. This approach is now successfully applied to the modelling of MXD near the M edges of RE and the L edges of TM ($3d^{10}4f^n \longrightarrow 3d^9 4f^{n+1}$ and $2p^6 3d^n \longrightarrow 2p^5 3d^{n+1}$ electronic transitions respectively) (see, e.g., Thole *et al* 1985b, van der Laan *et al* 1986, van der Laan 1987, 1990a, b, Goedkoop *et al* 1988b, c, Jo and Imada 1989, van der Laan and Thole 1990, 1991, 1992, Imada and Jo 1990, Carra and Altarelli 1990, Carra *et al* 1991, Sacchi *et al* 1991, Jo and Sawatzky 1991, Jo 1992). The atomic initial and final states should be calculated taking into account not only Coulomb and spin–orbit interactions in the open shell, but also a crystal field acting on the ion (see section 5.2). When the crystal field may be considered as a perturbation when compared with the Coulomb + spin–orbit interaction in the magnetic shell, the initial and final states may be treated as atomic ones. This case is realized when the 4f magnetic shell of RE is considered. Because both Coulomb and spin–orbit interactions are diagonal in J (the total angular momentum of the 4f shell), J is a good quantum number characterizing the initial and final states of the ion. On the other hand, the initial state of the ion may be represented as the almost pure Hund's rule ground state denoted as $|\alpha SLJM\rangle$ with $J = L - S$ if $n < 7$ (the 4f configuration of RE is considered) and $J = L + S$ otherwise. For the perfectly polarized ion $M = -J$, α denotes all of the other quantum numbers necessary to specify the state. The final state may be written as $|\alpha'J'M'\rangle$. It is characterized by a large core-hole–valence electron interaction, which results in the formation of a local bound state. The manifold of these states (their energies and wave functions) may be calculated using Hartree–Fock calculations (Thole *et al* 1985a). The dipole transition from a state $|\alpha SLJM\rangle$ to a final state $|\alpha' J'M'\rangle$ may be written according to the Wigner–Eckart theorem as

$$F_{\alpha' J'M',\alpha JM} = \begin{pmatrix} J' & 1 & J \\ -M' & q & M \end{pmatrix}^2 |\langle\alpha' J'\|C^1\|\alpha SLJ\rangle|^2.$$

The reduced dipole matrix element is the line strength in the absence of a magnetic field, $q = 0$ corresponds to light polarized in the direction of the field (z), and $q = \pm 1$ corresponds to right or left circularly polarized radiation. Since the magnetic splitting of the M' levels is much less than the experimental resolution, it could not be observed. So the temperature and polarization dependence of the intensity may be determined by taking the weighted average of the $3jm$-symbol:

$$\langle F_{\alpha' J'M',\alpha JM}\rangle = \langle A^q_{JJ'}\rangle \, |\langle\alpha' J'\|C^1\|\alpha SLJ\rangle|^2$$

where

$$\langle A^q_{JJ'}\rangle = Z^{-1} \sum_{MM'} \begin{pmatrix} J' & 1 & J \\ -M' & q & M \end{pmatrix}^2 \exp(-M/\Theta)$$

$$\Theta = \frac{kT}{g|\mu_B|H} \qquad Z = \sum_{M=-J}^{J} \exp(-M/\Theta).$$

To obtain a final x-ray absorption or MCXD spectrum, the calculated line strengths should be not only averaged as described above, but also convoluted with the linewidths associated with the core-hole lifetime. The latter is determined by all of the processes responsible for core-hole decay.

6.5 Sum rules

The importance of sum rules is well known in optical spectroscopy. A complete first-principles description of the x-ray magnetooptical effects is scarcely possible, because it requires the complete knowledge of the final states of the resonant multipole transitions involved. These states are strongly influenced by the crystal field, the hybridization, the interatomic exchange, and some other effects. Hence sum rules derived under very general assumptions provide a unique opportunity to obtain non-trivial ground-state properties without detailed investigation of the electronic structure of the sample. Thole *et al* (1992) showed that it is possible to measure the ground-state expectation value of the orbital angular momentum operator $\langle L_z\rangle$ using magnetic circular x-ray dichroism. This relationship may be written as follows (Wu *et al* 1992):

$$\frac{\Delta A_{j-} + \Delta A_{j+}}{A_t} = -\frac{1}{2\hbar} \frac{c(c+1) - l(l+1) - 2}{l(l+1)n} \langle L_z\rangle$$

where

$$\Delta A_{j\pm} = \int_{j\pm} (I^{+1} - I^{-1})\,\mathrm{d}E$$

$$A_t = \int_{j_+ + j_-} (I^{+1} + I^0 + I^{-1})\,\mathrm{d}E$$

are the integral intensities; j_+ and j_- denote the spin–orbit counterparts of the absorption edge (e.g. L_2 and L_3), $I^{0,\pm1}$ denote the absorbed intensities of the light for the various orientations of the photon spin relative to the magnetization direction—in particular, I^0 denotes the x-ray absorption intensity when the two directions are orthogonal to each other; c and l are the orbital angular momentum quantum numbers for the core and valence shell, and n is the number of holes in the valence shell in the ground state.

Later, Carra *et al* (1993) derived another sum rule which relates the MCXD signal to the ground-state expectation value of the total spin operator, $\langle S_z \rangle$, and the magnetic dipole operator:

$$\langle T_z \rangle = \left\langle \left[\sum_i s_i - 3\hat{r}_i (\hat{r}_i \cdot s_i) \right]_z \right\rangle.$$

These sum rules strongly aroused the interest of experimentalists in the field of magnetic surfaces (van der Laan *et al* 1992)[3]. This is no surprise, when we recall that one can make MCXD an extremely surface-sensitive technique. Some efforts have been made to confirm and generalize the sum rules (Chen *et al* 1995, Altarelli 1993, Sainctavit *et al* 1995, Ankudinov and Rehr 1995).

By and large, the sum rules significantly simplify the interpretation of the spectra and allow more, valuable information to be extracted from the MCXD.

6.6   Examples; MCXD and magnetic anisotropy

Consider now some typical experimental results obtained with the use of the MCXD technique. In the first example (Idzerda *et al* 1993) the magnetic structure of a Fe/Cr/Fe trilayer is considered. The absorption intensity was measured by collecting both the total electron yield and the Auger electron yield. The MCXD signal was measured for a Fe/Cr/Fe trilayer structure deposited on GaAs(001) at various stages of multilayer development. X-ray absorption spectra (XAS) and MCXD spectra of the L_3 and L_2 lines of the first deposited Fe film are shown in figure 6.3(a); they correspond to a perfectly magnetized

---

[3] It should be pointed out that in the experimental application of the sum rules one encounters certain difficulties (Wu *et al* 1992, Ankudinov and Rehr 1995). Among these are the arbitrariness as regards the choice of the integration range, and in the number of holes. Also, the derivation of the sum rules is based on non-relativistic quantum theory with relativistic corrections. Strange and Gyorffy (1995) derived a rule for interpreting MCXD experiments using an itinerant, rather than a localized approach, within a fully relativistic quantum-mechanical framework.

single-domain Fe sample. To investigate the magnetic orientation of the first monolayer (ML) of Cr, XAS and MCXD spectra of the lowest-coverage 0.25 ML Cr film were collected. They are shown in figure 6.3(b). It is immediately evident from the reversal of the Cr MCXD intensity that submonolayer coverages of Cr are aligned antiparallel to the first Fe layer. Additional deposition of Cr causes a continuous reduction in the MCXD signal. This makes it evident that subsequent Cr deposition gives rise to an antiferromagnetic structure in the Cr layer obtained. The XAS and MCXD spectra for 8 Å of second-layer Fe are shown in figure 6.3(c). The sign reversal of the MCXD signal indicates that the second Fe layer is ferromagnetic, and antiparallel to the first, thicker Fe film. This example shows the application of the element specificity of MCXD in probing the magnetic structures of multicomponent thin films. It also shows that MCXD is a powerful tool for studying even submonolayer coverages.

The MCXD technique can be used to determine the microscopic origin of the magnetic anisotropy. It was pointed out in section 5.7 that the sum rules can be used to investigate the spin–orbit interaction and magnetocrystalline anisotropy. Weller *et al* (1995) have shown that the orbital momentum anisotropy (as well as the spin momentum anisotropy) may be investigated with the aid of angle-dependent x-ray absorption spectroscopy (Stöhr and König 1995). They also confirm that the former can represent the origin of the magnetocrystalline energy anisotropy (MCA). The experiment was performed on a Au/Co staircase/Au sample[4]. The first Au layer was 28 nm thick. The Co staircase represented ten terraces with thicknesses from 3 to 12 atomic layers (AL). The structure was capped with a Au layer 9 AL thick. Such artificially made transition metal films are of great interest, because they exhibit perpendicular magnetic anisotropy. The magnetocrystalline anisotropy (MCA) energy is usually contributed by surfaces (interface anisotropy) and by the bulk crystal lattice (volume anisotropy). It is usually characterized through the measurement of the various phenomenological anisotropy constants, which can be compared with those obtained from microscopic calculations. Weller *et al* relate the perpendicular orientation of the magnetization direction in Au/CoAu sandwiches to a large anisotropy of the Co orbital magnetic moment. The latter was measured using high-field angle-dependent MCXD. In fact, Stöhr and König derived a new sum rule for a spin moment only. This rule is based on the angular average of MCXD intensities, and is valid for the 3d transition metals in an external magnetic field that is sufficiently strong to magnetically saturate the sample along all directions. Angle-dependent MCXD also allows independent measurements of the orbital and magnetic dipole moment anisotropies to be made. In an experiment, these values had been measured for varying Co thickness. The results have been used in the perturbation theory treatment of Bruno (1989). In accordance with the latter, the energy anisotropy in a uniaxial system caused by spin–orbit interaction is directly linked to the anisotropy of

---

[4] Our explanation now follows Weller *et al* (1995), and Stöhr and König (1995).

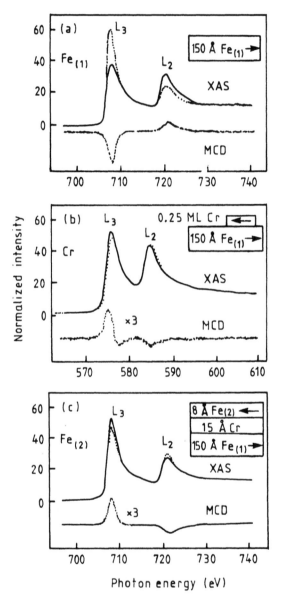

Figure 6.3. (a) XAS and MCXD spectra for the first Fe film, which is 150 Å thick. (b) The case with 0.25 ML of Cr deposited on top of the first Fe film. (c) The case with the second Fe film, 8 Å thick, on top of a 15 Å Cr interlayer. The solid (dashed) lines correspond to x-ray absorption with the spin direction of the incident photons parallel (antiparallel) to that of the majority electrons of the first Fe film. The ordinates of the MCXD spectra have been arbitrarily shifted (Idzerda *et al* 1993).

the orbital momentum, and for a more-than-half-filled d shell one obtains

$$\Delta E_{SO} = -\frac{G}{H}\frac{\lambda}{4\mu_B}\Delta m_{orb}$$

where the factor G/H is estimated to be about 0.2 for Co, η is the spin–orbit interaction constant, and Δm_{orb} is the anisotropy of the orbital contribution to the magnetic moment. Using the experimental value $\Delta m_{orb} = 0.12$ μ_B at a Co thickness of 4 AL, and $\eta = 0.05$ eV, one can obtain $\Delta E_{SO} = -3 \times 10^{-4}$ eV/atom.

Chapter 7

Domain structure

7.1 General notions

One of the most important properties of magnetically ordered materials is the existence of domain structures. Figures 7.1 and 7.2 illustrate typical domain structures observed in epitaxial iron garnet films.

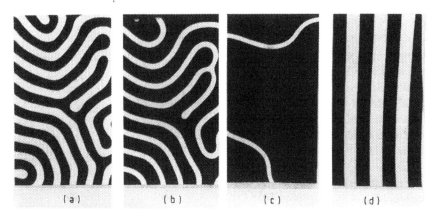

Figure 7.1. Maze (labyrinth) domain structure: $H = 0$ (a); $H = 0.7H_s$ (b); $H \leq H_s$ (c); and stripe domain structure (d).

The causes of the development of domain structures are well known, so we shall not dwell upon this issue or go into detail. Note only that a domain structure is most often formed because the division of a sample into domains reduces its magnetostatic energy, i.e. the energy of the magnetic poles that are formed at the surface. It immediately follows that the character of the domain structure strongly depends on the geometric form of the sample.

Regular domain structures (bubble lattices, mazes, stripes, and ring domains) are of interest in optoelectronics. They can be used for light control

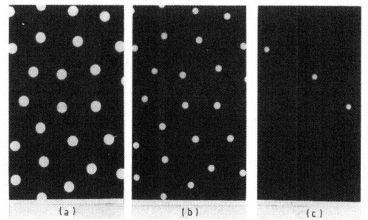

Figure 7.2. The bubble lattice: $H \geq H_2$ (a); $H_2 < H < H_0$ (b); and $H \approx H_0$ (c).

(modulation, deflection, the concentration of light beams), and for displaying and processing information.

There are a great variety of domain structures. The structures depend on the magnetic anisotropy and magnetization of the material, the sample shape, defects, the magnetic field and the temperature, the surface treatment, and the history of the sample. We shall mainly consider domain structures in magnetic films and plates with uniaxial magnetic anisotropy, assuming that the easy axis of magnetization is perpendicular to the plane of the film.

Films of this type have been studied in detail, and are of considerable practical interest. Such films contain domain structures with two kinds of domain, in which the magnetic moments are directed either along or oppositely to the z-axis (the z-axis is perpendicular to the film surface), the domains being separated by 180° domain walls. Such domain structures are usually in the form of a maze or stripe (figure 7.1) pattern.

The widths of the domains in this structure are determined by the relation (Malozemoff and Slonczewski 1979) (for $h > 10l$)

$$w \approx 2.7(lh)^{1/2}$$

where $l = (4\pi M_s^2)^{-1}\sigma_w$ is the parameter of the material which is referred to as the characteristic or magnetic length, σ_w is the energy per unit area of the domain wall, M_s is the saturation magnetization, and h is the slab (film) thickness. Typical values of h that are important for practical applications lie within the interval 0.1–10 μm. In thin films ($h < l$), the domain size w grows rapidly as the film thickness reduces.

In thick slabs ($h \geq 200l$), near the surface the domains branch, which leads to an additional lowering of the magnetostatic energy with a very small growth of the domain wall surface. The contours of the domain walls emerging at the film surface can be very complicated in this case.

Films with large uniaxial magnetic anisotropies $K_u > 2\pi M_s^2$, where K_u is the constant of uniaxial magnetic anisotropy, usually exhibit a maze domain structure. A stripe domain structure usually forms when the magnetization departs from the easy axis of magnetization. That can be observed in materials for which $K_u < 2\pi M_s^2$. In this case, the magnetic moment periodically changes its direction with respect to the film surface from one domain to another. The minimum of the total energy of the film is reached for a stripe structure whose domain walls do not contain magnetic poles. The magnetic moments can be made to tilt away from the easy axis, and a corresponding stripe domain structure can be obtained by applying a magnetic field perpendicular to the easy axis (in the plane of the film).

In a magnetic field that is directed along the easy axis, the domain structure is rearranged so that the domains having a magnetization direction that coincides with that of the external magnetic field expand, whereas the domains with the opposite orientation contract (see figures 7.1(b), 7.1(c)). On reaching some critical value H_s, which is referred to as the saturation field for a stripe or maze domain structure, the domains having magnetization oriented oppositely to the external field vanish.

In addition to such domain structures, in a certain interval of bias fields, which is limited above by the collapse field H_0, and below by the field of elliptical instability H_2, there can exist so-called cylindrical magnetic domains (bubbles) (figure 7.2). Bubbles are regions of cylindrical form, whose magnetization is oriented oppositely to the bias field. In this case the relationship

$$H_2 < H_s < H_0$$

holds true. Usually, the bubbles can be formed from maze domains, provided that a bias field $H < H_s$ and a pulsed magnetic field are simultaneously applied to the sample (the pulse duration is of the order of a microsecond). As a result, the unfavourably oriented domains contract inhomogeneously and the stripe domains are ruptured.

In zero bias field, hexagonal and irregular bubble lattices are stable. When a field which is directed oppositely to the bubble's magnetization is applied to the bubble lattice, the bubbles contract (see figures 7.2(b), 7.2(c)), and when the field reaches some critical value, every second bubble is likely to collapse. Further growth of the field leads to collapse of all of the bubbles in the field $H = H_0$. If the field direction coincides with the bubble magnetization orientation direction, they expand with increasing field to form a cellular structure, which disappears in fields substantially exceeding the collapse field.

An irregular lattice containing bubbles of different diameters emerges after a magnetic field exceeding the field of the uniaxial magnetic anisotropy (this is the field which has to be applied to the plane of the sample to remove the domain structure) is applied and then switched off in the plane of the film.

The domain types discussed do not cover the whole variety of domain structures. In epitaxial cobalt-substituted iron garnet films, domains are observed having almost rectangular shape. Under certain conditions, the stripe domains can be wound into spirals, and the bubbles can adopt elliptic or dumb-bell-like forms. If the width of the domains significantly exceeds the thickness of the film, the role of the demagnetizing fields becomes less important than that of the coercive force. In this case the domain might assume various configurations.

Crystals with anisotropy more complex than a uniaxial one (for instance, cubic anisotropy) may exhibit more complicated domain structures. For instance, there are domains with $M \parallel \langle 100 \rangle$ separated by 90° walls in iron, and domains with $M \parallel \langle 111 \rangle$ in yttrium iron garnet, nickel, and a number of other materials, where 71° and 109° walls are also possible. It is possible to completely close the magnetic flux in such crystals, with the result that there will be no magnetic poles on the surface. In this case the equilibrium dimensions of the domains are determined by the condition that the sum of the magnetoelastic and domain wall energies must be a minimum, because, in domains with non-collinearly oriented magnetic moments, various magnetoelastic stresses appear.

In speaking so far of magnetic structure, we have assumed that the magnetic material is ferromagnetic. Magnetooptics studies and employs materials with more complicated magnetic structure—for instance, ferrimagnets, weak ferromagnets, and antiferromagnets. Ferrimagnets and weak ferromagnets can be considered as ferromagnets with saturation magnetizations M_s until the intensity of an external magnetic field substantially affects the material's magnetization (for this, $H \ll 10^5$ Oe is usually enough).

In antiferromagnets, although at $H = 0$ the magnetization equals zero, and hence there are no magnetic charges on the surface of the sample, domains can exist. They differ from each other in the direction of the antiferromagnetism vector, as well as in the orientation of the magnetoelastic deformations. The domains in antiferromagnets can in many cases be observed by magnetooptical methods. That is possible if the domains are not separated by 180° walls. For the latter case, the domains are less readily observable, though in antiferromagnetic crystals with certain magnetic symmetries the magnetooptical contrast between such domains appears when a magnetic field is applied (for instance, in dysprosium orthoferrite at low temperatures, and in CoF_2 (Zvezdin and Kotov 1976b, Kharchenko and Belii 1980)).

7.2 The structure of domain walls

There is a transitional layer between domains—a domain wall (boundary)—in which the spin orientation gradually varies between the magnetization directions in the adjacent domains. In the Bloch domain wall the reorientation occurs in the plane of the wall. In this case the magnetization component that is normal to the plane of the wall remains equal to zero, and therefore there are no magnetostatic poles in the wall and the demagnetization energy of the wall is equal to zero. The thickness δ_w and the energy density σ_w of the 180° Bloch domain walls in uniaxial magnets are determined by the competition between the exchange energy and the magnetic anisotropy energy; they are defined as follows:

$$\delta_w = \pi(A/K_u)^{1/2} \qquad \sigma_w = 4(AK_u)^{1/2}$$

where A is the constant of non-uniform exchange (the exchange rigidity). For instance, for a typical iron garnet film with a width of the stripe domain $w = 3$ μm the values of the above-mentioned parameters are: $A = 2 \times 10^{-7}$ erg cm^{-1}; $K_u = 2 \times 10^4$ erg cm^{-3}; $\delta_w = 0.1$ μm; and $\sigma_w = 0.25$ erg cm^{-2}.

In the Néel domain wall, the spins are reoriented in the plane perpendicular to that of the wall, so magnetic poles occur and the energy of the Néel domain wall is higher than that of the Bloch domain wall. The difference in energies of such walls for materials with high quality factors $Q = (2\pi M_s^2)^{-1} K_u$ is small. In materials with $Q \ll 1$ the Néel domain walls may be energetically favourable only in sufficiently thin films.

Domain walls—in particular, Bloch domain walls—exhibit the property of 'chirality' (originating from the Greek $\chi\eta\iota\rho$ (*cheir*), meaning arm), which means that there exist Bloch domain walls that have equal energy but different directions of the spin rotation, from $+M$ to $-M$ (clockwise and anticlockwise). In the centre of the domain wall the directions of the spins in the 'left' and 'right' walls differ from each other by 180°. The domain wall can have several regions with different rotation directions of spins—i.e. with different polarities.

An intermediate region in the domain wall, which separates sections (subdomains) of different polarity, is referred to as the Bloch line. Although Bloch lines have been known of since the late 1950s, detailed research into them started only in the middle of the 1970s, and this was concerned with bubble-containing magnetic materials. Bloch lines turned out to affect the dynamic characteristics of the domain walls significantly, and to determine the direction of movement of magnetic bubbles in the magnetic field gradient, and the ultimate velocity of the domain walls.

Bloch lines are caused by various factors. If the easy axis lies in the plane of the film or slab, the formation of Bloch lines may be energetically favourable, since—depending on the direction of the spins—on both surfaces of the film in a narrow band (which is of width δ_w) there appear positive or negative magnetic poles. Their magnetostatic energy can be reduced through the formation of subdomains. That is the way in which the Bloch lines are formed in yttrium iron garnet slabs. In other cases Bloch lines appear when the domain wall moves at a sufficiently high velocity (dynamic conversion of the domain wall). The thickness Λ_L and energy e_L of the Bloch line are given by

$$\Lambda_L = (2\pi M_s^2)^{-1/2}\pi A^{1/2} \qquad e_L = Q^{-1/2} 8A.$$

From the topological viewpoint, a Bloch line is a magnetization whirl. If we choose an arbitrary sufficiently small loop around the Bloch line in the plane perpendicular to its axis, on passing along the loop the magnetization direction turns by ±360°. The two signs indicate the existence of two energetically equivalent Bloch lines with opposite topological charges ($v = \pm 1$). One Bloch line may have sections with opposite charges. The boundary between them is referred to as a Bloch point. The Bloch point is topologically a 'hedgehog' of magnetization, i.e. if we surround it with a small enough closed surface, we will observe all possible directions of spins on it. The energy of the Bloch point (for $Q \gg 1$) (Malozemoff and Slonczewski 1979) can be as high as

$$e_T = K^{-1/2} 2\pi A^{3/2} (\ln Q + 1.9)$$

and its characteristic dimensions are of the order of $\Lambda_L \Lambda_L \delta_w$.

7.3 Stripe domain structure

The shapes and sizes of the equilibrium domains in a magnetic material are determined from the condition that its free energy must be a minimum; the free energy is made up in general from the exchange energy, the magnetic anisotropy energy, the magnetostatic energy of the demagnetizing fields, and the magnetic energy in an external magnetic field. In particular, the formation of a domain structure without an external magnetic field leads to decreased magnetostatic energy. However, the formation of boundaries between the domains (the domain walls) requires additional energy, which is associated with the exchange and magnetic anisotropy energies. The sizes and configurations of the domains are determined by the balance of these two factors.

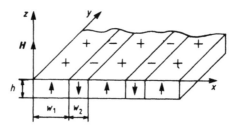

Figure 7.3. Stripe domain structure in the bias field H.

The free energy of a magnetic slab (figure 7.3) in an external magnetic field can be written as (Kooy and Enz 1960)

$$F = F_\sigma + F_H + F_M \qquad (7.1)$$

where F_σ is the energy density of the domain walls, F_H is the slab (magnetic) energy in an external magnetic field, and F_M is the magnetostatic energy, which arises from the scattering of the magnetic flux at the slab surface. In order to determine the domain widths w_1 and w_2 for an arbitrary magnetization M of the sample from expression (7.1) for the free energy, one should minimize the derivatives $\partial F/\partial(M/M_s)$ and $\partial F/\partial\tilde{h}$. After doing this, we have a set of two simultaneous equations:

$$(4\pi M_s)^{-1}(4\pi M - H) + \tilde{h}^{-1}\pi^{-2}(1+q^{1/2})^{-1}2q^{1/2}$$
$$\times \sum_{n=1}^{\infty} n^{-2}[1 - \exp(-2\pi n\tilde{h})]\sin[\pi n(1 + M/M_s)] = 0 \qquad (7.2)$$

$$(4\pi^2 h^2)^{-1}\sum_{n=1}^{\infty} n^{-3}[1 - (1 + 2\pi n\tilde{h})\exp(-2\pi n\tilde{h})]$$
$$\times \sin^2[(1/2)\pi n(1 + M/M_s)] - (32hqM_s^2)^{-1}\sigma_w(1 + q^{1/2}) = 0$$
$$(7.3)$$

where
$$q = 1 + K_u^{-1}2\pi M_s^2 \qquad \tilde{h} = (w_1 + w_2)^{-1}q^{1/2}h.$$

Using the set of equations (7.2) and (7.3), one can determine $\tilde{h}$ and M, and hence w_1 and w_2, for any given magnetic field intensity. For the particular case where $H = 0$, $M = 0$, $q = 1$, and $w_1 = w_2 = w = P/2$, from the set of equations (7.2) and (7.3) we have

$$l/h = \pi^{-3}h^{-2}P^2\sum_{n=1}^{\infty} n^{-3}[1 - (1 + P^{-1}2\pi nh)\exp(-P^{-1}2\pi nh)] \qquad (7.4)$$

where P is the period of the domain structure. The relationship (7.4) allows one to determine the value of l from the measurable P and h.

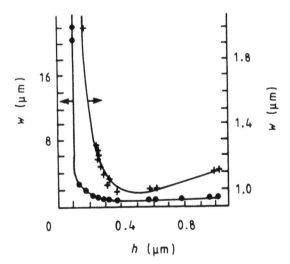

Figure 7.4. The stripe domain width versus the submicrometre film thickness (Kotov *et al* 1986a).

It is interesting to compare the above-cited theoretical expression with experiment. Figure 7.4 illustrates the width w of a stripe domain as a function of the thickness h of an epitaxial film of $Bi_{0.4}Sm_{0.2}Tm_{2.4}Fe_{4.3}Ga_{0.7}O_{12}$ with $4\pi M_s = 600$ G (Kotov *et al* 1986a). Here the crosses correspond to experimental points, and solid lines show the theoretical dependencies (according Kooy and Enz 1960).

Selecting P/h as a parameter, and using the set of equations (7.2) and (7.3), one can obtain an unambiguous correspondence between M/M_s and $(4\pi M_s)^{-1}H$, which provides the basis for determining the value of M_s from the measured values of M and H. The magnitude of the magnetization M is related to the domain structure parameters as follows:

$$M/M_s = (w_1 + w_2)^{-1}(w_1 - w_2). \tag{7.5}$$

From the experimental point of view it is convenient to determine the value of $M = M_s/2$, and then from this, using (7.5), find $w_1 = 3w_2$. It follows that $M = M_s/2$ when the width of a domain whose magnetic moment is oriented along the direction of the external magnetic field is three times that of a domain with the opposite magnetization direction. Measuring the magnitude of the bias field H for which the relationship $w_1 = 3w_2$ holds, and using the known value of P/h, we find $(4\pi M_s)^{-1}H$, from which we obtain the value of $4\pi M_s$. The bias field intensity can be found from the dependence of the rotation of the plane of polarization of the light on the bias field, because $M = M_s/2$ when $\Phi = \Phi_F h/2$ (figure 7.5).

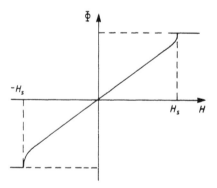

Figure 7.5. The angle of rotation of the plane of polarization of the light passed through a uniaxial film with a maze or stripe domain structure, as a function of the bias field.

Not only can the stripe domain structure be controlled by means of a bias field that is orthogonal to the film plane, but also the application of an in-plane field changes the width of the domains, and its rotation under specific conditions ensures rotation of the domain system.

7.4   The interaction of light with an isolated domain wall

Isolated domain walls have been investigated in many experimental studies. These walls are interesting from the practical viewpoint, since they can be used for constructing spatial light modulators, optical gates, etc. Most attractive are the high-frequency properties of the isolated wall.

How does an isolated wall form? An isolated plane 180° domain wall in a uniaxial material is unstable with respect to bending disturbances. The physical meaning of the instability can be understood via an analogy to a plane current wire (bus), which is based on the known Ampère conception of molecular currents. In this case the Ampère molecular current density along the domain wall is

$$j_s = 2cM_s.$$

Like a wire carrying a current, the domain wall tends to bend to reduce the current-induced magnetic field energy. The instability of an isolated wall can be suppressed by applying an inhomogeneous magnetic field dH_z/dx, where the z-axis coincides with the easy axis of magnetization, whereas the x-axis is normal to the wall plane.

Note that a tilt of the magnetic moments in domains with respect to the normal to the film or slab, which is caused by the deviation of the easy axis or by the external magnetic field, enhances the stability of the plane domain wall, because in this case the wall bending produces magnetic charges, whose magnetostatic energy makes this bending energetically unfavourable.

From the optical point of view, domain walls in magnetic materials constitute an optical inhomogeneity of the medium. Indeed, the permeability tensor of a material changes in the domain wall. A peculiarity of the domain optical inhomogeneity is that gyrotropic terms play the dominant role in changing the ε_{ik}-tensor in the wall, and in most cases it is the gyrotropic terms that determine the optical inhomogeneity.

An interesting manifestation of the inhomogeneity is the possibility of light waveguiding along the domain wall (Zvezdin and Kotov 1976a, 1977, 1988, Popkov 1977). Let us consider this effect in the geometry presented in figure 7.6, which illustrates two waveguide-type modes. One of the modes (A) is linearly polarized ($E_x \neq 0$, $E_y = E_z = 0$), and the other mode (B) is elliptically polarized ($E_x \neq 0, E_y \neq 0, E_z = 0$). The characteristic size of the mode localization is

$$\mathcal{L} = (2\pi g_0)^{-1} \varepsilon_{yy}^{1/2} \lambda.$$

For $g_0 = 10^{-2}, \lambda = 0.6\ \mu\text{m}$, and $(\varepsilon_{yy})^{1/2} = 2$, we have $\mathcal{L} = 20\ \mu\text{m}$. The domain wall can be considered as a non-reciprocal waveguide. If the A-mode propagates in the y-axis direction, then the B-mode can propagate only in the direction opposite to that of the y-axis. Changes in the propagation direction and reorientation of the magnetic moments in domains give rise to a transformation of the waveguide propagation mode into a radiation mode.

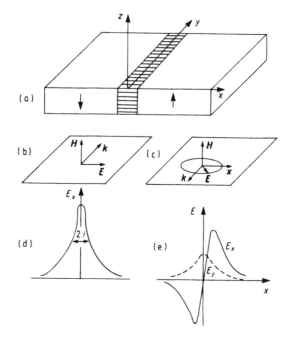

Figure 7.6. The position of a domain wall in a crystal (a), the geometry of the linearly (b) and elliptically (c) polarized modes, and the electric field distribution in the linearly (d) and elliptically (e) polarized modes.

Another interesting effect is the reflection of light from an isolated domain wall (Zvezdin and Kotov 1976a). Let the plane of incidence of an electromagnetic wave coincide with the xy-plane. For an s-polarized wave $(E_x = E_y = 0, E_z \neq 0)$, the reflection coefficient $R = 0$, whereas for a p-polarized wave $(E_x \neq 0, E_y \neq 0, E_z = 0)$,

$$R = (g_0^2 + \varepsilon_{xx}^2 \cos^2 \theta)^{-1} g_0^2$$

where θ is the angle of incidence of the wave on the domain wall (that is, the angle between the wave vector of the incident wave and the normal to the wall). For grazing oblique incidence $(\cos \theta \approx 10^{-2})$ and $g_0 \approx 10^{-2}$, $\varepsilon_{xx} = 4$, we have $R \approx 0.06$. Near the absorption band, the value of g_0 can reach large values $(g_0 \approx 10^{-1})$; in this case the reflection is more effective. A greater effect can be expected in the reflection from the stripe domain structure.

7.5 Observation of magnetic domains

The magnetooptical Faraday and Kerr effects, the phenomenon of magnetic circular dichroism (MCD), the Bitter powder pattern method, and Lorentz microscopy are used for magnetic domain observation. The Faraday effect and the phenomenon of MCD are used in studies of domain structures in thin magnetic films and slabs, which are transparent in the visible spectral region and have a magnetic moment component perpendicular to the sample surface. The domain structures in non-transparent magnetics are observed using the polar Kerr effect if there is a normal magnetization component, or the transverse Kerr effect if there is an in-plane magnetization. The Bitter powder pattern method produces a map of the domain walls in the case of in-plane magnetization. Lorentz microscopy is usually employed for studying the domains in thin metallic films.

7.5.1 Faraday effect observation of the domain structure

An individual magnetic domain is an isolated volume, which is magnetized up to saturation. When a linearly polarized light beam passes through the domain, the plane of polarization rotates by the angle

$$\Phi_1 = \Phi_F h \qquad \text{or} \qquad \Phi_2 = -\Phi_F h$$

depending on whether the M-vector is parallel or antiparallel to the direction of propagation of the light in the domain, i.e., domains with opposite magnetization rotate the plane of polarization of the light in opposite directions. In order to render the domain structure observable, the sample is placed between almost crossed polarizers and analysers.

The intensity of the light that has passed through the oppositely magnetized domains is given by the expressions

$$I_1 = rI_0[(1 - p)\sin^2(\beta + 2\Phi_F h) + p]\exp(-\alpha h) \qquad (7.6)$$

$$I_2 = rI_0[(1 - p)\sin^2\beta + p]\exp(-\alpha h) \qquad (7.7)$$

where I_0 is the intensity of the incident linearly polarized light, the factor $\exp(-\alpha h)$ takes into account the attenuation of the light intensity on passing through a layer of matter of thickness h, which has the optical absorption coefficient α, $\sin^2\beta$ determines the intensity of the light passed through the polarizer–analyser system, which is turned by an angle β with respect to the crossing position, the p-coefficient takes into account the non-ideality of the polarizers (the attenuation of the light intensity as a result of passing through the crossed polarizers), and the r-coefficient takes into account the reflection losses in the system. It follows from relations (7.6) and (7.7) that $I_1 > I_2$, i.e., while some domains look bright, the others are dark. The contrast of the domain structure is determined by the relation (Balbashov and Chervonenkis 1979)

$$K = I_1/I_2 = [(1 - p)\sin^2\beta + p]^{-1}[(1 - p)\sin^2(\beta + 2\Phi_F h) + p]. \qquad (7.8)$$

Figure 7.7 (curve 1) illustrates the dependence of K on the β-angle for the case where $\Phi_1 = 1°$ for $p = 2 \times 10^{-3}$. The maximum contrast is reached when the analyser is slightly turned from the crossing position; for lower Faraday rotation the contrast decreases. Analysis of expression (7.8) indicates that the contrast must depend significantly on the value of p. For instance, for $p = 2 \times 10^{-3}$ and $\Phi_1 = 6'$ the domain structure becomes practically invisible (figure 7.7, curve 2), whereas for $p = 10^{-5}$ and the same value of the angle Φ_1, the contrast is rather high (figure 7.7, curve 3). Values of p in the range 10^{-3}–10^{-4} are typical for dichroic polarizers; the value $p = 10^{-5}$ is reached when high-quality polarizing prisms are used.

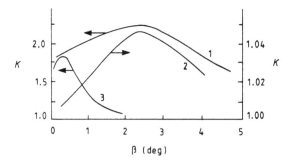

Figure 7.7. The angular dependence of the domain structure contrast K for polarizers turned from the crossing position (Kotov *et al* 1986b).

In principle, high-quality polarizing prisms ensure high contrast of the domain structure (figure 7.7, curve 3). In practice, however, achieving an attenuation of 10^{-5} requires beams that are almost parallel, which is difficult to arrange, and also decreases the optical microscope resolution. A slight depolarization—which results from multiple reflections from the epitaxial film–air interface at oblique incidence—and induced birefringence in the substrate and the elements of the optical path also contribute to the background and weaken the contrast. As a result, in many cases the utilization of high-quality polarizing prisms in the polarizing microscope does not ensure a significant improvement of the domain structure observation conditions.

The above estimates refer to the case of monochromatic radiation. Utilization of white light for illumination of an object only worsens the situation, because the dispersion of the Faraday rotation narrows the spectral interval in which the β-angle is optimal. Introduction of narrow-band interference filters does not resolve the problem, because in visual observations the diminished light intensity leads to lower contrast of the domain structure.

7.5.2 Optimization of the wavelength of the light source

The Faraday effect in magnets in the visible spectral region depends on the light wavelength; therefore, in order to make the methods of observation of domain structures using light microscopes as efficient as possible, one has to optimize the characteristics of the light source. In chapter 9, the spectral dependence of the Faraday effect for an epitaxial film of $Y_{2.6}Sm_{0.4}Fe_{3.8}Ga_{1.2}O_{12}$ is presented. This dependence is typical for all bismuth-free systems, provided that the lead content in the film is small enough. Figure 9.13 illustrates the Faraday effect dispersion in a film of $Bi_{0.5}Tm_{2.5}Fe_{3.9}Ga_{1.1}O_{12}$ (Burkov and Kotov 1975). The dependencies show that for bismuth-free compositions in the visible spectral region, the Faraday rotation maximum, which is approximately equal to 2 deg μm^{-1}, is reached at $\lambda = 0.435$ μm. For this specific Faraday rotation in real conditions, one may observe domain structures in epitaxial films of a thickness less than 0.03 μm.

For bismuth-substituted iron garnets, optimal observation conditions are reached for a 0.37 μm wavelength illuminator, where the specific Faraday rotation for the composition considered exceeds 10 deg μm^{-1}. One can expect that in this case domain structures in films of thickness less than 0.01 μm will be attainable for observation. The value of $\lambda = 0.37$ μm corresponds to invisible ultra-violet radiation. The most preferable wavelength values for visual observation of the domain structures in bismuth-substituted iron garnets are in the range 0.41–0.43 μm.

In the blue spectral region and for shorter wavelengths, the absorption coefficient of iron garnets exceeds 10^4–10^5 cm^{-1} (see chapter 9), thus limiting the maximum thickness of samples. For instance, for a wavelength of 0.44 μm the maximum sample thickness that still transmits enough light is 4 μm. Analysis of the characteristics of existing illuminators proves that to provide optimal conditions for domain structure observation in films of bismuth-substituted iron garnet, it is expedient to employ lasers with an operating wavelength of 0.44 μm. Figure 7.8(a) presents a photograph of the domain structure in an epitaxial iron garnet film of $(BiSmLuGd)_3(FeGaSc)_5O_{12}$, 0.04 μm thick, which was obtained by means of a laser microscope having an operating wavelength of 0.44 μm. The Faraday rotation of the film is equal to 6′; nevertheless the contrast is rather high. When taking the pictures, a $\lambda/4$ plate was used as an optical compensator to make the contrast higher (Kotov *et al* 1986b).

The utilization of a laser microscope ensures reliable observation of the domain structures in epitaxial bismuth-free iron garnet films having thicknesses up to 0.03 μm, and in bismuth-substituted iron garnet films with a thickness up to 0.01 μm (Kotov *et al* 1986b). Such thin films feature irregular domain structures with characteristic domain dimensions in some samples exceeding 100 μm.

7.6 Optical microscopy of domains

In addition to the reduced Faraday rotation, the contrast is limited by the optical resolution of the microscope. In the case of parallel beams the well-known Rayleigh criterion (Landsberg 1976) for two light point sources gives

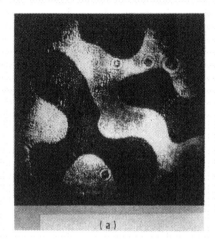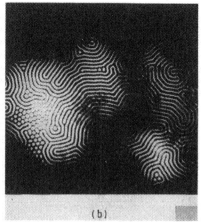

(a)   (b)

Figure 7.8. (a) Domain structure in an epitaxial iron garnet film of thickness $h = 0.04$ μm; the domain width is $w \approx 5$ μm (Kotov *et al* 1986b). (b) Domain structure in an epitaxial iron garnet film of thickness $h = 0.15$ μm; the domain width $w = 0.2$ μm (Starostin and Kotov 1982).

the following value for the ultimate resolution:

$$d \approx 1.22\lambda.$$

This value corresponds to the overlap of the first diffraction minimum of one of the sources with the maximum of the light intensity distribution in the image of the second source. The intensity drops in this case by 20%. In reality the naked eye can detect a little less difference in illumination.

Utilization of oblique incidence beams for illumination of an object almost doubles the ultimate resolution, so the real resolution of an optical microscope is

$$d \approx (A_1 + A_2)^{-1}\lambda$$

where A_1 is the condenser aperture, and A_2 is the objective aperture (Landsberg 1976). The resolution can be increased further by means of an immersion liquid, in which case the objective aperture may reach the value $A_2 = 1.5$.

The theory of phase diffraction gratings (Haskal 1970) provides a well-known expression for the angles of diffraction of the light:

$$P \sin \theta = n\lambda. \tag{7.9}$$

Here $P = 2w$ is the period of the stripe domain structure, w is the width of the stripe domain, and $n = 1, 2, \ldots$ is the diffraction order. Formula (7.9) is obviously valid for any diffraction grating. All of the information on a symmetric phase diffraction grating is contained in the data on the odd diffraction orders, the zero order of diffraction being a background noise. The domain structure is resolved by an optical microscope when the first diffraction order is within the objective aperture, i.e. if $P \sin\theta > \lambda$. Here $\sin\theta = A_2$ and we obtain the expression $P \approx \lambda/A_2$, corresponding to the case of illumination of an object with a parallel beam.

We should note that the object to be analysed here is the period of the domain structure P rather than the domain width w. Therefore, if we are interested in the ultimate resolution as regards the domain size, we should write $w \approx (2A_2)^{-1}\lambda$ for the condition for domain structure illumination with parallel beams. Taking into account the oblique-incidence beams, for the microscope resolution limit as regards the domain width we have

$$[2(A_1 + A_2)]^{-1}\lambda.$$

When using an immersion medium, assuming that $A_1 = 1.0$ and $A_2 = 1.5$, we have $w \approx \lambda/5$. Thus, by using a high-aperture objective, a condenser, and an immersion liquid at $\lambda = 0.44\ \mu m$, one can achieve an ultimate resolution as regards the domain size as high as $w \approx 0.1\ \mu m$.

Using a laser microscope with an operating wavelength of $\lambda = 0.44\ \mu m$ ensures observation of maze and bubble structures with a domain dimension of $0.2\ \mu m$ in an epitaxial film of $(CaSmLuGd)_3(FeGeGaSc)_5O_{12}$, $0.15\ \mu m$ thick (figure 7.8(b)) (Starostin and Kotov 1982, Kotov *et al* 1986b).

For the same epitaxial film, Starostin and Kotov observed isolated stripe domains just before the domain structures disappeared in the bias field. According to theoretical estimates, the width of the domain in that state has to be equal to $w = 0.1\ \mu m$. It is also possible to identify the moment of bubble collapse—that is, to observe domains with a diameter of about $0.1\ \mu m$. The fact that an optical microscope with a real resolution of $0.15\ \mu m$ ($A_1 = 0.2, A_2 = 1.35$) allows one to observe distinctly isolated domains of characteristic dimensions $0.1\ \mu m$ does not at all contradict the fundamental theoretical optical laws. In this case the problem is reduced to that of detecting a luminous source against the background noise, i.e. there is a formal analogy with observation of self-luminous celestial objects.

An optical microscope allows observation of isolated domains with dimensions less than the resolution limit, provided that the object features sufficient luminosity. In practice, the problem is reduced to suppression of the background noise, which is achieved to some extent by means of an optical compensator. In principle, the same problem can be resolved by the methods of spatial filtration of the optical signal.

A very significant factor that determines the background noise level, and, hence, the threshold of domain structure observation, is the occurrence of multiple reflections at the substrate–film and film–air interfaces. Deposition of antireflection coatings leads to significant lowering of the background noise. The effect of antireflection coating on contrast is discussed in more detail in chapter 13.

The above analysis of the ultimate possibilities for visual microscopy of domains, which is based on the Rayleigh diffraction limit, is of practical interest. There are sound mathematical reasons for suggesting that the resolution can exceed the classical diffraction limit for certain structures.

These notions are based on the fact that the Fourier representation $F(k_x, k_y)$ of an observed structure of finite dimensions is an analytical function. This means, in particular, that the entire $F(k_x, k_y)$ function can be defined by means of analytical continuation if it is defined for some spectral region (k_x, k_y). Therefore, if there is an opportunity to determine with high accuracy the Fourier representation $F(k_x, k_y)$ of an object in a particular spectral region (k_x, k_y), then the whole $F(k_x, k_y)$ function can be determined by means of known mathematical methods, and the form of an observed structure can be restored from the function.

This approach encourages expectations that for certain domain structures it will be possible to exceed the Rayleigh limit. In this case the practical problem of noise in $F(k_x, k_y)$ measurements is decisive as regards determining the ultimate resolution, and requires further thorough consideration.

The diffraction limit can also be surpassed by scanning near-field magnetooptical microscopy, which takes measurements of the Kerr effect at distances less than the wavelength of the light, λ. By scanning a small collecting aperture very close (about 10 nm) to the surface of the sample, one can obtain an optical image with spatial resolution as fine as $\lambda/40$ (Betzig and Trautmann 1992).

7.7 The magnetooptics of antiferromagnetic domains

Thermodynamic consideration of the antiferromagnetic ground state predicts the existence of antiferromagnetic domains, which differ as regards the direction of the antiferromagnetism vector L. In particular, such domains may appear in weak ferromagnets on transition from the weak ferromagnetic phase into the antiferromagnetic phase through the Morin point T_M—for instance, in $DyFeO_3$, for which $T_M \simeq 40$ K (see, e.g., Belov *et al* 1979). Let us consider this example in more detail.

It is easy to show that the special structure of the permeability tensor ε_{ik}, which is inherent in weak ferromagnets, allows antiferromagnetic domains to be observed by magnetooptical methods when the sample is placed into a magnetic field (Zvezdin and Kotov 1976b).

The ε_{ik}-tensor series expansion up to the term quadratic in L has the form

$$\varepsilon_{ik} = \varepsilon_{ik}^0 + ie_{ikl}G_l + \delta_{iklm}L_lL_m. \tag{7.10}$$

Here the second term on the right-hand side represents the usual Faraday effect, e_{ikl} is the Levi-Civita antisymmetric tensor, and $G_l = d_{lm}L_m$ is the gyrotropy vector (compare with (2.9)). The δ_{iklm}-tensor is of the same form as the magnetostriction tensor.

If the magnetic field H is applied along the z-axis of the crystal and the vector L is considered to be in the ab-plane, the ε_{ik}-tensor can be written as

$$\varepsilon_{ik} = \begin{pmatrix} \varepsilon_1 & b\sin 2\varphi + \mathrm{i}d\cos\varphi & 0 \\ b\sin 2\varphi - \mathrm{i}d\cos\varphi & \varepsilon_2 & 0 \\ 0 & 0 & \varepsilon_3 \end{pmatrix} \quad (7.11)$$

where φ is the angle that defines the position of the L-vector with respect to the a-axis (x-axis) of the crystal (figure 7.9(a)),

$$\varepsilon_1 = \varepsilon_{xx}^0 + b_{11}L_x^2 + b_{12}L_y^2$$
$$\varepsilon_2 = \varepsilon_{yy}^0 + b_{21}L_x^2 + b_{22}L_y^2$$
$$\varepsilon_3 = \varepsilon_{zz}^0 + b_{31}L_x^2 + b_{32}L_y^2$$

and $b = \frac{1}{2}b_{66}$ is one of the diagonal elements of the δ_{iklm}-tensor.

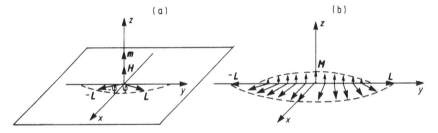

Figure 7.9. (a) The orientation of the antiferromagnetism vector L in antiferromagnetic domains separated by a 180° wall (the xz-plane) in an external magnetic field which is directed along the z-axis of the crystal. (b) The behaviour of the magnetic moment in the domain wall between antiferromagnetic domains (Zvezdin and Kotov 1976b).

Table 7.1. Propagation constants and eigenmodes of the electromagnetic wave in the crystal.

Propagation constant	Mode
$n_+^2 = \varepsilon_1 + (b^2\sin^2 2\varphi + d^2\cos^2\varphi)/(\varepsilon_1 - \varepsilon_2)$	$e_2/e_1 = (b\sin 2\varphi - \mathrm{i}d\cos\varphi)/(\varepsilon_1 - \varepsilon_2)$
$n_-^2 = \varepsilon_2 - (b^2\sin^2 2\varphi + d^2\cos^2\varphi)/(\varepsilon_1 - \varepsilon_2)$	$e_1/e_2 = (b\sin 2\varphi + \mathrm{i}d\cos\varphi)/(\varepsilon_1 - \varepsilon_2)$

Solving the problem of electromagnetic wave propagation in a medium with a permeability tensor of the same type as (7.11) (see sections 2.3 and 3.3), and using the condition $|\varepsilon_{xx}^0 - \varepsilon_{yy}^0| \gg |b|$, which is valid for dysprosium orthoferrite due to its high birefringence (this condition allows one to disregard the contribution of the δ_{iklm}-tensor to the diagonal components of the ε_{ik}-tensor), we can find propagation constants and eigenmodes of the electromagnetic wave in the crystal.

The appropriate values have the forms given in table 7.1.

It is evident that, in particular cases, for $\varphi = 0$ the solution describes the Faraday effect in ferromagnets in the presence of natural birefringence, whereas for $\varphi = \pi/2$ it describes the natural birefringence in an antiferromagnetic crystal.

Let us consider the problem of optical contrast between oppositely oriented antiferromagnetic domains. The intensity of light that passes through the polarizer–sample–analyser system is given by

$$I = \frac{c\varepsilon_0}{4\pi} |E \cdot n|^2. \tag{7.12}$$

Here the n-vector defines the position of the analyser with respect to a given system of coordinates. Let the analyser be rotated to an angle ψ with respect to the x-axis. Then

$$I = \frac{c\varepsilon_0}{4\pi} E_0^2 \left(1 + \frac{2\alpha'}{1 + |\alpha|^2} \sin 2\psi \sin^2 \frac{\varphi}{2} + \frac{\alpha''}{1 + |\alpha|^2} \sin 2\psi \sin \varphi \right). \tag{7.13}$$

Here

$$\varphi = \frac{2\pi(n_+ - n_-)}{\lambda} z \qquad \alpha' = \frac{b \sin 2\varphi(H, T)}{\varepsilon_1 - \varepsilon_2} \qquad \alpha'' = \frac{d \cos \varphi(H, T)}{\varepsilon_1 - \varepsilon_2}.$$

When deriving (7.13), we assumed that $|\alpha'|$, $|\alpha''| \ll 1$.

The orientation of the vector L in the antiferromagnetic domains is given by $\varphi = \pm\pi/2$. The magnetic field oriented along the x-axis rotates spins to the x-axis (see, e.g., Belov *et al* 1979), and, according to (7.13), induces the optical contrast between the antiferromagnetic domains. These domains were observed by Kharchenko and Belii (1980) in $DyFeO_3$ and CoF_2.

Another peculiarity of the antiferromagnetic phase in weak ferromagnets is the possibility of observing the wall between the antiferromagnetic domains, because in the wall region the antiferromagnetic vector L deviates from the b-axis of the crystal, which gives rise to a magnetic moment along the c-axis (figure 7.9(b)).

The variation of a magnetic moment in a wall between antiferromagnetic domains has the form $M_z = (d/A) \cos \varphi(x)$, which leads to there being non-zero off-diagonal components $\varepsilon_{xy} = id \cos \varphi(x)$ in the ε_{ik}-tensor in the absence of an external magnetic field. Thus, if light propagates along the z-axis (the c-axis of the crystal), the plane of polarization of the light rotates in the region of the domain wall, which leads to optical contrast, provided that the analyser is appropriately positioned.

PART 2

MAGNETOOPTICAL MATERIALS

Chapter 8

Dielectrics

8.1 Orthoferrites

8.1.1 Structure

The rare-earth orthoferrites (formula $REFeO_3$ and space group D_{2h}^{16} (P_{bnm})) have a distorted perovskite structure with only one type of Fe^{3+} ion octahedrally coordinated with O^{2-} ions. There are four Fe^{3+} ions, with $\bar{1}$ point group symmetry, per orthorhombic unit cell. The magnetization at room temperature lies along the crystalline c-axis (except for samarium orthoferrite, where it lies along the a-axis).

The basic magnetic interaction is antiferromagnetic between nearest-neighbour Fe^{3+} ions. A slight canting of the Fe^{3+} spins, due to antisymmetric exchange, produces a weak net magnetization $4\pi M$ of the order of 100 G at room temperature. There are two iron sublattices which are aligned nearly antiparallel to each other, leaving this small net magnetic moment. On the basis of magnetization data, the canting angle α is estimated to be about 0.5° for all of the orthoferrites.

The magnetic ordering temperature depends upon which rare-earth ions are present. The unit-cell volumes and Curie temperatures decrease gradually for the different rare-earth ions going across the lanthanide series from $LaFeO_3$ to $LuFeO_3$. As RE^{3+} varies from La^{3+} to Lu^{3+}, the unit-cell volume decreases from 243.1 to 218.8 Å^3, and the Curie temperature from 743 to 625 K. Rare-earth ions behave at room temperature like paramagnets in the effective field of the Fe^{3+} ions, and their magnetization can be neglected at room temperature.

8.1.2 Optical spectra

It is convenient to compare the orthoferrites with the closely related iron garnets, since the optical properties of the latter have been fairly well investigated. Figure 8.1 (curve 1) shows the absorption spectrum of $YFeO_3$ (Wood *et al* 1970).

This spectrum is typical for all orthoferrites and at the same time for rare-earth orthoferrites, in that there are many low rare-earth absorption lines in the transparent region.

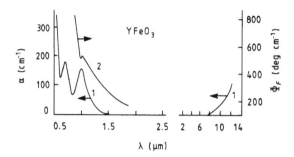

Figure 8.1. The optical absorption spectrum (curve 1) (Wood *et al* 1970) and Faraday rotation (curve 2) (Chetkin *et al* 1971) of $YFeO_3$.

Detailed information about the rare-earth absorption lines in orthoferrites is presented by Wood *et al* (1970). It can be seen that the absorption edge is situated in the visible region ($\alpha \sim 600$ cm^{-1} near 560 nm). At the same time two discrete broad absorption bands near 700 and 1000 nm are observed. There are two regions of relative transparency near 620 and 840 nm. Beyond 1300 nm, in the infra-red, the orthoferrites are intrinsically very highly transparent until the lattice absorption begins near 8 μm. The long-wavelength limit of the transparent region lies further out in the infra-red than for iron garnets, because the orthoferrites lack the higher-frequency vibration of the tetrahedral FeO_4 group present in iron garnets.

There are three principal spectra expected for the orthorhombic unit cell of orthoferrite. The two broad absorption bands near 1000 and 700 nm are due to the crystal-field transitions $^6A_1 \rightarrow {}^4T_1$ and $^6A_1 \rightarrow {}^4T_2$ of Fe^{3+} in approximately octahedral coordination (Wood and Remeika 1967, Wickersheim *et al* 1960), and the third band near 500 nm corresponds to the $^6A_1 \rightarrow {}^4E, {}^4A$ transition. When tetravalent ions such as Si^{4+} or Sn^{4+} are added to the melt during the growth of some rare-earth orthoferrites, a broad optical absorption occurs in the 'optical window'; the growth of absorption is also observed when orthoferrites are doped with divalent ions (Wood and Remeika 1967).

8.1.3 Faraday rotation

Because the orthoferrites are orthorhombic crystals, the magnetooptical behaviour is a combination of birefringence and Faraday rotation. In such a crystal, if the light propagates along the optical axis, the magnitude of Faraday rotation should be $\Phi_F \cos(\alpha)$, where α is the angle of the optical axis. The magnitude of the rotation is reduced by $\cos(\alpha)$, since the magnetization is still constrained to lie along the crystalline *c*-axis.

The spectral dependence of the Faraday rotation in YFeO$_3$ is shown in figure 8.1 (curve 2) (Chetkin *et al* 1971). It is interesting to note that the intrinsic Faraday rotation of the orthoferrites in the wavelength region from 600 to 1800 nm is greater then that in yttrium iron garnet by a factor of about 3 (Tabor *et al* 1970).

8.1.4 The complex polar Kerr effect

Kahn *et al* (1969) have measured the polar Kerr effects at room temperature between 1.8 and 5.5 eV for mirror-like growth surfaces of eleven single-crystal orthoferrites (RE = Sm, Eu, Gd, Tb, Dy, Y, Ho, Er, Tm, Yb, Lu); the results for EuFeO$_3$ are presented in figure 8.2. As has been noted by Kahn *et al* (1969), the results vary only slightly among the different rare earths.

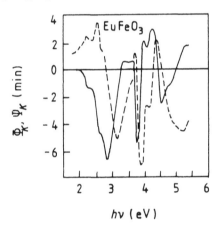

Figure 8.2. The spectral dependence of the polar Kerr effect for EuFeO$_3$ (Kahn *et al* 1969).

Examination of the spectra shows that, except the anomalous behaviour of SmFeO$_3$ above 4.35 eV and some low-energy details, the spectral features of the material investigated can be traced across the orthoferrite series from SmFeO$_3$ to LuFeO$_3$. While which rare earth is involved influences the spectra, it does not influence these common features. It is therefore apparent that the transitions responsible for these spectral features are associated with the Fe^{3+}–O^{2-} sublattices.

8.1.5 The nature of the magnetooptical activity of the orthoferrites

In most cases, when discussing the linear magnetooptic effects, it is assumed that only the non-diagonal element ε_{xy} is non-zero. In the most general case, as found for orthoferrites, ε_{yz} and ε_{zx} are non-zero, and the antisymmetric part of

ε may be represented by the pseudovector V (Kahn *et al* 1969):

$$V_l = (1/2) \sum_{jk} \varepsilon_{jkl} \varepsilon_{jk}$$

where ε_{jkl} is the alternating tensor and ε_{jk} is the jk-element of the dielectric tensor. For a system with non-interacting magnetic sublattices, it may take the form

$$V = (1/2) \sum_m \Gamma(m) S(m)$$

where the $S(m)$ is the spin of the mth sublattice. The magnitude and orientation of V depend on both the magnitude and orientations of $S(m)$, and the magnitudes and principal axes of $\Gamma(m)$. For the special case of orthoferrites, one can obtain (Kahn *et al* 1969)

$$V_z = \gamma S[\cos(\beta - \alpha)] \sin \beta = g(\alpha, \beta) \gamma S$$

where α is the angle between the x-axis and the spins $S(1)$ and $S(2)$, and β is the angle between the x-axis and the principal axes $P(1)$ and $P(2)$, in the xz-plane. This expression shows that the linear magnetooptical effects are proportional to the sine of the canting angle for $\beta = 90°$. Similarly, if the site symmetry of the magnetic ions were cubic, $\Gamma(m)$ would be a diagonal tensor, and V_z would be proportional to $\sin(\alpha)$. Anisotropy in $\Gamma(m)$ would enable the spin components $S_x(m)$ and $S_y(m)$, as well as $S_z(m)$, to contribute to the magnetooptical effects (Kahn *et al* 1969).

8.2 Manganites

(LaSr)MnO₃ and (BiLaSr)MnO₃.   The magnetooptical Kerr rotations of (LaSr)MnO₃ and (BiLaSr)MnO₃ are reported for wavelengths between 250 nm and 700 nm at 78 and 300 K (figure 8.3) (Popma and Kamminga 1975). The presence of bismuth enhances the Kerr rotation strongly at around 300 nm (for the composition $Bi_{0.3}La_{0.4}Sr_{0.3}MnO_3$ at 78 K, the rotation is 2.3° at 290 nm).

These manganites are ferromagnetic perovskites, with a Curie temperature T_c above room temperature for $0.1 < c_{Sr} < 0.6$ (Lawler *et al* 1994). The samples studied here were prepared by ceramic techniques. In the Kerr rotation spectra of the compound without bismuth (figure 8.3(a)), two dispersive transitions can be clearly observed at around 23 000 and 37 000 cm⁻¹, while some structure is also present at around 31 000 cm⁻¹. At 78 K a strong enhancement of the Kerr rotations on bismuth substitution is found (figure 8.3(a)), especially for the transition at about 37 000 cm⁻¹ (Popma and Kamminga 1975).

In the present compounds, two valences of manganese are present—Mn^{4+} and Mn^{3+}, surrounded by O^{2-} ions sited octahedrally. The ground states of these ions are $^4A_{2g}$ and ^5E_g, respectively. As noted by Popma and Kamminga, the situation is complicated by the itinerancy of the e electrons of Mn^{3+} in the compound without bismuth (resistivity at 300 K = 5×10^{-3} Ω cm) and by possible Jahn–Teller distortion around the Mn^{3+} ions in the compound containing bismuth (for $x = 0.3$, the resistivity at 300 K = 50 Ω cm).

The similarity of the characteristic frequencies in the spectra of these compounds with quite different conductivities suggests that the influence of the itinerancy of the e electrons upon the magnetooptical properties is of only minor importance. This may be due to the lack of first-order spin–orbit splitting of the orbitals with e symmetry.

Luminescence studies of Mn^{4+} ions surrounded octahedrally with oxygen ions show the spin-allowed crystal-field transitions to occur at 21 000 cm^{-1} ($^4A_{2g} \rightarrow {}^4T_{2g}$) and 25 000 cm^{-1} ($^4A_{2g} \rightarrow {}^4T_{1g}$) while the only spin-allowed d–d transition of Mn^{3+} occurs at around 20 000 cm^{-1} ($^5E_g \rightarrow {}^5T_{2g}$) (Popma and Kamminga 1975).

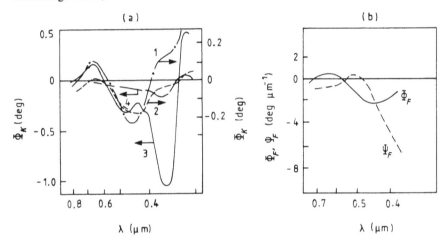

Figure 8.3. (a) The polar Kerr effect for $La_{0.7}Sr_{0.3}MnO_3$ (curves 1 and 2) and $Bi_{0.3}La_{0.4}Sr_{0.3}MnO_3$ (curves 3 and 4) at 300 K (dashed line) and 78 K (full line) (Popma and Kamminga 1975). (b) The spectral dependence of the Faraday rotation (Φ_F) and ellipticity (Ψ_F) for a film of $La_{0.82}Sr_{0.18}MnO_3$ (Cho *et al* 1990).

The Faraday rotation and ellipticity spectra of $La_{1-x}Sr_xMnO_3$ ($0 < x < 1$) films in the spectral range 400 to 750 nm (figure 8.3(b)) were studied by Cho *et al* (1990). The polycrystalline films of perovskite-type lanthanum manganese oxide were prepared by rf diode sputtering. Crystalline films were easily grown during sputtering at relatively low substrate temperatures (exceeding 170 °C). By annealing the films in an O_2 atmosphere above 900 °C, the saturation magnetization and magnetooptical properties were increased, and they approached the values of the bulk sample.

The Faraday rotation and ellipticity of the films deposited at 500 °C with pure O_2, and annealed in an O_2 atmosphere at 900 °C, reach -2.6×10^4 deg cm^{-1} at 2.9 eV and -5×10^4 deg cm^{-1} at 3.1 eV, respectively. Cho *et al* noted that the lanthanum manganese oxide films might be suitable for use as magnetooptical recording media—using the Kerr effect at short wavelength.

Other crystal-field transitions are spin forbidden, and as a consequence will have smaller oscillator strengths. The larger oscillator strengths in combination with the presence of spin–orbit splitting in the final T states mentioned above suggest that a transition of this kind is responsible for the dispersive transition in the Kerr spectra at around 25 000 cm^{-1}.

For Mn^{4+} at 30 000 cm^{-1}, the spin- and electric-dipole-allowed charge-transfer transitions ($O^{2-} + Mn^{n+} \rightarrow O^- + Mn^{(n-1)+}$) have been found, and such transitions are expected to occur for Mn^{3+} at higher frequencies because of its lower valence. These transitions have larger oscillator strengths than crystal-field transitions, and since the associated l–s splittings of the lowest charge-transfer transitions to the $t_{2g}(Mn)$ levels are comparable to those found in the crystal-field transitions, Cho *et al* attribute the features in the Kerr spectra of the material without bismuth above 30 000 cm^{-1} to these transitions.

The lowest internal bismuth transition $^1S_0(6s^2) \rightarrow {}^3P_1(6s6p)$ occurs in the energy region between 30 000 and 40 000 cm^{-1} in compounds where bismuth is coordinated with oxygen. Although spin forbidden, this transition is partly allowed by the admixture of $^1P_1$ character because of l–s coupling. Bi(6p) electrons have a large free-ion l–s coupling of 1700 cm^{-1}. Even if it is reduced by covalence, this value is large enough to produce considerable magnetooptical effects if the $^3P_1$ level is split in the exchange field. The increase of Φ_K may therefore be due to this transition.

However, these large Kerr rotations at around 37 000 cm^{-1} may also be attributed to an enhancement of the oxygen–manganese charge-transfer transitions by bismuth. The lowest charge transfer will be from an oxygen orbital which is π-bonding with the manganese ion ($O(\pi)$) to the $t_{2g}(Mn)$ orbital. Direct bismuth–manganese overlap will enhance the l–s splitting of the t_{2g} orbital. The hole left in $O(\pi)$ after charge transfer towards manganese may therefore feel the large l–s coupling of the Bi(6p) orbital because of the bismuth–oxygen covalence.

The small increase of Φ_K at around 23 000 cm^{-1} occurring because of bismuth substitution may be due to bismuth–manganese overlap. The $t_{2g}(Mn)$ orbitals are directed towards a bismuth position in the perovskite structure. Even a small admixture of Bi(6p) character into the $t_{2g}(Mn)$ orbitals may enhance its l–s splitting considerably. This admixture will also increase the oscillator strength of the manganese crystal-field transitions. Both of these effects will enhance Φ_K.

Popma and Kamminga (1975) suggested that the present results show that the enhancement of magnetooptical properties by diamagnetic Bi^{3+} ions occurs not only in bismuth-substituted iron garnets but also in $(LaBiSr)MnO_3$. The large increase of the Kerr rotation of the manganites studied at around 37 000 cm^{-1} is most probably due to an internal bismuth transition or to the admixture of Bi(6p) character into the oxygen–manganese charge transfer.

8.3 Spinel ferrites

8.3.1 Co-containing ferrimagnetic spinels

In cobalt-containing ferrimagnetic spinels, the Co^{2+} ion can be coordinated either tetrahedrally or octahedrally by oxygen atoms. The Curie temperature $T_c = 793$ K of cobalt-containing spinel ferrites is rather high. For tetrahedrally coordinated Co^{2+} ions, strong magnetooptical transitions are found in $CoFeRhO_4$, $CoFeCrO_4$, and $CoFeAlO_4$. These transitions are relatively narrow, and are centred at around 0.8 and 2.0 eV photon energy (1550 and 620 nm wavelength) (Ahrenkiel and Coburn 1975, Martens and Peeters 1983).

$CoFe_2O_4$. The polar Kerr rotation Φ_K and ellipticity Ψ_K of the magnetically saturated (111) surfaces of the single crystals $CoFe_2O_4$ are presented in figure 8.4(a) for the photon energy range from 0.65 to 4.5 eV (Martens and Peeters 1983). At the photon energies of the HeNe and AlGaAs lasers—1.96 and 1.55 eV—one observes Kerr rotations of 0.23° and 0.06°, respectively.

In $CoFe_2O_4$ the Co^{2+} ion is predominantly located on octahedrally arranged lattice sites. The broad magnetooptical transition found in $CoFe_2O_4$ is centred at around 2 eV (620 nm), and has been assigned to a $Co^{2+} \rightarrow Fe^{3+}$ charge-transfer transition at octahedral sites.

The Faraday rotation and spectral dependence of the absorption coefficients of thin polycrystalline $CoFe_2O_4$ films are given in figure 8.4(b). At 1.96 and 1.55 eV the Faraday rotations of 3.7×10^3 and 3.3×10^4 deg cm^{-1} were observed, respectively. The optical and magnetooptical properties of $CoFe_2O_4$ for the HeNe and AlGaAs laser wavelengths are given in table 8.1 (Martens and Peeters 1983).

Table 8.1. The optical and magnetooptical properties of $CoFe_2O_4$ (Martens and Peeters 1983), $CoRhFeO_4$, and $CoCrFeO_4$ (Ahrenkiel and Coburn 1975).

Material	λ (nm)	Φ_K (deg)	R	Φ_F (deg μm^{-1})	α (cm^{-1})
$CoFe_2O_4$	633	0.225	0.22	0.37	5×10^4
$CoFe_2O_4$	780	0.025	0.21	3.3	2×10^4
$CoRhFeO_4$	633			4	1.3×10^5
$CoCrFeO_4$	633			3.5	4.5×10^4

Crystalline Co spinel ferrite films were prepared by rf sputtering on glass substrates, and annealed at temperatures from 400 to 600 °C for 2 h in air by Zhang *et al* (1994). Anisotropy constants up to 2.2×10^5 erg cm^{-3}, a magnetic hysteresis loop coefficient of 0.94, and a Faraday rotation in the wavelength range from 600 to 800 nm were measured. A peak centred near 720 nm developed with increasing annealing temperature, reaching a maximum value of 3.8×10^4 deg cm^{-1} when $T_a = 600$ °C.

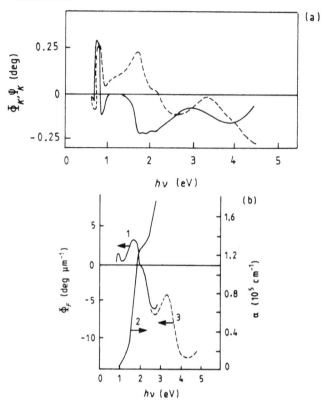

Figure 8.4. (a) The spectral dependence of the polar Kerr rotation (solid line) and ellipticity (dashed line); and (b) the Faraday rotation (solid line 1: experimental; dashed line 3: calculated) and absorption (line 2) for $CoFe_2O_4$ (Martens and Peeters 1983).

The MCD measurements on $Co_{0.4}Fe_{2.6}O_4$ films revealed two peaks near 560 nm (the transition $^4T_1(^4F) \rightarrow {}^4T_1(^4P)$ in an octahedron) and 650 nm (Edelman and Baurin 1978). Faraday rotation peaks of -2.5×10^4 deg cm^{-1} at 500 nm and 2×10^4 deg cm^{-1} at 740 nm were observed.

$CoRh_xFe_{2-x}O_4$. Large magnetooptical effects were reported for the compound $CoRhFeO_4$ ($T_c = 355$ K) at room temperature (Ahrenkiel and Coburn 1975). In figure 8.5, the dependences of the reflectance magnetic circular dichroism (RMCD) of $CoRhFeO_4$ and $CoCrFeO_4$ at 290 K, and $CoRh_{1.5}Fe_{0.5}O_4$ and $CoCr_2O_4$ at 80 K are plotted. The peak intensity for $CoRhFeO_4$ is about a factor of four larger at 80 K. The temperature dependence of the magnetooptical activity is related to the population of the higher spin states in the ground $^4A_2(F)$ manifold.

Ahrenkiel and Coburn (1975) have found that the Kerr rotation of $CoRhFeO_4$ attains $\Phi_K = -0.16°$ at 0.59 μm, crosses the zero level at 0.63 μm, and reaches $\Phi_K = 0.22°$ at 0.67 μm. It should be noted that zeros in Kerr

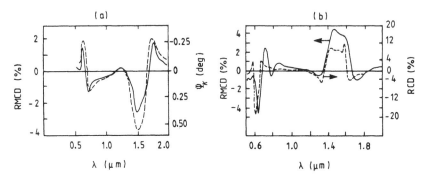

Figure 8.5. The spectral dependences of the reflectance magnetic circular dichroism (RMCD) of: $CoRhFeO_4$ (broken line) and $CoCrFeO_4$ (solid line) at 290 K (a); and $CoRh_{1.5}Fe_{0.5}O_4$ (solid line) and $CoCr_2O_4$ (broken line) at 80 K (b) (Ahrenkiel and Coburn 1975).

rotation correspond closely to peaks in the absorption band. The Kerr ellipticity may be shown to be related to the reflectance magnetic circular dichroism by the relation (Ahrenkiel and Coburn 1975)

$$\tan \Psi_K = -(1/4)(\delta R/R).$$

The magnetooptical activity of $Co^{2+}(T_d)$ ions is associated with the crystal-field transitions $^4A_2(F) \rightarrow ^4A_2(F)$ and $^4A_2(F) \rightarrow ^4T_1(P)$. The simple crystal-field model indicated that the allowed transitions to the various spin–orbit components of the $^4T_1$ state are highly circularly polarized. Obtaining high magnetooptical activity also required that only the lowest spin state, the $^4A_2(F)(3/2)$ level, is occupied at the operating temperature (Ahrenkiel and Coburn 1975).

It can be seen that the $^4A_2(F) \rightarrow ^4A_2(F)$ transition occurs at about the same energy as for $CoCr_2S_4$. However, the $^4A_2(F) \rightarrow ^4T_1(P)$ transition is blue-shifted by about 5000 cm^{-1} from that in the sulphide. This blue-shift is found for all of the oxides. A minor inflection in the spectrum is observed between 1000 and 1100 nm. This may be related to the well-known crystal-field transitions of Fe^{3+}. In the RMCD spectra of $CoRh_{0.5}Fe_{1.5}O_4$ near 620 nm at 290 K a ten-times-smaller amplitude of the RMCD was found as compared to that for $CoRhFeO_4$. This can be explained as an indication of a very low Co^{2+} (T_d) concentration in this compound (Ahrenkiel and Coburn 1975).

The measurements of the RMCD spectra of $CoRh_{1.5}Fe_{0.5}$ ($T_c = 160$ K) at 80 K have shown that the magnetooptical spectra are similar but the sign is inverted about the horizontal axis as compared to the spectra for $CoRhFeO_4$. It is known that the magnetization changes sign between $x = 1.2$ and $x = 1.5$ ($x =$ Rh content), and the a-sublattice magnetization predominates.

This means that the cobalt spin levels are reversed with respect to the external field. The decrease in the magnitude of the peaks in comparison to the case for $CoRhFeO_4$ at 80 K can be explained by some antiferromagnetic ordering in the cobalt sublattice (Ahrenkiel and Coburn 1975). The compound $CoRh_2O_4$ is antiferromagnetic, with the low-spin rhodium ion acting as a non-magnetic species.

$CoCr_xFe_{2-x}O_4$. This composition has been found to be normal over a range of x from about 0.5 to 2.0. The RMCD of $CoCrFeO_4$ ($T_c = 420$ K) at room temperature is shown in figure 8.5(a) (Ahrenkiel and Coburn 1975). Comparing the spectra of $CoRhFeO_4$ and $CoCrFeO_4$, it can see that they are almost identical, indicating again that cobalt is the optically active species or that the other host ions are relatively inactive. The blue-shift of the $^4A_2(F) \rightarrow {}^4T_1$ band for the chromium compound relative to that for the rhodium compound is 660 cm^{-1}, or about 30 nm. The absorption intensity at 630 nm for $CoCrFeO_4$ is 4.5×10^4 cm^{-1}, compared to 1.3×10^5 cm^{-1} for $CoRhFeO_4$. The calculated Faraday rotation at 630 nm for $CoRhFeO_4$ is about 3.5×10^4 deg cm^{-1}.

The substitution of Cr^{3+} ions into $Co_{0.4}Fe_{2.6}O_4$ films leads to major changes in the Faraday rotation and MCD spectra. For Cr^{3+} doping, the Faraday rotation peak near 500 nm with $\Phi_K = -2.5 \times 10^4$ deg cm^{-1} decreases down to 0.6×10^4 deg cm^{-1} at 530 nm, but another peak appears near 630 nm which attains 2.7×10^4 deg cm^{-1} for the compound $Co_{0.4}Fe_{1.55}Cr_{1.05}O_4$. At the same time, the peak near 740 nm also diminishes from 2×10^4 deg cm^{-1} for $Co_{0.4}Fe_{2.6}O_4$ to 1.3×10^4 deg cm^{-1} for $Co_{0.4}Fe_{1.55}Cr_{1.05}O_4$. In the MCD spectrum of this compound, two resolved maxima are clearly seen: a weak one at 560 nm and a strong one at 670 nm. The MCD maxima positions coincide with the points of inflection of the Faraday rotation spectra (Edelman and Baurin 1978).

$CoCr_2O_4$. This compound is ferrimagnetic, with a Curie temperature of 100 K. The spinel is normal, due to the strong preference of Cr^{3+} ions for octahedral coordination. The compound is ordered by a spiral spin arrangement of the chromium sublattice, and has a net negative moment, i.e., the a-sublattice magnetization predominates.

The RMCD of hot-pressed $CoCr_2O_4$ at 80 K is shown in figure 8.5(b) (Ahrenkiel and Coburn 1975). Characteristics of F- and P-band spectra occur at about 6700 cm^{-1} (1500 nm) and 15 000 cm^{-1} (600 nm), respectively. At 80 K, the intensity of the P band is about three times larger than that for $CoRhFeO_4$. The F-band intensity is similar for the two compounds. A plausible mechanism for the variation in P-band intensity between compounds is based on the fact that the oscillator strength is very sensitive to the crystal-field environment. The Faraday rotation at 1060 nm is calculated to be about 1.5×10^3 deg cm^{-1}.

When comparing optical and magnetooptical properties of oxides and sulphide spinels it should be noted that, making allowance for the blue-shift of the oxides, there is little doubt that the spectra have a common atomic origin. In fact, the spectral line shapes for $CoCr_2S_4$ and $CoRhFeO_4$ are almost identical. However, the intensities are different, and all of the oxide intensities are smaller by at least a factor of three. Thus, because of the smaller oscillator strengths, one should expect a smaller magnetooptical effect for the oxides than for the sulphides by about a factor of three (Ahrenkiel and Coburn 1975).

$Co_{1-x}Zn_xFe_2O_4$. Zn^{2+} ions tend to occupy the tetrahedral sites in a spinel structure, lower the magnetic moment at the tetrahedral sites, and increase the population of the Fe^{3+} ions at the octahedral sites. In the $Co_{1-x}Zn_xFe_2O_4$ system, the saturation magnetization M_s increased when a small amount of zinc was substituted in ($x < 0.4$), and then decreased with increasing zinc concentration ($x > 0.4$) (Suzuki *et al* 1988).

On substitution of zinc, an additional negative peak in the Faraday rotation spectra appeared near 550 nm, and the peaks near 750, 1600, and 1850 nm were lowered. It is expected that in CoZn ferrite the concentration of Co^{2+} ions at the tetrahedral sites will decrease with the increasing Zn^{2+} content. The measured changes in the peak heights near 1600 and 1850 nm are consistent with the view that these peaks are based on the crystal-field transition of Co^{2+} ions at the tetrahedral sites, since the concentration of Co^{2+} ions at the tetrahedral sites must be decreased by zinc substitution (Suzuki *et al* 1988).

$CoAl_xFe_{2-x}O_4$. Al^{3+} ions tend to occupy the octahedral sites, lower the magnetic moment at the octahedral sites, and increase the population of Co^{2+} ions at the tetrahedral sites. In the $CoAl_xFe_{2-x}O_4$ system, the saturation magnetization M_s decreases monotonically with increasing aluminium concentration (Suzuki *et al* 1988).

On aluminium substitution, a negative peak in the Faraday rotation spectra near 550 nm also appeared, but it had a positive background, unlike CoZn ferrite. The peaks near 750, 1600, and 1850 nm were shifted towards short wavelength. The peak near 750 nm was lowered with the increase in the aluminium concentration. These peaks near 1600 and 1850 nm in CoAl ferrite change differently to the peak near 750 nm. The heights of the peaks near 1600 and 1850 nm had a maximum when a small amount of aluminium was substituted in.

It is known that in CoAl ferrite the concentration of Co^{2+} ions at the tetrahedral sites increases with the Al^{3+} content. The measured changes in the peaks heights near 1600 and 1850 nm are consistent with the view that these peaks are based on the crystal-field transition of Co^{2+} ions at the tetrahedral site, since the concentration of Co^{2+} ions at the tetrahedral site must be increased by aluminium substitution. However, at the peak near 750 nm the Faraday rotation was decreased by both Al and Zn substitution. This indicates that some ions—not Co^{2+}—are concerned in the Faraday rotation at around 750 nm.

Suzuki *et al* (1988) suggested that the transition at around 2 eV (620 nm) could be a charge transfer from Co^{2+} to Fe^{2+}, both at octahedral sites. The absorption spectra of Co ferrite, CoZn ferrite, and CoAl ferrite thin films were studied by Suzuki *et al*. The absorption was lowered over the whole measured wavelength by substitution of zinc and aluminium.

The magnetooptical parameter $F = 2\Phi_F/\alpha$ was increased for CoZn ferrite compared with Co ferrite at 800 nm by a factor of 1.7, and decreased at 1600 nm.

At the same time F was increased for CoAl ferrite at 1600 nm by a factor of 1.4, and decreased at 800 nm (Suzuki *et al* 1988).

8.3.2 Spinel ferrites

$MgFe_2O_4$ and $Li_{0.5}Fe_{2.5}O_4$. The magnetooptical properties of these compounds are fully determined by the Fe^{3+} ions at the tetrahedral and octahedral positions. In magnesium ferrite the number of Fe^{3+} ions at octahedral sites is reduced with respect to that of Li ferrite (from 1.5 to 1.1). The same holds for the Curie temperature which is reduced from 943 to 713 K. The polar Kerr rotation and ellipticity spectra for $Li_{0.5}Fe_{2.5}O_4$ are shown in figure 8.7 (Višňovský *et al* 1986).

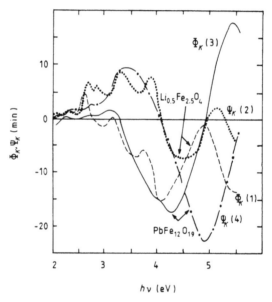

Figure 8.6. The spectral dependence of the polar Kerr rotation and ellipticity for $Li_{0.5}Fe_{2.5}O_4$ (curves 1 and 2) and $PbFe_{12}O_{19}$ (curves 3 and 4) crystals (Višňovský *et al* 1986).

8.3.3 Amorphous $NiFe_2-P_2O_5$ and $CoFe_2-P_2O_5$ films

The polar Kerr rotation in the spectral range 1.5–5 eV was measured in $NiFe_2-P_2O_5$ amorphous ferrite films prepared by vacuum evaporation (Hiratsuka and Sugimoto 1986). The Kerr rotation is weak (less than 0.15°), which indicates a photon energy dependence similar to that for crystalline $NiFe_2O_4$, except that the dispersion at 1.9 eV due to an electron charge-transfer transition is narrow in width compared with that observed for crystalline $NiFe_2O_4$. In the $CoFe_2-P_2O_5$ system, on optimizing the P_2O_5 vapour pressure, the Kerr rotation angle

$\Phi_K = 0.35°$ was attained. The films were obtained by the coevaporation of two targets: $CoFe_2$ and P_2O_5.

8.4 Hexagonal ferrites

8.4.1 Magnetoplumbite

$PbFe_{12}O_{19}$. The polar Kerr rotation and ellipticity spectrum for hexagonal $PbFe_{12}O_{19}$ (magnetoplumbite) at photon energies between 2 and 5.6 eV (figure 8.6) revealed some peaks with amplitudes of a few minutes of arc between 2 and 3.5 eV, and two intense peaks with $\Phi_K = -0.3°$ at 4.3 eV and $\Phi_K = -0.3°$ at 5.4 eV (Višňovský *et al* 1986).

The polar Kerr ellipticity spectra shown in figure 8.6 reveal some peaks between 2 and 4 eV, and one very intense peak at 4.9 eV with amplitude $\Phi_K = -0.4°$. The magnetooptical features shown in this figure provide us with evidence for attributing the peaks in the Kerr rotation and ellipticity in the spectral range 4–6 eV to one magnetooptical transition at 4.9 eV usually called the paramagnetic-type magnetooptical transition (Višňovský *et al* 1986).

8.4.2 Barium hexaferrite

$BaFe_{12}O_{19}$. The hexagonal ferrite $BaFe_{12}O_{19}$ shows rather high Faraday rotation in the visible region (Nakamura *et al* 1987, Krumme *et al* 1989), and perpendicular magnetic anisotropy (Coey 1978, Masterson *et al* 1993). The spectral dependencies of the Faraday and Kerr rotation, and ellipticity of $BaFe_{12}O_{19}$ are shown in figure 8.7(a) (Krumme *et al* 1989).

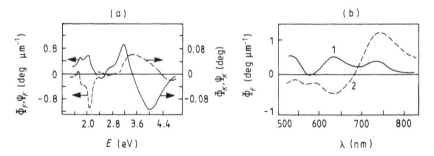

Figure 8.7. (a) The spectral dependence of the Faraday and Kerr rotation (solid lines), and Faraday and Kerr ellipticity (dashed lines) for $BaFe_{12}O_{19}$ at 295 K (Krumme *et al* 1989). (b) The Faraday rotation for $BaFe_{12}O_{19}$ (solid line 1) and $BaFe_{10.42}Co_{0.78}Ti_{0.8}O_{12}$ (dashed line 2) (Nakamura *et al* 1987).

$BaFe_{12-x-y}Co_xTi_yO_{12}$. It was found that the Faraday rotation in $BaFe_{12}O_{19}$ in the near infra-red and at 800 nm was enhanced by Co substitution (Nakamura *et al* 1987, Šimša *et al* 1992, Masterson *et al* 1993, 1995). There are two magnetooptical features in Co-substituted barium hexaferrite in the near infra-red between 0.7 and 1.0 eV, and at 1.7 eV. The Faraday rotations of $BaFe_{12}O_{19}$ and $BaFe_{10.4}Co_{0.8}Ti_{0.8}O_{12}$ in the spectral range 830–520 nm are shown in figure 8.7(b) (Nakamura *et al* 1987).

Šimša *et al* (1992) relate the first three peaks in the Faraday rotation spectra at 0.72, 0.86, and 1.0 eV to two paramagnetic transitions at 0.75 and 1.07 eV, and one diamagnetic transition at 0.86 eV. The paramagnetic lines are related to the $^4A_2(F) \rightarrow {}^4T_1(F)$ transition of the Co^{2+} ions at tetrahedral positions, while the diamagnetic transition at 0.86 eV is due to Co^{3+} ions at octahedral sites.

Barium ferrite films in which Co, Cr, Mn, and Ni are selectively introduced to substitute for between 5 and 20 at.% of the Fe were prepared by Carey *et al* (1994a, b). For most samples, the saturation Kerr rotation at 670 nm was enhanced compared with the rotation produced by a barium ferrite film, but there was only a slight improvement at 825 nm for Mn substitution. A 5 at.% Co substitution provides the maximum polar Kerr rotation at 670 nm of 0.14°, and at the same wavelength 5 at.% Mn gives −0.12° and at 825 nm 0.09°.

8.5   Chalcogenide spinels

CdCr$_2$S$_4$. The ferromagnetic semiconductor $CdCr_2S_4$ has the cubic spinel structure with $T_c = 221°C$, and is rather transparent in the spectral range 0.9–14 μm. The absorption spectrum (curve 1) at 300 K of hot-pressed $CdCr_2S_4$ in the spectral range 0.8–50 μm (Moser *et al* 1971), and the Faraday rotation (curve 2) at 4.2 K near the absorption edge (Ahrenkiel *et al* 1971) are given in figure 8.8.

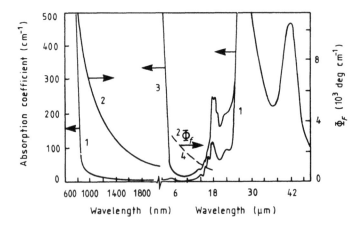

Figure 8.8. The absorption spectra at 300 K of CdCr$_2$S$_4$ (curve 1) (Moser *et al* 1971) and of CoCr$_2$S$_4$ (curve 3) (Carnall *et al* 1972), and the Faraday rotations for CdCr$_2$S$_4$ (curve 2) at 4.2 K near the absorption edge (Ahrenkiel *et al* 1971), and CoCr$_2$S$_4$ (curve 4, $2\Phi_F$) at 80 K in the transparency band (Carnall *et al* 1972).

CoCr₂S₄. $CoCr_2S_4$ is a ferrimagnetic semiconductor with $T_c = 221$ K. The compound is a normal spinel, with the cobalt ion occupying the a-site of the spinel lattice. The absorption coefficient at room temperature in the spectral range 3–40 μm, and the Faraday rotation at 80 K in the spectral range 6–16 μm and for an applied field of $H = 7$ kOe (curve 2) for $CoCr_2S_4$ are shown in figure 8.8 (Carnall *et al* 1972).

The polar Kerr rotation and ellipticity for hot-pressed $CoCr_2S_4$ at 80 K are shown in figure 8.9 (Ahrenkiel and Coburn 1973). The strong dispersion-like peaks centred at about 1.0 and 1.7 μm are associated with the $^4A_2(F) \rightarrow {}^4T_1(F)$ crystal-field transition of Co^{2+} ions in tetrahedral positions.

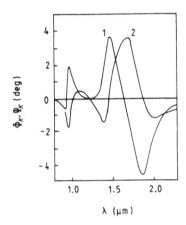

Figure 8.9. The polar Kerr rotation (curve 1) and ellipticity (curve 2) for $CoCr_2S_4$ single crystal at 80 K and in a saturating field (Ahrenkiel and Coburn 1973).

CuCr₂Se₄. The spinel-type chalcogenide $CuCr_2Se_4$ with metallic-type conductivity is a new material, with a pronounced magnetooptical Kerr effect at room temperature. The compound shows the Curie temperature $T_c = 430$ K. The polar Kerr rotation Φ_K was measured between 0.55 and 5.0 eV, and the corresponding Kerr ellipticity Ψ_K was measured between 0.55 and 3 eV. The results are shown in figure 8.10 (Brandle *et al* 1990). A pronounced structure with values up to $-1.11°$ in Φ_K at 0.81 eV and $-1.19°$ in Ψ_K at 0.96 eV has been discovered.

Measurements of the near-normal-incidence reflectivity R show the presence of a steep decrease in reflectivity from more than 80% at 0.03 eV down to a minimum of 24% at 1.24 eV. The minimum in the reflectivity occurs at an energy somewhat higher than the screened plasma energy E_p. This fact, and the resonant-like shape of the magnetooptical anomaly led Brandle *et al* to suggest that the large magnetooptical Kerr effect in this compound is due to the step plasma edge in the reflectivity.

The plasma energy, screened by the optical inter-band transition, is defined as the energy at which the so-called plasma edge occurs in the reflectivity.

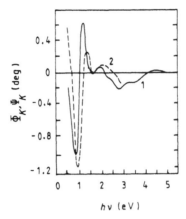

Figure 8.10. The polar Kerr rotation (curve 1) and ellipticity (curve 2) for $CuCr_2Se_4$ single crystal at room temperature in an applied field of 20 kOe (Brandle *et al* 1990).

Assuming for $CuCr_2Se_4$ a carrier concentration corresponding to one hole per formula unit at room temperature, and a contribution of inter-band transitions similar to that for $CdCr_2Se_4$, Brandle *et al* predicted the screened plasma edge energy $E_p = 1.1$ eV. It should be noted that the minimum in the reflectivity occurs at an energy somewhat higher than the screened plasma energy E_p (Brandle *et al* 1990).

$CuCr_2Se_4$ has a maximum figure of merit of 0.84° at 0.88 eV, which is actually slightly higher than the value of 0.83° ($\Phi_K = 1.27°$) obtained at 1.57 eV for PtMnSb. An interesting feature of plasma edge enhancement of the magnetooptical Kerr effect is the possibility of fine tuning the screened plasma energy E_p, particularly by variation of the carrier concentration. In the case of bromide-containing copper chromium selenide, a shift of the reflectivity minimum towards a lower photon energy as expected in a simple ionic model has already been found (Brandle *et al* 1990).

8.6   Chromium trihalides

8.6.1   Properties of $CrCl_3$, $CrBr_3$, and CrI_3

These compounds are moderately transparent over a wide spectral range, and a very large Faraday rotation may be observed at low temperature. Exhaustive information about the structural, optical, and magnetooptical properties of ferromagnetic chromium trihalides is given by Dillon *et al* (1966) in their review.

$CrCl_3$ and $CrBr_3$ have the same crystal structure at low temperature: a covalently bonded 'sandwich' in which a hexagonal net of Cr^{3+} ions lies between two close-packed layers of halogen ions. While the bonding within the sandwiches is predominantly covalent, successive sandwiches are held together only by weak van der Waals bonds. It is supposed that chromium iodide has the same structure.

The point symmetry for the Cr^{3+} site is C_3, but the leading term in the expansion of the crystal field is cubic, and this is followed by a trigonal term. The crystal-field parameters of $CrCl_3$ and $CrBr_3$ are given in table 8.2 (Wood *et al* 1963, Dillon *et al* 1966).

Table 8.2. The crystal-field parameters of $CrCl_3$ and $CrBr_3$ (Wood *et al* 1963, Dillon *et al* 1966).

	Dq (cm^{-1})	B (cm^{-1})	C (cm^{-1})	C/B
$CrCl_3$	1370	550	3400	6.3
$CrBr_3$	1340	370	3700	10

The main feature of chromium trihalide structure is that the chromium ion is surrounded by a very nearly regular octahedron of halogen ions. Chromium tribromide and chromium triiodide are true ferromagnets; on the other hand chromium trichloride is a metamagnet. In the latter compound, below T_c the spin sublattices are arranged antiferromagnetically, but in relatively small fields— a few thousand oersted—they become ferromagnetically arranged. Table 8.2 gives detailed information on the crystal-field parameters of magnetic chromium trichalides (Dillon *et al* 1966).

8.6.2 Optical absorption

In the optical spectra of chromium trihalide, many of the lower-lying sharp lines and broad bands typical of the Cr^{3+} ion in a crystal environment with octahedral symmetry are observed. The absorption spectra of chromium trihalides measured at 1.5 K in a saturating field are shown in figure 8.11 (Dillon *et al* 1966). The positions of the bands and lines are consistent with the results predicted by crystal-field theory. The high-energy end of each absorption curve is determined by a strong absorption band edge.

8.6.3 Magnetooptical properties

The Faraday rotation data for chromium trihalides measured at 1.5 K in a saturating field by Dillon *et al* (1966) are shown in figure 8.11. For the tribromide a dispersion-like structure associated with the two quartet absorption peaks is easily seen. The trichloride Faraday rotation reveals a dispersion at the $^4T_1$ band. Narrower peaks corresponding to the various doublet transitions are much more prominent here than for the tribromide.

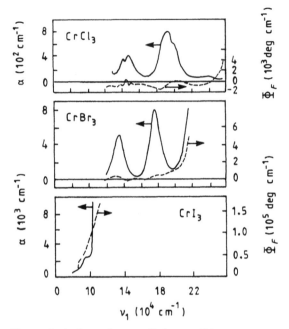

Figure 8.11. The optical absorption coefficient (solid curves) and specific Faraday rotation (dashed lines) for $CrCl_3$, $CrBr_3$, and CrI_3 at 1.5 K in a saturating magnetic field (Dillon *et al* 1966).

For $CrBr_3$ and CrI_3, differences between the positions of the absorption edges for the two senses of circular polarization were found (Dillon *et al* 1966). The measured differences are ≈ 0 cm^{-1} for $CrCl_3$, ≈ 200 cm^{-1} for $CrBr_3$, and ≈ 90 cm^{-1} for CrI_3. It is interesting to note that the $^2T_2$ line in $CrBr_3$ appears only for the sense of polarization for which the edge is apparently higher in energy.

A general expression for the complex Faraday rotation has been derived that is based on the Heitler–London model for the unpaired spins of chromium ions. From comparing theoretical results based on the calculation of the spin–orbit coupling constant of the final state for several electron configurations, and the magnitude of the trigonal splitting of the final state, with the observed difference between the absorption band edges for the two senses of circular polarization, Dillon *et al* (1966) attribute the absorption band edge of the chromium trihalides to the electron-transfer transition from a $t_{1u}(\sigma)$ orbital to an e_g^* orbital (an electron on a Br$^-$ orbital is promoted into an orbital of the Cr^{3+} ion).

As noted by Dillon *et al* (1966), the Faraday rotation at the foot of the edge would seem to arise from the splitting of a much weaker electron-transfer transition from a $t_{1u}^n(\pi)$ or $t_{2u}^n(\pi)$ orbital to an e_g^* orbital. The large splitting of this transition is due to the spin–orbit interaction which arises from the itinerancy of the p_π hole among the halogen ions.

8.7   Europium oxide and monochalcogenides

EuO, EuS, EuTe, and EuSe. EuO and EuS are ferromagnetic crystals, EuSe is a metamagnet which becomes ferromagnetic in easily accessible fields, and EuTe is a simple antiferromagnet. The Curie temperatures are 69, 16.6, and 7 K for EuO, EuS, and EuSe, respectively. The compounds have the exceedingly simple crystal structure of rock-salt. The Néel temperature of EuTe is 10 K (Suits 1972).

The magnetooptical properties of europium compounds can be interpreted on the basis of an energy level diagram in which the valence band consists of $Se(^4P)$ states and the conduction band consists of $Eu^{2+}(5d, 6S)$ states, while the lower edge consists mostly $Eu^{2+}(6S)$ levels (Dillon 1968). Below the edge of the conduction band there are a number of magnetic excitons of 5d character in which a hole in the 4f band and an electron in the conduction band are bound together.

Divalent europium has seven electrons half-filling its 4f shell. These levels are localized on the Eu^{2+} ions and lie above a filled valence band. The magnetooptical transitions are identified as being between the 4f levels and an optically active 5d-like magnetic exciton. These excited states are split by the spin–orbit interaction, and therein lies the origin of the large magnetooptical rotations.

8.7.1   EuS and doped EuS films

Figure 8.12(a) displays the Faraday rotation and magnetic circular dichroism spectra for EuS at magnetic saturation ($H = 100$ kOe) at $T = 2.5$ K measured on thin evaporated films—of 100 to 200 nm thickness (Schoenes 1987). The magnetic circular dichroism (MCD) has been obtained by Kramers–Kronig inversion of the Faraday rotation spectrum. The direct measurement of the MCD gave excellent agreement with the computed spectrum.

The first remarkable property exhibited in figure 8.12 is the size of the magnetooptical effects. The maximum Faraday rotation of 2.5×10^6 deg cm^{-1} was observed at 4.4 eV. A second maximum, at 2.15 eV, reaches 2×10^6 deg cm^{-1}.

Gambino *et al* (1992) have measured the magnetooptical properties of EuS–Gd and EuS–Tb films at $T = 5$ K. The general formula for these materials is $Eu_{1-x}RE_xS_{1-x}[\ ]_y$, where $y \simeq x + 0.1$ and [ ] denotes the vacancies on the sulphur sites. Because the trivalent rare-earth metal is added without a corresponding amount of sulphur, we can expect vacancies on the sulphur sites. As Tb or Gd is added, the sulphur deficiency increases as expected: one S vacancy per added RE atom. Analysis of the S/RE ratio shows that even undoped films of EuS are about 10% deficient in sulphur. When EuS is codeposited with Tb, the resulting films have the rock-salt structure, indicating solid-solution formation at least up to $x = 0.2$.

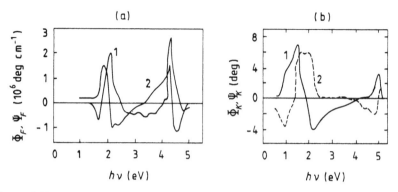

Figure 8.12. (a) The specific Faraday rotation (curve 1) and the magnetic circular dichroism (curve 2) for EuS films at 2.5 K and $H = 100$ kOe. (b) The polar Kerr rotation (curve 1) and the ellipticity (curve 2) for EuO single crystal at 10 K and $H = 40$ kOe (Schoenes 1987).

The Kerr rotation spectra of undoped EuS have a number of well-defined features: a large positive peak at 1.6 eV ($+4°$), a very large negative peak at 1.8 eV ($-8°$), and two smaller peaks at 2.4 and 2.9 eV ($+2°$). There are a number of other features at higher energy. With doping, the peaks at 1.6 and 1.8 eV are generally suppressed, and the peak at 2.4 eV grows significantly to typically $+6°$. A new feature appears between 3 and 4 eV ($-6°$), which broadens and shifts to higher energy at higher Tb concentration. The $x = 0.4$ sample is somewhat exceptional in that it shows an enhancement of both the 1.8 eV peak ($-12°$) and the 2.4 eV peak ($+9°$).

The Gd-doped samples show the same suppression of the peaks at 1.6 and 1.8 eV. The peaks at 2.4 eV are only enhanced in the $x = 0.1$ sample. The negative peak between 3 and 4 eV observed for the Tb-doped samples is absent for the Gd-doped samples. The Faraday spectrum of EuS shows a peak at 2.1 eV ($+7 \times 10^5$ deg cm^{-1}) and negative peaks at 1.7 and 3.5 eV. The Tb-doped samples show similar spectral features, but scaled down by a factor of four. As noted by Gambino *et al* (1992), the enhancement of the Kerr rotation peak at 2.4 eV for doped samples is probably related to the low reflectivity observed in this energy range, which indicates low values of the index $n(\omega)$ and the absorption coefficient $k(\omega)$. This would also explain why the Faraday spectrum is largely unchanged; this property is less sensitive to the magnitude of the refractive index n than the Kerr effect.

A two-phase material—EuS that contains cobalt precipitates ($\cong 100$ Å) (Gambino and Fumagalli 1994)—is described in chapter 12.

8.7.2   EuO

The second example showing high magnetooptical activity that we consider is EuO. The Kerr rotation and ellipticity for EuO single crystal at $T = 10$ K and $H = 40$ kOe are shown in figure 8.12(b) (Wang *et al* 1986). A maximum Kerr rotation of $7.1°$ at 1.4 eV was observed. A qualitative comparison with the Faraday rotation measurements for EuS shows similarities of Φ_K with Ψ_F, and Ψ_K with Φ_F. Nevertheless, there is also a remarkable difference between the two sets of spectra—namely, the relative weakness of the high-energy structure in Φ_K and Ψ_K. In this case, $\delta E_{so} \delta E_h$ is somewhat smaller due to the larger bandwidth f_0 of the 5d(t_{2g}) and (e_g) subbands. For EuO, Wang *et al* have obtained a crystal-field splitting of the 5d states of $10Dq = 3.7$ eV, an exchange splitting of $\delta E_{fd} = 0.59$ eV, and a spin–orbit coupling constant of $\zeta = 0.055$ eV (Schoenes 1987).

8.8 Ferric borate and ferric fluoride

FeBO$_3$ and FeF$_3$ represent two materials which can be described as ferromagnetic crystals that are transparent at room temperature (Wolfe *et al* 1970). These materials—ferric borate and ferric fluoride—are both green, and are weak ferromagnets at room temperature. FeBO$_3$ has the rhombohedral calcite structure. The crystal structure of FeF$_3$ is rhombohedral with a bi-molecular unit cell. Some magnetic, optical, and magnetooptical properties of FeBO$_3$ and FeF$_3$ are presented in table 8.3 (Wolfe *et al* 1970).

Table 8.3. Properties of FeBO$_3$ and FeF$_3$ (Wolfe *et al* 1970). (All of the optical and magnetooptical parameters shown were measured at 525 nm.)

Parameters	FeBO$_3$	FeF$_3$
T_c (K)	348	363
$4\pi M_s$ (G)	115	40
Absorption edge (nm) ($\alpha = 1000$ cm^{-1})	450	244
Absorption minimum (cm^{-1})	39	4.4
Φ_F (deg cm^{-1})	2300	180
Φ_F/α (deg dB^{-1})	14	9.5
n	2.1	1.54
Birefringence (deg cm^{-1})	-4×10^3	8×10^4

In both materials, the (111) plane is the easy plane of magnetization, and the threefold [111] axis is the hard axis. All of the iron spins in each material are parallel to the (111) plane, with a small amount of canting away from the antiparallel configuration. Some magnetic and magnetooptical parameters of the materials discussed are presented in table 8.3 (Wolfe *et al* 1970). In both materials Fe^{3+} ions lie in distorted octahedral environments, and are birefringent.

The Faraday rotation, birefringence, and optical absorption spectra of FeBO$_3$ (Kurtzig *et al* 1969) and FeF$_3$ (Kurtzig and Guggenheim 1970) are

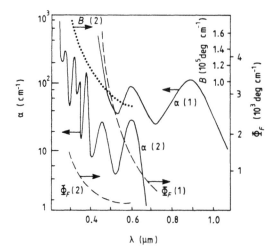

Figure 8.13. The optical absorption coefficient (α), Faraday rotation (Φ_F), and birefringence (B) for $FeBO_3$ (1) (Kurtzig *et al* 1969) and for FeF_3 (2) (Kurtzig and Guggenheim 1970).

shown in figure 8.13. The absorption spectrum of each material consists of a series of absorption bands and a fundamental absorption edge. In $FeBO_3$ there are two pronounced broad absorption bands; at the same time, in FeF_3, six absorption peaks are distinguishable at room temperature. At low temperature, in both materials, each of these broad absorption peaks is split into several well-resolved peaks. It should be noted that, due to natural birefringence, in $FeBO_3$ and FeF_3, the maximum achievable rotation for green light propagating parallel to the magnetization would be only 0.7° and 0.3°, respectively. The main disadvantage of these materials is that while in these uniaxial crystals, for light propagating along the [111] axis, there is no birefringence, with the magnetization lying in the (111) plane, there is no Faraday effect either.

Chapter 9

Ferrimagnetic garnets

9.1 Magnetic and structural properties

The unit cell of garnet is cubic with space group O_h^{10} ($Ia3d$). The magnetic unit cell contains eight formula units of $RE_3Fe_5O_{12}$ with 24 Fe^{3+} ions in the (d) (tetrahedral) sites, 16 Fe^{3+} in the [a] (octahedral) sites, and 24 rare-earth ions RE^{3+} in the {c} (dodecahedral) sites. The ions O^{2-} form a close-packed structure; the voids between the oxygen ions are filled by the rare-earth and iron ions.

There are three types of void in the garnet structure—dodecahedral, octahedral, and tetrahedral. In the dodecahedral position the rare-earth ion is bonded to eight oxygen ions, in the octahedral position the iron ion is bonded to six oxygen ions, and in the tetrahedral position the iron ion is bonded to four oxygen ions. One formula unit of the garnet contains three dodecahedral, two octahedral, and three tetrahedral positions.

For one volume unit (cm^3) of the garnet there are 5.0×10^{22} oxygen ions, 2.1×10^{22} iron ions, and 1.3×10^{22} rare-earth ions. The rare-earth cations occupy the dodecahedral positions {RE^{3+}}, and the iron cations occupy the octahedral [Fe^{3+}] and tetrahedral (Fe^{3+}) positions. The presence of the iron ions in the octahedral and tetrahedral positions gives rise to complicated optical and magnetooptical spectra as compared, for example, with those for the orthoferrites and other materials, containing iron ions only in the octahedral positions.

In the iron garnet the [Fe^{3+}] ions have the local point symmetry $\bar{3}$, and the (Fe^{3+}) ions have the local point symmetry $\bar{4}$. While the distance Fe^{3+}–O^{2-} in oxygen polyhedra is constant, there is some distortion of the polyhedra. The Fe^{3+} ions placed in the octahedral positions form an octahedral magnetic sublattice, and the Fe^{3+} ions placed in the tetrahedral positions form a tetrahedral magnetic sublattice.

The main superexchange interaction is between the tetrahedral and octahedral iron ions. It is known that such an interaction gives rise to antiparallel ordering of the magnetic moments in the octahedral and tetrahedral iron sublattices. There are also superexchange interactions between the ions belonging the same sublattice, but such intra-sublattice interactions are only one tenth as strong as the inter-sublattice interactions.

When magnetic rare-earth ions are placed in dodecahedral positions they form a third magnetic sublattice—the dodecahedral magnetic sublattice. The superexchange interaction between the rare-earth and iron ions is the smallest among the inter-sublattice interactions. The main exchange interaction between the rare-earth and iron ions is the interaction between the dodecahedral rare-earth lattice and the tetrahedral iron lattice.

It can be said that the rare-earth ions are in the molecular field formed by the iron ions; the magnitude of such a field is of the order of 100–300 kOe at room temperature, depending on the kind of rare-earth ion involved (Krinchik and Chetkin 1961). It should be noted that the effective magnetic field related to the superexchange between the octahedral and tetrahedral sublattices is nearly 2 MOe.

Magnetic materials that possess two or more inequivalent magnetic sublattices are called ferrimagnetics. The saturation magnetization M_s of a rare-earth iron garnet can be expressed as the sum of the magnetizations of each individual sublattice:

$$M_s = M_a + M_d + M_c$$

where M_a, M_d, and M_c are the saturation magnetizations of the octahedral, tetrahedral, and dodecahedral sublattices, respectively. In the yttrium iron garnet there are two magnetic sublattices because the Y^{3+} ion is non-magnetic.

9.2   Yttrium iron garnets

9.2.1   Optical properties

The interpretation of the optical transitions which take place in iron garnet is very difficult. For this reason, in discussing the optical properties of the iron garnet it is convenient to consider the optical spectra for wavelengths longer than 440 nm separately from those for wavelengths less than this.

The iron garnets are highly transparent in the near-infra-red range 1.3–5.5 μm (figure 9.1) (Wood and Remeika 1967); optical absorption coefficients as low as 0.03 cm^{-1} are attained. At wavelengths longer than 5.5 μm there is an increase in the optical absorption due to the lattice vibrations (phonon absorption). The increase of the optical absorption at wavelengths less than 1.5 μm is related to the intrinsic electron transition in octahedral [Fe^{3+}] ions, with the centre of the absorption band near 900 nm.

The ground state of free Fe^{3+} ions is ^6S. By Hund's rule, in this state all five electrons of the d shell have parallel spins, so the spin of the Fe^{3+} ion is equal to $S = 5/2$ and the angular momentum $L = 0$. The first excited state of the Fe^{3+} ion (the ^4G term) has $S = 3/2$ and $L = 4$, the next excited state (the ^4F term) has $S = 3/2$ and $L = 3$, then $S = 3/2$ and $L = 2$ (the ^4D term), and $S = 3/2$ and $L = 1$ (the ^4P term) follow.

The crystal field changes the energy of the excited states and partly removes the degeneracy of the states. Figure 9.2 shows the splitting of the terms for the Fe^{3+} ion in the tetrahedral and octahedral crystal fields (Wood and Remeika 1967). The Dq-factor characterizes the value of the crystal field: as can be seen from figure 9.2 the value of the crystal field in the octahedral position of the garnet is nearly twice that in the tetrahedral position. From the practical point of view, the most interesting feature is the splitting of the lowest term ^4G, which gives, in a crystal field, three sublevels: $^4T_1$, $^4T_2$, and $^4A_1$, ^4E. If the local distortion is taken into account, the states $^4T_1$ and $^4T_2$ split into sublevels and the degeneracy of the states $^4A_1$ and ^4E is removed. The ground state ^6S of the Fe^{3+} ion does not split in octahedral and tetrahedral crystal fields.

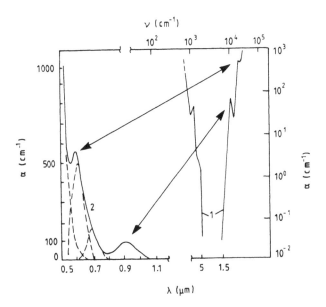

Figure 9.1. The optical absorption of $Y_3Fe_5O_{12}$ at low frequency due to lattice vibrations, and at high frequency due to octahedral Fe^{3+} crystal-field transitions (curves 1), and a detailed view of the absorption edge (curve 2) (Wood and Remeika 1967).

As can be seen from figure 9.2, because the crystal field of the octahedral ion is expected to be larger than that of the tetrahedral ion, the lowest-lying line in the spectrum is expected to arise from the transition $^6A_{1g} \rightarrow {}^4T_{1g}$ of the octahedral Fe^{3+} ion. The next transition will be $^6A_{1g} \rightarrow {}^4T_{2g}$ at the same position, and the transitions $^6A_1 \rightarrow {}^4T_1$ and $^6A_1 \rightarrow {}^4T_2$ in the tetrahedral position will be the next in energy.

The absorption edge. The optical properties of the iron garnet near the absorption edge (1.0–1.3 μm) are very important from the practical point of view, because iron garnet is widely used for developing non-reciprocal and other magnetooptical devices for wavelengths of 1.06, 1.15, 1.3, and 1.55 μm.

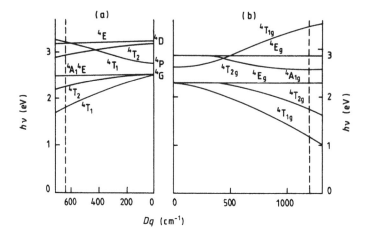

Figure 9.2. The crystal-field computation for the octahedral (right) and tetrahedral (left) sites of $Y_3Fe_5O_{12}$. The best fit with experiment lies on the vertical line for each case (Wood and Remeika 1967).

The spectral and temperature dependences of the absorption edge of $Y_3Fe_5O_{12}$ (YIG) in the spectral range between 1.0 and 1.2 μm were studied by Wood and Remeika (1967). The spectral dependence between 1.1 and 1.9 μm of the absorption coefficient of high-purity YIG, which has a low absorption level in the transparent window, is given in chapter 13 (Nakano *et al* 1984).

The temperature dependence of the absorption level at 1.06 μm is related to the broadening of the absorption band $^6A_{1g} \rightarrow {}^4T_{1g}$ with temperature increase. It should be noted that iron garnet is potentially very promising as regards practical use at wavelengths of 1.15 and 1.3 μm at room temperature; also, if the temperature is reduced to 77 K it may be suitable for use in the development of non-reciprocal optical devices (optical isolators) for the wavelength of 1.06 μm.

The absorption spectrum between 440 and 1100 nm. In this spectral range the energy position of the absorption lines of YIG can be predicted rather well on the basis of crystal-field considerations. Comparison of the energy positions measured for the absorption lines (figure 9.1) with the calculated ones (figure 9.2) for Fe^{3+} ions at octahedral and tetrahedral sites of the iron garnet allows us to identify the positions of the different crystal-field transitions.

Crystal-field theory is also useful for the interpretation of intensities, as well as the energies of the transitions. Specifically, one expects a tetrahedrally coordinated ion to have lines at least one order of magnitude more intense than the corresponding lines for an octahedrally coordinated ion. However, the absolute magnitudes are very appreciably larger than expected for spin-forbidden transitions. Thus, while the intensity ratios support the crystal-field assignment, the absolute intensities require that some perturbation of the energy level system be present.

In reality, the absorption spectra of YIG are more complicated than is predicted from simple crystal-field theory, because there are local distortions of the octahedral and tetrahedral sites, and the spin–orbit and exchange interactions. The splitting of crystal-field transitions resolved using low-temperature optical and magnetooptical measurements, and the magnetic circular dichroism data reveal a rather complicated energy level structure (Scott *et al* 1975).

Studies of the iron garnet compositions in which the iron sublattice has been diluted with either Ga^{3+} or Sc^{3+} have shown (figure 9.3) that there is, at all energies, a general trend of reduction in the absorption with iron dilution in either the tetrahedral or octahedral sublattices. In the former case the tetrahedral iron is removed, while in the latter case the octahedral iron is removed.

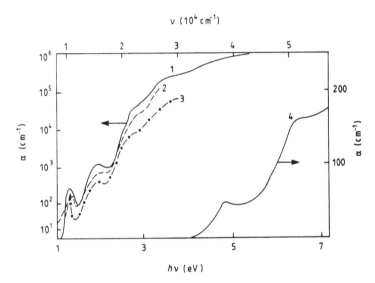

Figure 9.3. Absorption spectra of $Y_3Fe_5O_{12}$ (curve 1), $(YGdTm)_3Fe_{4.3}Ga_{0.7}O_{12}$ (curve 2), $Yb_3Fe_{3.8}Sc_{1.2}O_{12}$ (curve 3) (Wemple *et al* 1974), and Al_2O_3 doped with 0.005 at.% of Fe^{3+} (curve 4) (Scott *et al* 1974).

As has been shown by Hansen and Krumme (1984), when either tetrahedral or octahedral positions are diluted, the absorption coefficient changes according to

$$\alpha(y) = \alpha_0(1 - y_a/2)(1 - y_d/3)$$

where y_a and y_d are the contents of diluting ions in the octahedral and tetrahedral positions, and $y = y_a + y_d$.

The absorption at short wavelengths. The absorption spectra of some iron garnets at room temperature obtained from direct absorption measurements on films of various thicknesses and from Kramers–Kronig analysis of reflectance data are shown in figure 9.3 (Wemple *et al* 1974). The main feature of the absorption spectrum at high levels of absorption is the existence of absorption bands with oscillator strengths of the order of 10^{-2}. At energies above 3.2 eV the absorption coefficient of YIG increases rapidly, and approaches 5×10^5 cm^{-1} above 5 eV.

It is more difficult to account for the absorption spectrum corresponding to the region between 2.4×10^4 cm^{-1} and 3.6×10^4 cm^{-1} of intermediate absorption. It is understood that this is not the O^{2-}–Fe^{3+} charge-transfer region (Scott *et al* 1974, Wemple *et al* 1974). It would seem most likely that the absorption is due to a combination of bi-exciton excitations—that is, simultaneous crystal-field transitions of two neighbouring ions (Scott *et al* 1974), and charge-transfer excitations where the charge is transferred between two Fe^{3+} ions making a Fe^{4+}–Fe^{2+} pair.

In order to understand the nature of the absorption spectrum of YIG at high photon energies, it is very useful to discuss the absorption spectra of a matrix containing small numbers of Fe^{3+} ions in octahedral and tetrahedral sites. The absorption spectrum of Al_2O_3 containing 0.005 at.% of the octahedrally coordinated Fe^{3+} ions is shown in figure 9.3 (curve 4). The oscillator strength of the first absorption band is 0.1, while the second one attains 0.9. Such oscillator strengths are possible only for allowed electrical dipole transitions. It is thought that these transitions are charge-transfer transitions between the oxygen band and the octahedral Fe^{3+} band (Scott *et al* 1974).

The spectral positions and assignments, and the oscillator strengths of the optical transitions in YIG at 77 K are given in table 9.1 (Scott *et al* 1974). It should be noted that the three most intense transitions given in table 9.1 do not correlate with the crystal-field transition assignment. The result obtained by Wood and Remeika (1967) that, when the concentration of iron ions in garnet exceeds a few per cent, the absorption coefficient is very roughly proportional to the square of the fraction of iron ions present, means that the intensity of the transitions is affected by the presence of other iron ions in the neighbourhood of a given ion, and the resulting absorption level is related to an Fe–O–Fe molecule rather than an individual ion.

As to the origin of the strong bands at 21 640, 23 110, 25 600, 27 400 cm^{-1}, Scott *et al* (1974) suggest that they may arise from excitation of double excitons, i.e., transitions where two Fe^{3+} ions on the same sublattice or different sublattices are simultaneously excited to the spin-quartet levels, e.g.

$$^6A_{1g}(^6S) \rightarrow {}^4T_{1g}(^4G) + {}^6A_{1g}(^6S) \rightarrow {}^4T_{1g}(^4G).$$

It can be expected that these transitions will occur at the sums of the energies of the two transitions involved, and will have oscillator strengths many times those of the single-exciton (magnon-assisted) bands.

Such a description would account for the following facts.

(i)   These bands do not occur below 21 000 cm^{-1}.

(ii) Those which can be studied while varying the Fe^{3+} concentration have concentration-dependent oscillator strengths.

Table 9.1. Parameters of the optical transitions in YIG at 77 K (Scott *et al* 1974).

λ (nm)	Transition	Fe^{3+}	Oscillator strength
900	$^6A_{1g}(^6S) \rightarrow \, ^4T_{1g}(^4G)$	Octahedral	2×10^{-5}
690	$^6A_{1g}(^6S) \rightarrow \, ^4T_{2g}(^4G)$	Octahedral	2×10^{-5}
620	$^6A_{1g}(^6S) \rightarrow \, ^4T_1(^4G)$	Tetrahedral	8×10^{-5}
510	$^6A_{1g}(^6S) \rightarrow \, ^4T_2(^4G)$	Tetrahedral	1.6×10^{-4}
480	$^6A_{1g}(^6S) \rightarrow \, ^4E, ^4A_1(^4G)$	Tetrahedral	3×10^{-5}
470	$^6A_{1g}(^6S) \rightarrow \, ^4E_g, ^4A_{1g}(^4G)$	Octahedral	2×10^{-5}
440	$^6A_{1g}(^6S) \rightarrow \, ^4T_{2g}(^4D)$	Octahedral	1×10^{-4}
430			2×10^{-3}
410	$^6A_{1g}(^6S) \rightarrow \, ^4T_2(^4D)$	Tetrahedral	6×10^{-5}
390			4×10^{-3}
365			1×10^{-2}

(iii) The oscillator strengths are much larger than those of the single excitons. Taking sums of transition energies in YIG, and comparing these energies with the energy positions of the intense absorption bands, Scott *et al* (1974) have shown that there is reasonable agreement with the energies of the observed bands (see table 9.2).

On the basis of optical and magnetooptical measurements, Scott *et al* (1975) assumed that the most attractive assignments of the absorption bands of intermediate strength in the region of the spectrum from 20 000 to 28 000 cm^{-1} are to bi-exciton transitions except for the line at 23 250 cm^{-1}, which is tentatively assigned to a tetrahedral-iron–octahedral-iron charge-transfer transition.

Table 9.2. Probable double-exciton energies (E_{cal}) for YIG compared with band energies (E_{ex}) obtained from Gaussian fittings for YIG and $Y_3Ga_5O_{12}:Fe^{3+}$ spectra (Scott *et al* 1974).

E_{cal} (10^3 cm^{-1})	Transition	E_{ex} (10^3 cm^{-1})	Linewidth, 2Γ (cm^{-1})
20.8, 22.0, 23.2	$^4T_{1g}(^4G) + \, ^4T_{1g}(^4G)$	21.6, 23.1	1000, 1800
25.0, 25.6, 26.2	$^4T_{1g}(^4G) + \, ^4T_{2g}(^4G)$	25.6	2700
26.5, 27.3, 27.1	$^4T_{1g}(^4G) + \, ^4T_{1g}(^4G)$	27.4	2500
27.9, 27.7, 28.5, 29.1	$^4T_{2g}(^4G) + \, ^4T_{2g}(^4G)$	29.0	2200

Wemple *et al* (1974) summarize the results of investigation of the optical properties of YIG films as follows.

(i)   A series of Fe^{3+} crystal-field transitions lying between 1.4 and 3.4 eV involve localized d electron states that can be associated with either octahedral or tetrahedral iron. The strengths of all of the crystal-field transitions decrease with dilution of tetrahedral iron or octahedral iron sublattices.

(ii) Optical transitions having strengths more than a factor of 10 above the crystal-field transitions for tetrahedral iron are observed at 2.9 and 3.2 eV for both the full and diluted lattices. Wemple *et al* suggested that these transitions involve charge-transfer excitons of the configuration $2P^5 3d^6$. Results for the Ga^{3+}- and Sc^{3+}-substituted systems indicate that these features cannot be separately identified with either the octahedral or tetrahedral sublattices. They suggest that exchange-coupled Fe–O–Fe complexes are involved, in keeping with an earlier suggestion of Wood and Remeika (1967).

(iii) Strong optical absorptions above 3.4 eV have been associated with 2p → 3d charge-transfer bands for which the final states are itinerant one-electron band states.

(iv) Very strong semiconductor-like optical transitions occur above 8 eV. These transitions, which make the major contribution to the refractive index, almost certainly involve 2p–4s transitions between the oxygen 2p band and the $Fe^{3+}(4S)$ band.

Another model of the absorption mechanism for the spectral range above 2.5 eV is based on the assumption that intense absorption bands are related to iron–iron charge-transfer transitions between octahedral and tetrahedral pairs (Blazey 1974). The possible charge-transfer transitions of this type are given in table 9.3 (Scott and Page 1977a, b).

Table 9.3. Parameters of possible charge-transfer optical transitions in YIG (Scott and Page 1977a, b).

Transition	Energy (cm^{-1})	Intensity
$(Fe^{3+})_{T_2} + h\nu \rightarrow [Fe^{2+}]_{T_{2g}}$	23 100	Low
$[Fe^{3+}]_{E_g} + h\nu \rightarrow [Fe^{2+}]_E$	25 600	Low
$[Fe^{3+}]_{E_g} + h\nu \rightarrow (Fe^{2+})_{T_2}$	29 100	High
$(Fe^{3+})_{T_2} + h\nu \rightarrow [Fe^{2+}]_{E_g}$	33 000	High
$(Fe^{3+})_E + h\nu \rightarrow [Fe^{2+}]_{T_{2g}}$	30 100	Medium
$[Fe^{3+}]_{T_{2g}} + h\nu \rightarrow (Fe^{2+})_E$	38 600	Medium
$(Fe^{3+})_E + h\nu \rightarrow [Fe^{2+}]_{E_g}$	40 100	Low
$[Fe^{3+}]_{T_{2g}} + h\nu \rightarrow (Fe^{2+})_{T_2}$	42 100	Low

9.2.2   Magnetooptical properties

9.2.2.1   Faraday rotation and magnetic circular dichroism

The first investigation of the magnetooptical properties of magnetic garnets was made by Dillon (1958). Faraday rotation measurements on some iron garnets in the near infra-red were made by Krinchik and Chetkin (1961). It was found that the Fe^{3+} ions on different crystal sites (octahedral and tetrahedral) give different contributions, of opposite sign, to the dispersive rotation, the octahedral ions giving the greater, positive, contribution.

For example, at 633 nm and $T = 295$ K the contribution of the octahedral sublattice to the Faraday rotation is $\Phi_{F_a} = 8670$ deg cm^{-1}, while the tetrahedral one contributes $\Phi_{F_d} = -7840$ deg cm^{-1}, and the resulting rotation is only $\Phi_F = 830$ deg cm^{-1} (Hansen and Krumme 1984). The Faraday rotation spectrum between 0.9 and 2.6 eV is shown in figure 9.4 (Wettling *et al* 1973).

The maximum Faraday rotation attained in the visible region, at 435 nm, is 2.8×10^4 deg cm^{-1}. The spectral dependence of the Faraday rotation of sputtered $Gd_3Fe_5O_{12}$ films at shorter wavelengths is shown in figure 9.4 (Mee 1967). In the ultra-violet region, the Faraday rotation changes sign and attains 10^5 deg cm^{-1} at 300 nm. At this wavelength, rare-earth ions give some contribution to the value of the Faraday rotation, and there are some differences between the magnetooptical spectra of $Y_3Fe_5O_{12}$ and $Gd_3Fe_5O_{12}$.

Concentrational and temperature dependencies of the Faraday rotation. The introduction of diamagnetic and paramagnetic ions instead of iron ions in iron garnet, and changing one rare-earth ion to another leads to a change in the absolute value and temperature dependence of the Faraday rotation. When the tetrahedral sublattice is diluted, at first—while the dilution does not exceed 0.7 formula units (at room temperature)—there is an increase in the Faraday rotation, because the negative contribution of the tetrahedral sublattice diminishes. When the dilution increases further, there is a decrease of the Faraday rotation due to the diminishing of the exchange interaction.

The temperature dependencies of the Faraday rotation and ellipticity of $Y_3Fe_{5-x}Ga_xO_{12}$ at 633 nm for different values of x are shown in figure 9.5 (Hansen and Krumme 1984).

The dependence of the Faraday rotation in $Y_3Fe_{2-x}U_xFe_{3-x}V_yO_{12}$ (at 633 nm and $T = 295$ K) on diluting the iron sublattices, where U is the diluting ion in the octahedral sublattice and x is the concentration, while V and y are the diluting ion and concentration for the tetrahedral sublattice, can be represented by the formula

$$\Phi_F(x, y) = 8670\left(1 - \frac{x}{2}\right) - 7835\left(1 - \frac{y}{3}\right)$$

where $\Phi_F(x, y)$ is in deg cm^{-1} (Hansen and Krumme 1984).

The temperature dependence of the Faraday rotation at 633 nm in

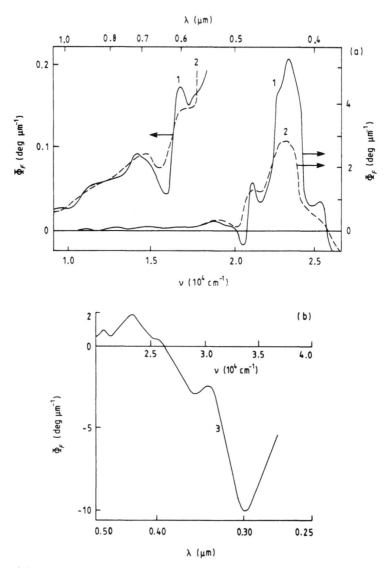

Figure 9.4. The spectral dependence of the Faraday rotation for YIG at 20 K (curve 1) and 300 K (curve 2) (Wettling *et al* 1973), and the room temperature dependence of the Faraday rotation for $Gd_3Fe_5O_{12}$ (curve 3) at a shorter wavelength (Mee 1967).

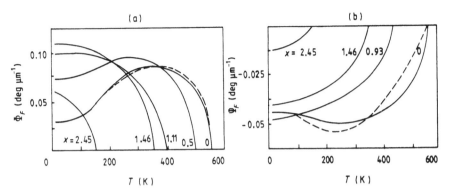

Figure 9.5. The temperature dependence of the Faraday rotation (a) and ellipticity (b) at 633 nm for $Y_3Fe_{5-x}Ga_xO_{12}$. Solid lines: experimental results; dashed lines: theoretical calculations including cubic terms (Hansen and Krumme 1984).

$Y_3Fe_{5-x}Ga_xO_{12}$ can be represented by the formula

$$\Phi_F(T) = 484M_a(T) - 312M_d(T)$$

where M_a and M_d are the sublattice magnetizations for the octahedral and tetrahedral sublattices, respectively (Hansen and Krumme 1984).

The polar Kerr effect. The polar Kerr rotation spectra of $Y_3Fe_5O_{12}$, $Y_3Fe_{4.3}Al_{0.7}O_{12}$, and $Y_3Fe_{4.5}In_{0.5}O_{12}$, and the ellipticity spectrum of $Y_3Fe_5O_{12}$ are shown in figure 9.6 (Višňovský *et al* 1981, Thuy *et al* 1981). The most remarkable features of the YIG spectra are the strong peak of ellipticity at 4.7 eV, and the two peaks of rotation near 4.1 eV and 5.0 eV. These features can be related to the very intense magnetooptical transition of paramagnetic form near 4.6 eV.

9.3 Rare-earth iron garnets

9.3.1 Optical properties

The optical properties of rare-earth iron garnets are mainly determined by the iron atoms, and therefore the spectral peculiarities are common to all iron garnets. The presence of the rare-earth ions in the dodecahedral sublattice is exhibited in the following way: in the transparent band and in the background of bands of relatively low optical absorption there appear narrow peaks caused by transitions inside the partially filled 4f shells of the rare-earth ions; the intensities of the peaks are of the order of $\alpha \approx 100$ cm^{-1}.

Among the trivalent rare-earth ions, only La^{3+} and Lu^{3+} ions have no characteristic transitions for the 4f electrons, because in the former the 4f electrons are absent and in the latter the 4f shell is full. Gd^{3+} ions exhibit weak absorption bands near $\lambda \approx 0.33$ μm that can be observed only for non-magnetic

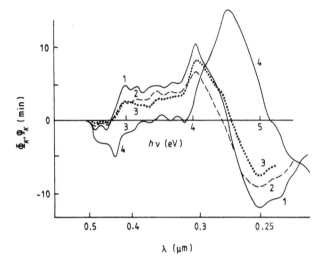

Figure 9.6. Spectral dependencies of the polar Kerr rotation spectra of $Y_3Fe_5O_{12}$ (curve 1), $Y_3Fe_{4.3}Al_{0.7}O_{12}$ (curve 2), and $Y_3Fe_{4.5}In_{0.5}O_{12}$ (curve 3), and the ellipticity spectrum of $Y_3Fe_5O_{12}$ (curve 4) (Višňovský *et al* 1981, Thuy *et al* 1981).

garnets. Y^{3+}, Ca^{2+}, Si^{4+}, Ge^{4+}, Sc^{3+}, Ga^{3+}, and Al^{3+} ions, which are often incorporated into various garnet sublattices, also do not produce absorption lines in the transparent region and in the visible spectral band.

A crystalline field of dodecahedral symmetry leads to a slight shift in position of the 4f transition for the rare-earth ions, because the 4f shell is screened from the influence of the crystalline field by the outer electrons. Typical absorption spectra for a number of rare-earth iron garnets in the transparent band are shown in figure 9.7.

At the same time the presence of the rare-earth ions leads to some increase in the optical absorption in the spectral band below 0.3 μm, where there are allowed electric dipole transitions from the 4f to the 5d state with the oscillator strength $f \sim 10^{-3}$–10^{-2} (Suits 1972). As the transitions in the rare-earth ions lie deep in the absorption band of the iron garnets, where the absorption coefficient is higher than 4×10^5 cm^{-1}, their absorption bands will only insignificantly increase the absorption in this region.

The shift in position of the first transition $^6A_{1g}(^6S) \rightarrow {}^4T_{1g}(^4G)$ of the octahedral Fe^{3+} complex due to the rare-earth-induced change of the lattice parameter of the iron garnet is extremely important from the point of view of applied magnetooptics. The shift is mainly caused by transformation of the crystalline field in which the Fe^{3+} is located, due to the change in the distance between the iron and the oxygen ions. The change in the position of the transition leads to changes in the optical absorption at the wavelengths 1.06 and 1.15 μm, which are used in optical communication systems.

For magnetooptical modulators and non-reciprocal elements that operate in these wavelength regions, it is advisable to use iron garnets with large unit-cell parameters. It should also be taken into account that the presence of small amounts of Fe^{2+} and Fe^{4+} significantly affects the optical absorption in this region; therefore to minimize the optical absorption it is necessary to thoroughly control the charge compensation and growth parameters during the single-crystal growth and the film epitaxy (Wood and Remeika 1967).

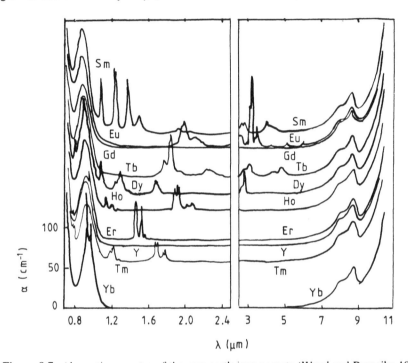

Figure 9.7. Absorption spectra of the rare-earth iron garnets (Wood and Remeika 1967).

9.3.2 Magnetooptical properties

The intense electric dipole transitions between 4f and 5d levels of the rare-earth ions, and the charge-transfer transitions from oxygen to rare-earth ions, which both correspond to the ultra-violet spectral band, substantially affect the magnetooptical properties of a number of rare-earth iron garnets in both the infra-red and the visible spectral regions. This effect is especially noticeable at low temperatures, because with the decrease of the temperature, the magnetization of the rare-earth sublattice grows considerably faster than the magnetization of the octahedral or tetrahedral sublattices.

The rare-earth iron garnets have magnetic compensation points, and all odd magnetooptical effects change their sign when crossing that point. The Faraday effect changes its sign because above the compensation point magnetization of the tetrahedral lattice orients along the direction of the external magnetic field, while below the compensation point the octahedral lattice orients along the field direction. As the rare-earth sublattice is antiferromagnetically bound to the tetrahedral sublattice, the direction of its magnetization also changes after crossing this point.

Figure 9.8(a) shows the spectral dependencies of the Faraday effect of the rare-earth iron garnets in the visible spectral band (Hansen and Krumme 1984). In table 9.4 the Faraday rotations for the rare-earth iron garnets $RE_3Fe_5O_{12}$ at $\lambda = 1.064$ μm, $T = 300$ K are summarized (Wemple *et al* 1974). The temperature dependencies of the Faraday rotation of some rare-earth garnets are presented in figure 9.8(b) (Hansen and Krumme 1984).

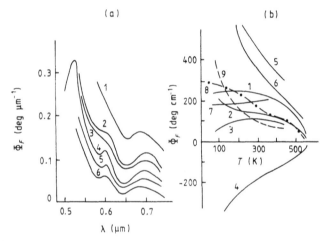

Figure 9.8. (a) The dispersion of the Faraday effect for the rare-earth iron garnets $RE_3Fe_5O_{12}$: RE = Tb (curve 1), Gd (curve 2), Dy (curve 3), Y (curve 4), Eu (curve 5), and Er (curve 6). (b) The temperature dependence of the Faraday rotation at $\lambda = 1152$ nm for $Y_3Fe_5O_{12}$ (curve 1), $Er_3Fe_5O_{12}$ (curve 2), $Tm_3Fe_5O_{12}$ (curve 3), $Sm_3Fe_5O_{12}$ (curve 4), $Tb_3Fe_5O_{12}$ (curve 5), $Dy_3Fe_5O_{12}$ (curve 6), $Lu_3Fe_5O_{12}$ (curve 7), $Eu_3Fe_5O_{12}$ (curve 8), and $Ho_3Fe_5O_{12}$ (curve 9) (Hansen and Krumme 1984).

9.3.3 Faraday rotation in the transparent band

Magnetooptical properties of the rare-earth iron garnets in the transparent region require special consideration, because in this band, along with the gyroelectric contribution to the magnetooptical effect, which is associated with allowed electric dipole transitions in the UV spectral band, there is a significant gyromagnetic contribution, which is caused by ferromagnetic and exchange resonance phenomena with characteristic wavenumbers of the order of 0.3 and 30 cm^{-1} (Krinchik and Chetkin 1960, 1961, Krinchik 1985).

Table 9.4. The Faraday rotation for rare-earth iron garnets, for $\lambda = 1.06$ μm and $T = 300$ K (Wemple *et al* 1973).

Garnet	Φ_F (deg cm^{-1})	Garnet	Φ_F (deg cm^{-1})
$Y_3Fe_5O_{12}$	+280	$Ho_3Fe_5O_{12}$	+135
$Y_2Pr_1Fe_5O_{12}$	−400	$Er_3Fe_5O_{12}$	+120
$Sm_3Fe_5O_{12}$	+15	$Tm_3Fe_5O_{12}$	+115
$Eu_3Fe_5O_{12}$	+167	$Yb_3Fe_5O_{12}$	+12
$Gd_3Fe_5O_{12}$	+65	$[Lu_3Fe_5O_{12}]^*$	+200
$Tb_3Fe_5O_{12}$	+535	$[Nd_3Fe_5O_{12}]^*$	−840
$Dy_3Fe_5O_{12}$	+310		

* In the brackets the hypothetical single-crystal compositions are given.

The Faraday rotation spectra for a number of rare-earth iron garnets are shown in figure 9.9 (Krinchik and Chetkin 1961, Krinchik 1985). A characteristic feature of the spectra presented is the very low dispersion of the Faraday rotation in the long-wavelength region of the transparent band.

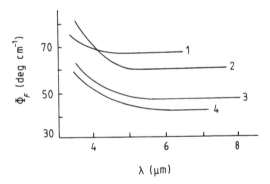

Figure 9.9. The Faraday effect for $Y_3Fe_5O_{12}$ at $T = 77$ K (curve 1) and for $Y_3Fe_5O_{12}$ (curve 2), $Er_3Fe_5O_{12}$ (curve 3), and $Ho_3Fe_5O_{12}$ (curve 4) at $T = 290$ K, where $H = 2.4$ kOe (Krinchik and Chetkin 1961).

When analysing the dispersion of the Faraday effect in the infra-red spectral region, it is convenient to divide the effect into two contributions: gyroelectric and gyromagnetic (Krinchik and Chetkin 1961, Krinchik 1985):

$$\Phi_F = \Phi_E + \Phi_M \qquad (9.1)$$

where

$$\Phi_E = \frac{\pi\omega}{c\sqrt{\varepsilon}}[\chi_e^- + \chi_e^+]\frac{\varepsilon + 2}{9} \qquad (9.2)$$

$$\Phi_M = \frac{\pi \omega \sqrt{\varepsilon}}{c} [\chi_m^- + \chi_m^+]. \tag{9.3}$$

The rotation of the plane of polarization of the infra-red light, which is independent of the radiation frequency, results from the precession of the magnetization vector induced by the magnetic field of the light wave, i.e., it is a consequence of the magnetic spin resonance. The characteristic frequencies ω_0 of this resonance correspond to the rf spectral region; therefore observation of the Faraday effect in the IR spectral region corresponds to the condition $\omega \gg \omega_0$. The Faraday effect in the case under consideration is determined by the value of the off-diagonal component of the magnetic permeability tensor.

If $\omega \gg \omega_0$

$$\tilde{M} = \frac{4\pi \gamma}{\omega} M_s \tag{9.4}$$

where M_s is the saturation magnetization of the ferromagnet, and $\gamma = e/mc$. The non-zero value of the off-diagonal component of the magnetic permeability tensor results in a Faraday rotation of the plane of polarization of the light wave, the specific Faraday rotation being equal to

$$\Phi_M = \frac{\sqrt{\varepsilon}}{2c} \omega M. \tag{9.5}$$

Making the substitution $M = \tilde{M}$ yields

$$\Phi_M = \frac{2\pi \sqrt{\varepsilon}}{c} \gamma M_s. \tag{9.6}$$

As follows from formula (9.6), the gyromagnetic contribution to the Faraday rotation in the IR spectral region is independent of the wavelength of the light.

In ferrimagnets the phenomenon of exchange resonance can arise—the spin magnetic resonance in the effective field that is caused by exchange interaction between inequivalent magnetic sublattices (Krinchik 1985). The intensity of the exchange resonance is proportional to $(\gamma_1 - \gamma_2)^2$, where $\gamma_1 = g_1 e/mc$ and $\gamma_2 = g_2 e/mc$—the gyromagnetic ratios for the ions in the two different magnetic lattices. The frequency of the exchange resonance

$$\omega_{ex} = \lambda(\gamma_2 M_1 - \gamma_1 M_2) \tag{9.7}$$

where λ is the molecular-field coefficient, and M_1 and M_2 are the sublattice magnetizations of the ferrimagnet. When $H_{ex} = \lambda M = 10^6$ Oe, the value of $\omega_{ex} = 3 \times 10^{12}$ Hz—that is, the resonance frequency—lies in the far-infra-red spectral region.

Near the magnetic compensation point, $M_1 \simeq M_2$, and ω_{ex} decreases and shifts towards the rf band region, which is convenient for measurements. For ferrimagnetic material with two sublattices, in the frequency interval $\omega_{res} \ll$

$\omega \ll \omega_{ex}$, the expression for specific the gyromagnetic Faraday rotation has the following form:

$$\Phi_M = \frac{2\pi \sqrt{\varepsilon}}{c} \gamma_{eff} M \tag{9.8}$$

where

$$\gamma_{eff} = \frac{M_1 - M_2}{M_1/\gamma_1 - M_2/\gamma_2} \tag{9.9}$$

(the ratio of the total magnetic moment to the mechanical moment). For frequencies $\omega \gg \omega_{res}, \omega_{ex}$, in the presence of exchange resonance the following formula is valid:

$$\Phi_M = \frac{2\pi \sqrt{\varepsilon}}{c} (\gamma_1 M_1 - \gamma_2 M_2). \tag{9.10}$$

The physical sense of the expressions obtained is that, while they share a common ferromagnetic resonance, the two magnetic sublattices of a ferrimagnet, which are associated with the exchange interaction, precess as a single entity; in the exchange type of precession, the two sublattices contribute independently to the Faraday effect, with the sign of the contribution corresponding to the orientation of the sublattice magnetic moment.

9.3.4   The Faraday effect in a strong magnetic field

9.3.4.1   Field dependence

The dependence of magnetooptical phenomena exhibited by the rare-earth iron garnets on the magnetic field has been intensively studied by many authors. Figure 9.10 shows the field dependencies of the Faraday rotation of some rare-earth iron garnets at three temperatures: 290, 77, and 4.2 K (Valiev *et al* 1983). It can be seen that saturation of the Faraday effect is reached in weak fields (the measurements were carried out along the axis of easy magnetization).

As the field grows further, a strong change of the Faraday rotation is observed. Note that the main characteristics of the Faraday rotation—the value of the spontaneous Faraday effect Φ_F, which is obtained by extrapolation from the strong-field region to the zero-field region, and the variation of the Faraday effect with the field, $d\Phi/dH$—vary substantially from one iron garnet to another, and have different temperature dependencies for different rare-earth garnets (figure 9.11) (Valiev *et al* 1983).

Let us consider briefly the physical nature of these dependencies. As we mentioned earlier, the specific Faraday rotation in the rare-earth iron garnets can be represented as

$$\Phi_F = \pm\Phi_{Fe} \mp \Phi_{RE} \tag{9.11}$$

where Φ_{Fe} is the Faraday rotation caused by the octahedral and the tetrahedral sublattices, and Φ_{RE} is that caused by the rare-earth sublattice.

Above the magnetic compensation point ($T > T_{com}$), the magnetization of the rare-earth sublattice is antiparallel to the field, and therefore in formula (9.11)

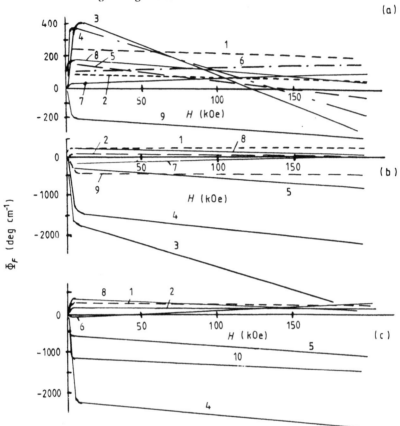

Figure 9.10. The field dependence of the Faraday effect for some rare-earth iron garnets: Y (curve 1), Gd (curve 2), Tb (curve 3), Dy (curve 4), Er (curve 5), Tm (curve 6), Yb (curve 7), Eu (curve 8), Sm (curve 9), and Ho (curve 10) at the temperatures 290 K (a), 77 K (b), and 4.2 K (c) (Valiev *et al* 1983).

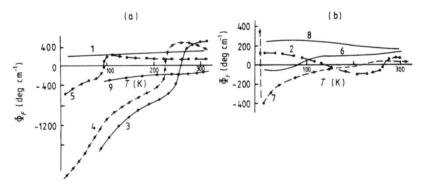

Figure 9.11. The temperature dependences of the Faraday effect for some rare-earth iron garnets (the key to the curves is the same as for the previous figure) (Valiev *et al* 1983).

the lower signs should be chosen, whereas below T_{com} the upper ones should be chosen. For $T > T_{com}$ the increase of the external field leads to a reduction of the magnetic moment of the rare-earth sublattice, and when the external field is equal to the molecular field H_m^0, which affects the main multiplet, the magnetization of the rare-earth sublattice vanishes. Does the Faraday rotation of the rare-earth sublattice also vanish? Experience shows that it does not. Indeed, as seen from table 9.5, the critical field H_{cr}, at which the effect vanishes, is substantially different from the molecular field (assuming that the value of Φ_{Fe} is the same as for yttrium iron garnet). This indicates the significant role of the mixing mechanism (and, probably, of the diamagnetic component) in the Faraday rotation of the rare-earth sublattice of the iron garnets considered.

The dependencies $\Phi_F(H)$, which are shown in figure 9.10, are described by the formula

$$\Phi_R = C M_R(H_{eff}^0) + D H_{eff}^1 \tag{9.12}$$

where

$$H_{eff}^0 = H \pm H_m^0 = H \pm [2(g_J - 1)/g_J]H_{ex}. \tag{9.13}$$

Table 9.5. The molecular (H_m^0) and critical (H_{cr}) fields of rare-earth iron garnets.

Garnet	H_m^0 (kOe)	H_{cr} (kOe)
$Gd_3Fe_5O_{12}$	300	1000*
$Tb_3Fe_5O_{12}$	160	55
$Dy_3Fe_5O_{12}$	120	55
$Er_3Fe_5O_{12}$	80	-150*
$Tm_3Fe_5O_{12}$	90	270*
$Yb_3Fe_5O_{12}$	105	400*

* These values are obtained by extrapolation of the experimental dependencies Φ_F to $H = H_m^0$.

$H_{eff}^1 = H \pm 2H_{ex}$, where H_{ex} is the exchange field affecting the rare-earth ion, and g_J is the Landé factor of the main multiplet. The difference between the effective fields H_{eff}^0 and H_{eff}^1 explains the observed difference between the fields H_m^0 and H_{cr}. The coefficients C and D are described in chapter 4. We should note that these coefficients agree with the corresponding coefficients of the rare-earth garnet gallates and aluminates.

9.4 Fe^{2+} and Fe^{4+} ions in garnets

In iron garnets, Fe^{2+} and Fe^{4+} ions always exist simultaneously in low concentration. However, at equilibrium concentrations near 10^{-4} formula units they do not affect the optical and magnetooptical properties in any observable way. When iron garnets are doped with non-trivalent ions, the concentrations of Fe^{2+} and Fe^{4+} ions can reach ~ 0.1 formula units, and in this case the optical and magnetooptical properties of the iron garnets can change drastically, in particular in the infra-red spectral region.

For example, in yttrium iron garnet at the wavelength 1.3 μm, the absorption coefficient can be less than 0.1 cm^{-1}, provided that optimal growth conditions were maintained; however, in the samples doped with uncompensated Ca^{2+} ions—which leads to the appearance of the Fe^{4+} ions in the iron garnet— or with Ge^{4+} ions—which leads to the appearance of the Fe^{2+} ions in the iron garnet—at this wavelength the optical absorption coefficient can reach the level of 100 cm^{-1} and even 10^3 cm^{-1} (Lucari *et al* 1977, Antonini *et al* 1980).

Doping of the iron garnets with doubly positive ions, e.g. Ca^{2+} or Mg^{2+}, creates p-type conductivity; doping with the four-valent ions or by five-valent ions, say Ge^{4+}, Si^{4+}, Zr^{4+}, or Nb^{5+}, leads to the appearance of conductivity of n type.

9.4.1 Optical properties

The growth of the optical absorption due to the presence of two- and four-valent iron ions in the iron garnet results from different mechanisms. First of all, each type of ion has its own absorption bands in the infra-red spectral region. The energy levels of the Fe^{2+} ion in the octahedral environment and of the Fe^{4+} ion in the tetrahedral surroundings, if we take into account the level splitting in the cubic and in the trigonal crystalline field, the spin–orbital interaction, the exchange field, and local perturbations, have a complicated structure of the ground and the excited states. In both cases one can expect the formation of intra-ion transitions with energies of about 10^4 cm^{-1} and a number of other transitions of lower energies.

Naturally, in the single-ion spectra of these ions a lot of transitions of higher energy will be present, but they will lie deep inside the absorption band of the iron garnet, while the low-frequency transitions occupy the transparency region of the iron garnets. The most spectacular structures in the absorption spectra lie in the 0.9 μm region for the Fe^{2+} ion, and in the 0.95 μm region for the Fe^{4+} ion (Antonini *et al* 1980).

In the absorption spectrum of a Sn^{4+}-doped sample with a concentration $x = 0.088$ of the formula unit, Lucari *et al* (1977) observed structure near 2 μm. Measurements at liquid nitrogen temperatures on Zr^{4+}-doped single crystals of yttrium iron garnet revealed four weak structures in the absorption spectra at 1.10, 1.60, 1.95, and 2.15 μm (Lucari *et al* 1980).

Absorption tails in the infra-red spectral region in the samples that contain fairly high concentrations of bivalent and tetravalent iron ions are usually associated with the jumping mechanism of Verwey conductivity, which is responsible for the relatively high conductivity of heavily doped iron garnets (Nassau 1968). According to this mechanism, an electron can jump from an Fe^{2+} ion to one of the neighbouring Fe^{3+} ions and from an Fe^{3+} ion to an Fe^{4+} ion as a result of the absorption of a photon from the infra-red band.

Nassau (1968) proposed the following empirical relationships between the absorption coefficient at the wavelength $\lambda = 1.2$ μm and the concentration of Fe^{2+} and Fe^{4+} ions:

$$\alpha_{1.2}^{Fe^{2+}} = 950[Fe^{2+}]$$

and

$$\alpha_{1.2}^{Fe^{4+}} = 6500[Fe^{4+}].$$

For the iron garnets containing substantial numbers of bivalent and tetravalent iron ions, a significant increase of the optical absorption is also observed in the visible spectral region. There is no unambiguous explanation of the mechanisms responsible for the observed growth of the absorption as yet. Increase of the absorption in the absorption band of the iron garnet can result from the increase of the oscillator strength of the transitions associated with the Fe^{3+} ions, due to the additional perturbation of the local symmetry of the octahedral and tetrahedral complexes of the trivalent iron ions; the increase of the oscillator strength and the growth of the magnetooptical activity of these transitions can be associated with the admixing of the main and excited states of the Fe^{2+} and Fe^{4+} ions with the appropriate states of the Fe^{3+} ions. New transitions involving charge transfer between the Fe^{2+} or Fe^{4+} ions and ions that are located in different sites in the iron garnet are also likely to appear, first of all with the participation of the Fe^{3+} or Pb^{2+} and Pb^{4+} ions (Scott *et al* 1975).

The value of the optical absorption in the samples doped with bivalent and tetravalent ions can vary significantly at the same doping level, depending on the peculiarities of the technological process of the sample preparation, on the melt composition, and on a number of other factors. For example, in 0.1-formula-unit calcium-doped epitaxial films, the absorption spectrum is barely different from that of the undoped sample. At the same time, at a doping level of 0.34 formula units an abrupt growth of the absorption coefficient over a wide spectral region was observed by Lucari *et al* (1977).

This sharp difference between the spectra of the two samples is likely to be associated with the fact that for the sample with $x = 0.10$, compensation of the Ca^{2+} ions was achieved without the formation of Fe^{4+} ions—for instance, by formation of Pt^{4+} and Pb^{4+} ions, and O^- centres. It is well known that, beginning from the photon energy $h\nu = 2.8$ eV, which corresponds to the abrupt growth of the optical absorption, the iron garnets exhibit photoconductivity. At higher levels of doping with Ca^{2+} ions (beginning from $x = 0.1$), the photoconductivity edge shifts toward the long-wavelength spectral region, the shift being a function of the calcium concentration squared. In the absorption spectrum of the sample with $x = 0.34$, two spectral peculiarities are clearly manifested in the regions near 1.35 eV and 2.05 eV (Thavendrarajiah *et al* 1990).

The optical absorption in iron garnets containing bivalent and tetravalent iron ions can be significantly varied by annealing in reducing or oxidizing atmospheres. The consequences of the annealing of the calcium-doped iron garnet films (containing Fe^{4+} ions) were studied by Antonini *et al* (1980). The initial film had $\alpha \sim 1800$ cm^{-1} at the wavelength 1.3 μm. Annealing in the oxygen atmosphere led to a slight increase of the absorption coefficient to the level of $\alpha \sim 2100$ cm^{-1} ($\lambda = 1.3$ μm). After successive annealings in the nitrogen atmosphere, the absorption coefficient was reduced to a rather low level.

It should be noted here that further annealing resulted in the optical absorption growing again; additional experiments have shown that the samples always had p-type conductivity and that the concentration of the Fe^{4+} ions correlated with the optical absorption: first the Fe^{4+}-ion concentration and the value of optical absorption decreased, reaching a minimum after the fifth annealing; then, during the following annealings, they grew again.

Today, annealing is widely used for the formation of bismuth-substituted iron garnets with record-low optical losses in a desired spectral region, usually at the wavelengths 0.8 μm or 1.3 μm; according to the type of doping and its concentration, to the growth peculiarities, to the melt composition, and to other technological factors, one should carefully select the annealing atmosphere, and the duration and temperature of the annealing. A number of particular examples are considered in chapters 13, 15, and 16.

The effects of a reducing treatment on the optical absorption loss in the near-infra-red region were studied for Ca-free and Ca-doped Bi-substituted iron garnet films by Yokoyama *et al* (1987). $(BiGd)_3(FeAlGa)_5O_{12}$ films were grown by the conventional liquid-phase-epitaxy technique and annealed in a hydrogen (H_2) atmosphere between 150 and 400 °C, in steps of 50 °C, for 22 hours. The absorption loss at the wavelength $\lambda = 800$ nm reached minimum values of 250–300 dB cm^{-1} on annealing at 250 °C for both Ca-free films and Ca-doped films.

The decrease in the absorption loss at $\lambda = 800$ nm caused by this annealing is mainly ascribed to the reduction of Pb^{4+} to Pb^{2+} on the dodecahedral sites in the garnets. The wavelength dependence of the difference in absorption loss between as-grown films and films annealed in H_2 at 250 °C has a peak at around $\lambda = 560$ nm. This peak is thought to be due to a charge-transfer transition from Fe^{3+} to Pb^{4+}.

The absorption loss at $\lambda = 1.15$ μm decreased for both Ca-free and Ca-doped films on annealing at 250 °C. The values reached $\sim$2 dB cm^{-1} for Ca-doped films, and $\sim$9 dB cm^{-1} for Ca-free films. It is thought that the absorption loss of $\sim$9 dB cm^{-1} for the Ca-free films is due to Fe^{2+} which exists in the as-grown state. The increase in the absorption loss on annealing above 250 °C may be due to Fe^{2+} ions created by the reduction of $Fe^{3+} \rightarrow Fe^{2+}$.

At low temperatures (below 150 K), the presence of the Fe^{2+} and Fe^{4+} ions in the iron garnets leads to very interesting photo-induced phenomena. For example, in Ca^{2+}-doped iron garnets, at temperatures below 150 K, the hole is localized on the Fe^{4+} ion. At higher temperatures, delocalized magnetic polarons appear. In this case, the hole is relatively strongly bound to the Ca^{2+} ion, but the polaron has a sufficiently large radius for the hole to rapidly jump from one Fe^{3+} ion to another in the vicinity of the Ca^{2+} ion, never becoming localized at any single iron-ion site. The binding energy of the hole and the Ca^{2+} ion is estimated to be $E_a = 0.41$ eV in the temperature interval 150–300 K.

As the d shell of the Fe^{4+} ion is less than half-full, the superexchange interaction between the tetrahedral Fe^{4+} ions and octahedral Fe^{3+} ions changes its sign with respect to the case in which both the tetrahedral and octahedral sites are occupied by the trivalent iron ions; as a result the magnetic moment of the Fe^{4+} ion is oriented parallel to that of the octahedral sublattice. At low temperatures and for larger Fe^{4+}-ion contents, these magnetic moments substantially affect the resulting magnetic moment of the iron garnet. The temperature dependence of the Fe^{4+}-ion concentration can be described by the formula

$$n = C_0\left[1 - \exp\left(\frac{T_0}{T}\right)\right]$$

where $T_0 = 25$ K, and C_0 is the concentration of the Ca^{2+} ions not compensated by other dopants, i.e., of the ions that give rise to the tetravalent iron ions. Abnormal magnetic anisotropy and magnetostriction, and a number of other unusual properties, are also intrinsic to the iron garnets that contain Fe^{2+} ions in sufficient concentrations at low temperatures.

9.4.2 Magnetooptical properties

Studying the magnetooptical properties of the iron garnets that contain Fe^{2+} and Fe^{4+} ions, first of all one should take into consideration the fact that doping of the iron garnet films with Ge^{4+}, Si^{4+}, and other more-than-trivalent ions as a rule results in an increased content of lead ions, which strongly affect the magnetooptical properties of the iron garnets. Incorporation of Ca^{2+} and other bivalent ions usually reduces the lead content in the film and thus can also substantially change the Faraday rotation.

Nevertheless the study of the magnetic circular dichroism (MCD) spectra allows one to discover more detailed information about the transitions associated with the Fe^{2+} and Fe^{4+} ions, as compared to that found from common optical spectra.

MCD spectra of the samples of Sn^{4+}- and Zr^{4+}-doped iron garnets were studied by Lucari *et al* (1977) and Antonini *et al* (1980). Both ions are tetravalent, and they preferentially occupy the octahedral sites. While in the samples doped with Sn^{4+} and Zr^{4+} ions, the MCD peaks correspond to wavelengths of 1.15, 1.7, and 2.0 μm, in the Si^{4+}-doped sample only one peak is observed, at a wavelength of 1.15 μm. For the Fe^{2+} ion in a silicon-doped yttrium iron garnet, the calculated parameter of the spin–orbital bond, which was determined from the measured MCD spectra, turned out to be $\lambda = 132$ cm^{-1} (Lucari *et al* 1980).

An even more interesting result is observed for the samples doped with Nb^{5+} ions. While for the samples doped with Si^{4+}, Sn^{4+}, and Zr^{4+} there is an MCD peak at 1.15 μm, and the sign of the effect is positive, for the Nb^{5+}-doped samples the sign of the MCD in the same region is negative, and the value of the MCD varies monotonically (increases in its absolute value) with the decrease of the light wavelength in the 1.1–1.6 μm spectral region (Antonini *et al* 1980).

One has to pay attention to the difference between the MCD spectra for the samples doped with Si^{4+} and Zr^{4+} ions on one hand and with Nb^{5+} ions on the other in the 1.1–1.6 μm spectral region. According to Lucari *et al* (1977), the main reason for these peculiarities of the MCD spectra of the Nb^{5+}-doped samples is that, in this case, the Fe^{2+} ions preferentially occupy tetrahedral rather than octahedral sites, and these are usually occupied by the dopants Si^{4+} and other ions.

It was also noted (Lucari *et al* 1980) that while the MCD spectra of Si^{4+}- and Ge^{4+}-doped iron garnets exhibit no magnetooptical peculiarities in the spectral region above 1.6 μm, in the samples doped with Sn^{4+} and Zr^{4+} ions the spectrum is far more complicated. It should be noted here that the Si^{4+} and Ge^{4+} ions occupy tetrahedral sites, and the Sn^{4+} and Zr^{4+} ions occupy octahedral sites in the iron garnets. Lucari *et al* (1980) suggested that the negative MCD peaks for the Sn^{4+}- and Zr^{4+}-doped samples can be associated with Fe^{2+} ions located at the tetrahedral sites.

9.5   Cobalt-substituted iron garnets

9.5.1   Optical properties

Optical measurements by Wood and Remeika (1967) on single-crystal aluminium and gallium garnets doped with Co and/or Si^{4+} and Ca^{2+} provided evidence that both Co^{3+} and Co^{2+} can be substituted in both tetrahedral and octahedral sites, the large optical response arising from tetrahedral Co, because the lack of inversion symmetry of the tetrahedral sites relaxes the selection rules opposing d–d transitions.

Incorporation of cobalt ions into the iron garnet (both bivalent and tetravalent) also results in increased optical absorption in the visible and near-infra-red spectral regions in comparison with the case for yttrium iron garnet. But as the growth of the absorption in the visible spectral region is associated not only with the cobalt ions, but also with the presence of the Fe^{2+} and Fe^{4+} ions, it is very difficult to estimate their contribution to the optical absorption.

Depending on the formation conditions and on the presence of ions of different valences in the iron garnets, the ratio of Co^{2+} to Co^{3+} charge states can be changed, but it turns out that counterdoping with quadrivalent dopants enhances the Co^{2+} content, but is not able to substantially decrease the number of Co^{3+} ions which are incorporated.

The absorption line of the lowest tetrahedral Co^{2+} ion in the garnet must be associated with the $^6A_2(^4F) \rightarrow {}^4T_1(^4P)$ transition. Toriumi *et al* (1987) stated that when Co^{2+} ions are incorporated into the iron garnet, two absorption bands appear at 0.65 μm (1.91 eV) and at 1.5 μm (0.83 eV).

For the tetrahedral Co^{3+} ion, the lower absorption band must be associated with the $^5E(^5D) \rightarrow {}^5T_2(^5D)$ transition. When Co^{3+} ions are incorporated into an iron garnet, two absorption bands appear, one at 0.7 μm (1.77 eV), and another at 1.32 μm (0.94 eV). Daval *et al* (1987) assumed, on the basis of the results of Faraday rotation and MCD measurements, that transitions related to Co^{2+} ions were situated at 0.7 and 1.6 μm, and that the Co^{3+} transition is near 1.3 μm.

The optical and magnetooptical properties of cobalt-substituted iron garnet have been studied by Egashira and Manabe (1972), Itoh *et al* (1985), Ferrand *et al* (1987), Saito *et al* (1987b), Toriumi *et al* (1987), Daval *et al* (1987), Šimša (1990) and many others.

9.5.2 Magnetooptical properties

The spectral dependencies of the Faraday rotation (solid lines) and MCD (dashed lines) for epitaxial films of composition $Y_3Fe_5O_{12}$:Co^{3+} and $Y_3Fe_5O_{12}$:Co^{2+}–Zr^{4+} are presented in figures 9.12(a) and 9.12(b), respectively (Daval *et al* 1987). It should be noted that both samples contain both Co^{2+} and Co^{3+} ions; the full amount of Co was near 0.1 formula units, resulting in some difficulties in interpretation of the results of the measurements. The experiments in which the Faraday rotation and MCD were measured simultaneously are the most suitable for use in characterizing the optical transitions involved.

The MCD of the Co^{2+} transitions at 0.7 and 1.6 μm, and of the Co^{3+} transition at 1.3 μm, are clearly seen to be peaked at the transition wavelength, while the Faraday rotation curves have rather complicated forms. All of the above transitions are thus seen to have paramagnetic character. The magnetooptical anomaly in the Faraday rotation near 1.5 μm is in fact related to the superposition of the contributions of two transitions at 1.3 μm (from Co^{3+}) and 1.6 μm (from Co^{2+}), which looks like one transition at 1.45 μm with diamagnetic dispersion of the Faraday rotation.

The annealing experiments performed in hydrogen atmospheres with $Y_3Fe_5O_{12}$:Co^{2+}–Zr^{4+} samples demonstrated a halving of the MCD peak near 1.3 μm but no change in the value of the MCD at 1.6 μm, showing that these peaks are related to different transitions (Daval *et al* 1987).

9.6 Bismuth-substituted iron garnets

9.6.1 Optical properties

While doping of iron garnets with bismuth greatly increases their magnetooptic activity, growth of the bismuth content in an epitaxial film is usually accompanied by higher absorption in the visible spectral region. The deter-

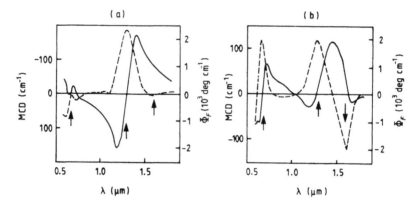

Figure 9.12.   The dispersion of the Faraday effect (solid lines) and magnetic circular dichroism (dashed lines) for the compositions $Y_3Fe_5O_{12}:Co^{3+}$ (a) and $Y_3Fe_5O_{12}:Co^{2+}-Zr^{4+}$ (b), respectively (Daval *et al* 1987).

mination of the contribution of the Bi^{3+} ions to the optical absorption is a rather difficult task. For technological reasons, epitaxial growth of films with a high content of bismuth is impossible without increasing the lead-ion content, which substantially affects the optical absorption. A study of the absorption spectra of the bismuth-substituted iron garnets deep in the absorption band allows one to distinguish the contribution associated with the presence of the bismuth ions in the iron garnets. As shown by Kaneko *et al* (1987), incorporation of the Bi^{3+} ions up to 1.6 formula units into the iron garnet does not result in growth of the optical absorption near $\lambda = 0.8$ μm if lead ions do not enter the film simultaneously.

When measuring the optical transmission of iron garnets with high bismuth concentrations, one should ensure that the samples are magnetized to the saturation point, and that all existing domains are removed. Otherwise the measured value of the transmission will be substantially lowered due to the diffraction of light by the domain structure, and therefore the optical absorption will seem higher.

9.6.2 Magnetic properties

Incorporation of the Bi^{3+} ions into the iron garnet structure results in significant changes of magnetooptical properties of the material; in particular, the Faraday rotation in the visible spectral region changes its sign. Besides this, the bismuth ions also affect the magnetic properties of the iron garnets; for example, at room temperature the saturation magnetization M_s grows; growth of the Curie temperature, and increases of the cubic magnetic anisotropy and magnetostriction coefficients were also observed. For most of the iron garnet compositions, incorporation of bismuth leads to the appearance of substantial uniaxial magnetic anisotropy (Hansen and Krumme 1984).

It is important to note that the nature of the growth-induced magnetic anisotropy in the bismuth-substituted iron garnets is different from that in the case that is observed for epitaxial films of iron garnets with two inequivalent rare-earth elements (the Sm–Lu system, and the like). As a result, the temperature dependencies of the growth-induced magnetic anisotropy constant, K_u^g, in these films are also different; this difference can have very important consequences in applications.

In epitaxial films the growth-induced magnetic anisotropy constant is mainly determined by the rare-earth ions, and the dependence $K_u^g(T)$ is mainly determined by the magnetization of the rare-earth sublattice, while in the bismuth-substituted films it is mainly determined by the magnetization of the tetrahedral sublattice ($K_u^g \approx M_d^2 - M_d^3$). The reason for the difference is that in the first case the uniaxial magnetic anisotropy is caused by the ordering of the rare-earth ions and is determined in the long run by their single-ion anisotropy, while in the second case it depends on distortions of the local symmetry of the iron-ion environment (Hansen and Krumme 1984).

The magnetic properties of the iron garnets with bismuth ions incorporated change for several reasons. The growth of the unit-cell parameter results in some reduction of the magnetic moment per unit volume. On the other hand, the presence of the bismuth ions affects the indirect exchange interaction between the octahedral and tetrahedral sublattices that leads to the higher Curie temperature. To explain the rise of the Curie temperature for the bismuth-substituted iron garnets, two models were proposed; they can be loosely called 'structural' and 'electron' models (Hansen and Krumme 1984).

The first (structural) model presumes that the distortion of the local symmetry of the different crystalline sites results in changes of the bond angles between the two iron ions that interact with each other via the oxygen ion. The indirect exchange reaches its maximum when all three ions are located in a single line. Although incorporation of the bismuth ion does lead to some increase of the bond angle, this mechanism cannot entirely explain the observed phenomenon. The electron model suggests that mixing of the wave functions of the bismuth and oxygen ions results in an increase of the indirect exchange interaction, and growth of the spin–orbital interaction of the iron ions. Incorporation of lead and vanadium ions into the iron garnet also leads to a slight increase of the Curie temperature (Hansen and Krumme 1984).

9.6.3 Magnetooptical properties

Faraday rotation and magnetic circular dichroism. Figures 9.13 and 9.14 (Burkov and Kotov 1975, 1983, Burkov *et al* 1986) show the dispersion characteristics of the Faraday rotation and of the magnetic circular dichroism (MCD) deep in the absorption band of the bismuth-substituted iron garnets; corresponding spectral characteristics for iron garnet diluted with gallium ions are also shown there for comparison. The Faraday effect spectra measured for two samples with the same gallium content but with different concentrations of bismuth allow one to discern the partial contribution associated with incorporation of the Bi^{3+} ions into the iron garnet. In the difference spectrum, one can easily see an asymmetric curve with extremes at $\lambda = 0.47$ and $0.375 \ \mu$m.

The shape of this difference curve looks like that of a dispersion characteristic of the diamagnetic type with one wing absent. The spectral characteristics of the MCD for gallium-substituted iron garnet deep inside the absorption band are shown in figure 9.14. The spectral dependence of the Faraday effect in the long-wavelength spectral region is presented in the inset in figure 9.13 (Hibiya *et al* 1985).

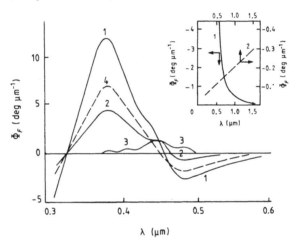

Figure 9.13.   The dispersion of the Faraday effect for epitaxial films of composition $Bi_{0.5}Tm_{2.5}Fe_{3.9}Ga_{1.1}O_{12}$ (curve 1), $Bi_{0.25}Tm_{2.75}Fe_{3.9}Ga_{1.1}O_{12}$ (curve 2), and $Y_{2.6}Sm_{0.4}Fe_{3.8}Ga_{1.2}O_{12}$ (curve 3), and the difference spectrum curve 1 − curve 2 (curve 4) (Burkov and Kotov 1975). Inset: the dispersion of the Faraday effect for the film of composition $Bi_{1.56}Gd_{2.5}(FeAlGa)_5O_{12}$ (curve 1), and the dependence of the Faraday rotation on the concentration x of bismuth (curve 2) for $Bi_xGd_{3-x}(FeAlGa)_5O_{12}$, with $\lambda = 1.3 \ \mu$m (Hibiya *et al* 1985).

In the MCD difference spectrum, the presence of a dispersion curve of diamagnetic type with the band centre at $\lambda = 0.37 \ \mu$m is clearly manifested. Analysis of the MCD and Faraday effect spectra leads us to infer that in the bismuth-substituted iron garnets there is a new magnetooptical peculiarity centred at $\lambda = 0.37 \ \mu$m. It can be anticipated that the Faraday rotation at $\lambda = 0.37 \ \mu$m for the iron garnet $Bi_{2.3}(YLu)_{0.7}Fe_5O_{12}$ (Hansen *et al* 1984) will exceed 60 deg μm^{-1}.

The Kerr effect. The spectral dependencies of the polar Kerr effect were rather rigorously studied by Wittekoek *et al* (1975). Figure 9.15 shows the spectral dependencies of the rotation and ellipticity. The rotation spectrum has

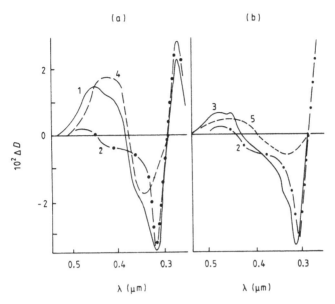

Figure 9.14. The spectral dependences of the magnetic circular dichroism for epitaxial films of $Bi_{0.5}Tm_{2.5}Fe_{3.9}Ga_{1.1}O_{12}$ (curve 1), $Y_{2.6}Sm_{0.4}Fe_{3.8}Ga_{1.2}O_{12}$ (curve 2), and $(YLu)_{2.8}Pb_{0.2}Fe_{3.8}Ga_{1.2}O_{12}$ (curve 3), plus the difference spectrum curve 1 − curve 2 (curve 4), and the difference spectrum curve 3 − curve 2 (curve 5) (Burkov and Kotov 1983).

a maximum at $\lambda = 0.46$ μm and a minimum at $\lambda = 0.32$ μm, and crosses the zero point at $\lambda = 0.366$ μm. These singular points are in good agreement with the MCD data (extrema at $\lambda = 0.48$ μm and $\lambda = 0.32$ μm, and a zero-point transition at $\lambda = 0.37$ μm). In the same figure the rotation and ellipticity difference spectra for the Kerr effect (dashed line 4) are presented; these were obtained by subtraction of the dispersion dependence of the yttrium iron garnet from the corresponding dependency of the bismuth-substituted iron garnet. The difference spectrum reveals a magnetooptic peculiarity of diamagnetic type with the band centre at $\lambda = 0.375$ μm.

From this analysis of the spectral dependencies of the Faraday effect, the MCD, and the polar Kerr effect rotation and ellipticity, we can conclude that the main feature of the magnetooptical spectra for bismuth-substituted iron garnets is an intense magnetooptically active transition centred at 0.37 μm. Some asymmetry in the form of the curve could be related to a paramagnetic-type transition near 0.45 μm.

The microscopical origin of the Faraday rotation of bismuth-substituted iron garnets was investigated theoretically by Kahn *et al* (1969), Scott *et al* (1975), Shinagawa (1982), and Šimša *et al* (1984), who assumed that the large Faraday rotation is caused by the increase of the spin–orbit interaction caused by the

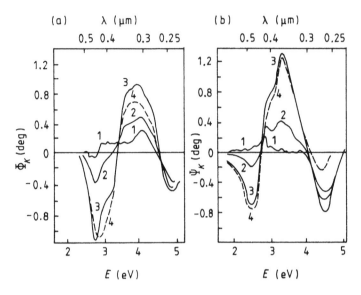

Figure 9.15. The spectral dependence of the rotation (a) and ellipticity (b) of the Kerr effect for polycrystalline samples of $Bi_xY_{3-x}Fe_5O_{12}$: curve 1: $x = 0.00$; curve 2: $x = 0.25$; curve 3: $x = 1.00$; curve 4: the difference spectrum, curve 3 − curve 1 (Wittekoek *et al* 1975).

formation of the molecular orbit between the 3d orbital in Fe^{3+} and the 2p orbital in O^{2-} mixed with the 6p orbital in Bi^{3+}, which has a large spin–orbit interaction coefficient.

Using a model based on the assumption of four π-type charge-transfer transitions (of diamagnetic type) from 2p in O^{2-} to 3d in Fe^{3+}, such as $t_{1u} \rightarrow t_{2g}^*$ (the a site), $t_{1u} \rightarrow e^*$ (the d site), $t_2 \rightarrow e^*$ (the d site), and $t_{2u} \rightarrow t_{2g}^*$ (the a site) (Shinagawa 1982), and adding the contributions of four paramagnetic-type transitions, Matsumoto *et al* (1992) have obtained rather good agreement between theory and experiment for Faraday rotation spectra of bismuth-substituted iron garnet over the spectral range 1.5–3.4 eV.

Recently Dionne and Allen (1994), in an attempt to support the theory that the enhanced magnetooptical effect in bismuth-substituted iron garnets originates from covalent interactions of Bi^{3+} and Fe^{3+} ions, discussed a self-consistent approximation to a two-level bonding–antibonding hybrid formed from the excited ^4P term of Fe^{3+} and the excited ^3P term of Bi^{3+}. They assumed that the anomalous magnetooptical properties of bismuth-substituted iron garnets are related to two transitions, at 2.6 and 3.15 eV, at the tetrahedral and octahedral sites of iron, and found that more than 30% of the hybrid eigenfunctions of the upper states can arise from the Bi^{3+} orbital term, and that the multiplet splittings are of the order of 0.5 eV.

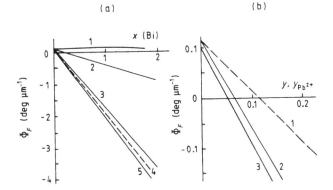

Figure 9.16. (a) The concentration dependences of the Faraday effect for films of $Y_3Fe_{5-x}Ga_xO_{12}$ (curve 1), $Y_{3-x}Pr_xFe_5O_{12}$ (curve 2), $Y_{3-x}Pb_x^{2+}M_x^{4+}Fe_{5-x}O_{12}$ (curve 3), $Bi_xGd_{3-x}Fe_5O_{12}$ (curve 4), and $Bi_xY_{3-x}Fe_5O_{12}$ (curve 5); $\lambda = 0.63$ μm. (b) The dependence of the Faraday rotation for $Pb_xY_{3-x}Fe_5O_{12}$, with $\lambda = 0.63$ μm, on the concentration of Pb^{2+}, $y_{Pb^{2+}}$ (solid lines). Dashed line 1: full content of Pb, y; solid lines (lines 2 and 3 correspond to different compositions of the melt): content of Pb^{2+} (Hansen and Krumme 1984).

9.6.4 Concentration and temperature dependencies

The magnetooptical properties of iron garnets vary linearly with the concentration of the bismuth ions incorporated. The dependence of the Faraday effect for iron garnets of different compositions on the bismuth concentration is shown in figure 9.16(a) (Hansen and Krumme 1984). The largest value of Φ_F achieved so far for epitaxial films at the wavelength $\lambda = 0.63$ μm has been reached for the composition $Bi_{2.3}(YLu)_{0.7}Fe_5O_{12}$, and is $\Phi_F = 4.8 \times 10^4$ deg cm^{-1} at 295 K (Hansen *et al* 1984). Dilution of the tetrahedral and octahedral sublattices decreases the magnetooptical effects. For example, the Faraday effect in $Bi_xY_{3-x}Fe_{5-y}Al_yO_{12}$ changes according to the formula

$$\Phi_F(x, y) = \Phi_F(x)\left(1 - \frac{1}{2}y_a\right)\left(1 - \frac{1}{3}y_d\right)$$

where y_a and y_d are the concentrations of the Al^{3+} ions in the octahedral and tetrahedral sites, and $y = y_a + y_d$ (Hansen and Krumme 1984).

The temperature and concentration dependencies of the Faraday effect in the bismuth-substituted iron garnets $Bi_xY_{3-x}Fe_5O_{12}$ are shown in figure 9.17 (Hansen and Krumme 1984).

9.6.5 The magnetooptical figure of merit

Applied magnetooptics requires materials with minimal optical absorption and maximum Faraday rotation. In order to compare magnetooptical materials, a

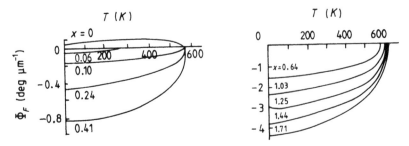

Figure 9.17. The temperature and concentration dependence of the Faraday effect for $Bi_xY_{3-x}Fe_5O_{12}$, with $\lambda = 0.63$ μm (Hansen and Krumme 1984).

new parameter has been introduced—the magnetooptical figure of merit, which is defined as the ratio of twice the specific Faraday rotation to the absorption coefficient of the material:

$$F_1 = 2|\Phi_F|/\alpha.$$

The magnetooptical figure of merit is expressed in degrees if the absorption coefficient α is expressed in cm^{-1}, and the Faraday rotation in deg cm^{-1}, or in deg dB^{-1} if an optical loss parameter, which is measured in dB cm^{-1}, is used instead of the absorption coefficient. In some cases the ratio $|\Phi_F|/\alpha$ is also called the magnetooptical figure of merit. The figure of merit determines the optimal thickness of the magnetooptical layer when it is used in magnetooptical modulators, switches, transparencies, and deflectors. The explicit meaning of the F_1-parameter becomes obvious after analysis of the formulae presented in chapter 13.

Typical spectral characteristics of F_1 for the bismuth-substituted iron garnets contain three maxima at $\lambda = 0.56, 0.78$, and 1.15 μm; these values are determined by the spectral dependence of the absorption coefficient of the iron garnet. In the region where $\lambda > 1.0$ μm, despite the Faraday rotation diminishing with the growth of the wavelength, the magnetooptical figure of merit grows sharply, because here the absorption coefficient drops exponentially in the transparent region. Consequently in the transparent region the magnetooptical figure of merit can exceed $1000°$.

9.7   Lead-substituted iron garnets

9.7.1   Optical properties

Incorporation of the lead ions into dodecahedral sites of the iron garnet structure results in a significant increase of the optical absorption in the visible spectral band (figure 9.18). It is very surprising that lead enters the garnet structure at the level of 0.4 formula units in the absence of ions that provide charge compensation.

To explain this fact, Wittekoek *et al* (1975) and Scott and Page (1977a, b) proposed a self-compensation model, suggesting that in the garnet structure the lead ions form $Pb^{2+}-Pb^{4+}$ pairs. The presence of the lead ions must give rise to a number of optical transitions, with charge transfer between the lead and iron ions. The positions and types of these additional transitions are summarized in table 9.6 (Hansen *et al* 1983).

Additional absorption in the visible spectral band for lead-substituted iron garnets should be attributed to the charge-transfer transition from Pb^{2+} to Pb^{4+} at $\lambda = 0.56\ \mu m$ (Scott and Page 1977a, b, Hansen *et al* 1983). Additional absorption near $\lambda = 0.8\ \mu m$ is also usually associated with $Fe^{3+} \rightarrow Pb^{4+}$ charge-transfer transitions (Hansen *et al* 1983, Yokoyama *et al* 1987).

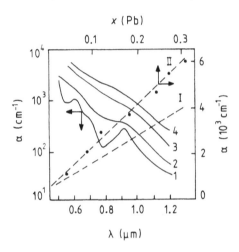

Figure 9.18. The effect of the growth temperature on the optical absorption of the epitaxial film $Y_3Fe_5O_{12}:Pb^{2+}$, Si^{4+}. $T_g = 955$ °C (curve 1), 873 °C (curve 2), 824 °C (curve 3), and 742 °C (curve 4) (Larsen and Robertson 1974). Also shown are the dependences of the absorption coefficients on the lead concentrations in: $Y_{3-x}Pb_xFe_5O_{12}$ (Hansen *et al* 1983) (I), with $\lambda = 0.63\ \mu m$; and $RE_{3-x}Pb_xFe_5O_{12}$ (II), with $\lambda = 0.56\ \mu m$ (Scott and Page 1977a, b).

9.7.2 Magnetooptical properties

Incorporation of the divalent lead ions, like that of trivalent bismuth ions, into the iron garnet structure results in unusual changes in the magnetooptical properties of the material; the effect caused by the influence of the lead ions in terms of divalent ions is comparable with the similar effect of the bismuth ions.

Figure 9.14(b) shows the spectral characteristics of the MCD for two samples with different lead contents (Burkov and Kotov 1983). Dashed line 5, which is the difference of curves 3 and 2, represents the contribution of a magnetooptical transition associated with the presence of the lead ions in the iron

Table 9.6. Calculated positions of the optical transitions with participation of lead ions (Hansen *et al* 1983). (p is the hole on the oxygen ion.)

Transition type	Wavelength (μm)
$Pb^{2+} + Pb^{4+} + \hbar\omega \longrightarrow Pb^{3+} + Pb^{3+}$	0.56
	0.35
$Pb^{2+} + (Fe^{3+}) + \hbar\omega \longrightarrow Pb^{3+} + (Fe^{2+})$	0.42
	0.38
$Pb^{4+} + [Fe^{3+}] + \hbar\omega \longrightarrow Pb^{3+} + p$	0.51
	0.31
$Pb^{4+} + (Fe^{3+}) + \hbar\omega \longrightarrow Pb^{3+} + p$	0.51
	0.37
$O^{2-} + Pb^{4+} + \hbar\omega \longrightarrow Pb^{3+} + p$	0.31
$^1S_0 \longrightarrow {}^3P_1\{Pb^{2+}\}$	0.28
$Pb^{2+} \longrightarrow$ conduction band	0.25
Valence band $\longrightarrow Pb^{4+}$	0.37

garnet. Comparing the difference spectrum with a similar one for the bismuth-substituted iron garnet (figure 9.14(a)), we can conclude that incorporation of both Bi^{3+} and Pb^{2+} ions into the iron garnet structure leads to enhanced magnetooptical activity of the same transition. The dependence of the Faraday rotation of the iron garnet on the lead content at the wavelength $\lambda = 0.63$ μm is presented in figure 9.16(b) (Hansen and Krumme 1984).

9.8   Cerium-substituted iron garnets

9.8.1   Optical properties

Cerium ions in different crystals usually exist in trivalent and tetravalent states. In garnets, both trivalent and tetravalent cerium has been observed (Alex *et al* 1990). The polyvalence of cerium can lead to the appearance of different absorption lines for the cerium-containing materials, including those arising from the charge-transfer transitions between the cerium ions of different valences. The optical properties of the cerium-substituted iron garnets have not yet been thoroughly studied.

Gomi *et al* (1988) presented absorption spectra for iron garnet films of composition $Ce_xY_{3-x}Fe_5O_{12}$ ($0 \leq x \leq 0.7$) that were epitaxially sputtered onto heated gadolinium–gallium garnet (GGG) substrates (figure 9.20 (curves 3, 4, and 5)). Taking into account the magnetooptical spectra, which are discussed below, the increased absorption in the spectral interval 1–2 eV can be associated with two transitions near 1.4 and 1.9 eV.

For $Ce_1Y_2Fe_5O_{12}$ near 0.8 μm, the additional absorption as compared with that of yttrium iron garnet reaches approximately 10^3 cm^{-1}. Taking into account the results of Alex *et al* (1990), this can suggest that in the films studied, Ce^{3+} and Ce^{4+} ions were simultaneously present. Therefore we can expect Fe^{2+} ions, which significantly affect the value of the optical absorption in the 1–2 eV region, to appear in the films.

Because of this, all of the available experimental data taken together still do not provide a sound basis for drawing any definite conclusions as regards the nature of magnetooptically active transitions in cerium-substituted iron garnets at 0.9 and 0.45 μm. Gomi *et al* (1988) associated these transitions with intra-ion transitions among Ce^{3+} ions, while Alex *et al* (1990) considered the possibility of charge-transfer transitions between the Fe^{2+} and Ce^{3+} ions; a significant role— in magnetooptical spectra too—can be played by charge-transfer transitions between Ce^{3+} and Ce^{4+} ions, and between Ce^{4+} and Fe^{2+} ions.

Cerium-substituted iron garnets are very interesting as a medium for magnetooptical memory—say in magnetooptical disks—being superior in their characteristics to bismuth-substituted iron garnets in the 0.8 μm spectral region. The absence of experimental data on optical absorption in the regions near 1.3 and 1.55 μm renders it impossible to judge the possibilities for this material as regards the development of non-reciprocal and other magnetooptical devices for use in the infra-red spectral band.

9.8.2 Magnetooptical properties

Incorporation of cerium ions (Ce^{3+}) into the iron garnet structure results in dramatic changes in the magnetooptical properties, in both the infra-red and visible spectral regions (Gomi *et al* 1988, 1990, Alex *et al* 1990, Gomi and Abe 1994). From the studies of the magnetooptical properties of glasses doped with various trivalent rare-earth ions, it is known that the Ce^{3+} ions yield the greatest specific contribution to the Faraday effect. This can be associated first of all with the longest-wavelength magnetooptically active transitions of cerium among all of the trivalent rare-earth ions. Similarly it could be expected that, in the iron garnets too, the Ce^{3+} ions, as compared with other rare-earth ions, will strongly modify the magnetooptical properties of the matrix.

A tremendous technological breakthrough in the formation of cerium-substituted iron garnets was achieved as a result of the development of rf sputtering of epitaxial iron garnet films onto a heated garnet substrate. Using this technique, it proved possible to form films with cerium concentrations up to one formula unit (Gomi *et al* 1990).

The study of these films has shown that incorporation of the Ce^{3+} ions into the iron garnets results in substantial changes in the magnetooptical properties of the material; for example, for $Ce_1Y_2Fe_5O_{12}$ the Faraday rotation near 1.6 eV (0.8 μm) reaches 3.3 deg μm^{-1}, which in absolute value exceeds fourfold the Faraday rotation for the bismuth-substituted iron garnet $Bi_1Y_2Fe_5O_{12}$ in this spectral region (Gomi *et al* 1990).

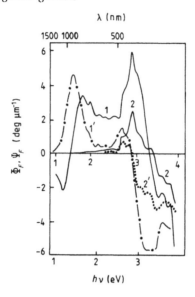

Figure 9.19. Spectral characteristics of the Faraday rotation (curves 1 and 2) and ellipticity (curves 1' and 2'), for $Ce_1Y_2Fe_5O_{12}$ (curves 1 and 1') and $Y_3Fe_5O_{12}$ (curves 2 and 2') iron garnets (Gomi *et al* 1990).

The Faraday rotation (curve 1) and ellipticity (curve 1') for $Ce_1Y_2Fe_5O_{12}$ iron garnet are shown in figure 9.19, and the corresponding characteristics for yttrium-substituted iron garnet (curves 2 and 2') are also shown there for comparison. Difference spectra of the Faraday rotation (curve 1) and ellipticity (curve 2) for these compositions, which are presented in figure 9.20, make the contribution of cerium ions clearly evident (Gomi *et al* 1990).

Comparison of the Faraday rotation and ellipticity spectra led to the conclusion that incorporation of Ce^{3+} ions into the iron garnets results in the appearance of two magnetooptical abnormalities of comparable intensity but opposite sign at 1.4 and 3 eV. These unique magnetooptical properties of the cerium-substituted iron garnets make them a most promising material as regards use in applied magnetooptics.

The transition at 1.4 eV indicates the paramagnetic form of the dispersion curve. There are not enough experimental data available for us to come to an unambiguous conclusion regarding the form and position of the high-frequency magnetooptically active transition. One of the possible interpretations implies the existence of a transition with a paramagnetic shape of the dispersion curve at 3.1 eV. On the other hand, because of the strong asymmetry in the short-wavelength wing of the difference ellipticity spectrum, and the absence of experimental data for the energy region above 4 eV, it is still possible to assume the existence of a magnetooptical peculiarity of diamagnetic form with a transition at 3.5 eV, which lies in the same region as in the bismuth- and

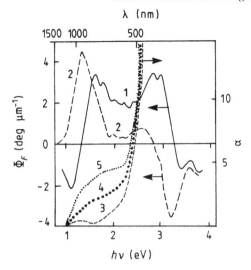

Figure 9.20. The specific contribution of cerium ions to the Faraday rotation (curve 1) and ellipticity (curve 2) for $Ce_1Y_2Fe_5O_{12}$, and the absorption spectrum of $Ce_xY_{3-x}Fe_5O_{12}$ films at $T = 295$ K without correction for the reflection: $x = 0$ (curve 3); $x = 0.3$ (curve 4); and $x = 0.7$ (curve 5) (Gomi *et al* 1988, 1990).

lead-substituted iron garnets.

It is also possible that in the high-frequency spectral region, two magnetooptical peculiarities are superimposed, one of which is associated with the second transition of the Ce^{3+} ion, and another which is caused by the so-called 'mixing' mechanism, which usually operates in bismuth- and lead-substituted iron garnets.

Cerium-substituted iron garnets are the most interesting as regards applications in the wavelength region $\lambda > 0.6\ \mu$m, where their magnetooptical characteristics are better than the corresponding ones of the bismuth-substituted iron garnets. At the same time, as there are no reliable data on the optical absorption in the infra-red spectral band, it is difficult to reach final conclusions as to the advantages of cerium- or bismuth-substituted iron garnets in different spectral regions.

Gomi *et al* (1991) proposed a Ce^{3+} (4f)–Fe^{3+}(tetrahedral) charge-transfer model to explain the origin of the strong enhancement of the magnetooptical properties of cerium-substituted iron garnets. Investigation of the Faraday rotation and ellipticity of the rf-sputtered films with the composition $Y_2Ce_1Fe_{5-x}M_xO_{12}$ (M = Al, In; $x = 0$–5) showed that with the amount of substitution increasing, the Faraday rotation and ellipticity for the films at 1.4 eV reduced at nearly the same rates for the ions of Al and In (Gomi and Abe 1994). Gomi and Abe found, from the analysis using molecular-field theory, that these reductions are in proportion to the magnetic moment of the tetrahedral iron sublattice, which supports the proposed charge-transfer model.

Chapter 10

Metals and alloys

It is well known that the magnetooptical properties of metals and alloys are strongly dependent on the difference in population near the Fermi level of spin-up and spin-down electrons, and on the oscillator strength of the transitions involved. Because of this strong dependence, for metallic systems it is difficult to establish a direct connection between magnetooptical effects and microscopical origins. Many investigations in this area have dealt with the Kerr effect in thin metallic films, because erasable magnetooptical recording is feasible with thin metallic films of rare-earth–transition metal composition.

10.1 Transition metals and alloys

10.1.1 Iron, cobalt, and nickel

At the start of this section devoted to the magnetooptical properties of metals and alloys, it is necessary first of all to discuss simple systems such as Fe, Co, and Ni. The results of measurements of the polar Kerr effect for pure Fe, Co, and Ni are shown in figure 1.4 (Krinchik and Artem'ev 1968, van Engen 1983, Carey et al 1983, Buschow 1988). The common feature of all of the dependencies is the existence of the negative peak of the rotation near 1.2–1.6 eV for all of these metals, and the low-energy sign reversal for Fe and Co occurring near 0.4 eV; then the Kerr rotation attains a maximum of $\Phi_K = 0.8°$ near 0.2 eV, and then slowly falls to 0.04° at 0.1 eV for Co, and attains a maximum of $\Phi_K = 0.03°$ near 0.2 eV and falls to 0.025° at 0.1 eV for Fe.

The low-energy sign reversal of Φ_K for Ni occurs near 0.9 eV, and then the Kerr rotation attains a maximum of $\Phi_K = 0.17°$ near 0.4 eV and slowly decreases to 0.02° near 0.1 eV. Comparison of the results obtained by different authors for Ni shows that in most cases there is good agreement as regards the shape of the energy dependence when the results were obtained from direct measurements. The fact that the absolute values of the Kerr rotation for curve 1 (for Ni) in figure 1.4 are lower than those for other samples can be explained by the fact that the sample was not completely saturated.

10.1.2 Alloys of transition metals

Fe–Co. The results of many magnetooptical investigations on Fe–Co alloys have been presented in a review (Buschow 1988). The concentration dependence of the real and imaginary parts of the dielectric tensors obtained for polycrystalline thin films shows a sharp increase in the absolute value of both contributions in the intermediate-concentration region. This can be explained as a manifestation of the Fe–Co atomic order–disorder transition.

The main effect is not symmetric around the equiatomic position, but occurs at approximately 45 at.% of Co. It is interesting to note that comparisons of the concentration dependence of the magnetooptical properties of the various Fe–Co alloys with the corresponding saturation magnetization have shown that the popular view of the magnetooptical intensity as being proportional to the magnetization is a misconception.

Co–Cr. The polar Kerr rotation, ellipticity, and reflectivity for sputtered $Co_{1-x}Cr_x$ were studied by Tsutsumi *et al* (1983). The intensity of the polar Kerr rotation is much reduced in the Co–Cr alloys as compared to Co metal. The polar Kerr rotation angle Φ_K at 633 nm and the saturation magnetization decrease monotonically with increasing Cr content from $\Phi_K = 0.32°$ and $M_s = 1400$ emu cm^{-3} at $x = 0$ to $\Phi_K = 0$ and $M_s = 0$ at $x = 0.3$, and the reflectivity slowly decreases from 66% at $x = 0$ to 58% at $x = 0.3$.

Fe–V. Voloshinskaya and Fedorov (1973) measured the equatorial Kerr effect and the polar Kerr effect in Fe–V alloys. It was found that increasing the V concentration strongly reduces the absolute value of the Kerr rotation. The result that for 22% of V the Kerr effect no longer shows the sign reversal of Φ_K observed near 0.3 eV (3.7 μm) for Fe and more Fe-rich materials might indicate that alloying Fe with increasing amounts of V leads ultimately to a sign reversal of the spin polarization of the conduction electrons.

10.1.3 Iron-based alloys and compounds

Iron-based alloys. The Kerr rotations measured for various materials, including Fe–Al, Fe–Ga, Fe–Si, Fe–Ge, Fe–V, Fe–Co, Fe–Pd, and Fe–Pt, at 633 and 830 nm, are listed in table 10.1, together with the corresponding room temperature magnetization values (σ_r (A m^2 kg^{-1})) (Buschow *et al* 1983, Buschow 1988).

Fe–Pt and Fe–Pd. The magnetooptical properties of Fe–Pt and Fe–Pd have been studied by Buschow *et al* (1983) and van Engen (1983). The results of the investigations are shown in figure 10.1(a), where they can be compared with the spectrum obtained for pure Fe metal. The following features become apparent. The location of the first minimum in Φ_K (near 1.2 eV) is not much affected by alloying, even for compositions in which half of the Fe has been replaced by Pt or Pd. There is a considerable increase in intensity in the high-energy region of the spectra. In these alloys the enhancement of the polar Kerr rotation in the high-energy range is stronger for Pt than for Pd.

Table 10.1. The Kerr rotations for several Fe alloys (Buschow 1988).

Alloy	σ_r (A m^2 kg^{-1})	Φ_K (deg) $\lambda = 633$ nm	Φ_K (deg) $\lambda = 830$ nm
Fe	213	−0.41	−0.53
Fe$_{90}$Al$_{10}$	203	−0.42	−0.53
Fe$_{70}$Al$_{30}$	98	−0.34	−0.41
Fe$_{80}$Ga$_{20}$	167	−0.45	−0.56
Fe$_{60}$Ga$_{40}$	101	−0.34	−0.44
Fe$_{80}$Si$_{20}$	160	−0.36	−0.42
Fe$_{85}$Ge$_{15}$	169	−0.44	−0.55
Fe$_{80}$V$_{20}$	156	−0.27	−0.36
Fe$_3$Co	234	−0.42	−0.46
FeCo	230	−0.51	−0.60
FeCo$_3$	200	−0.48	−0.58
Fe$_{75}$Pd$_{25}$	167	−0.33	−0.37
Fe$_{50}$Pd$_{50}$	96	−0.14	−0.19
Fe$_{85}$Pt$_{15}$	172	−0.53	−0.62
Fe$_{65}$Pt$_{35}$	107	−0.52	−0.68

It should be noted that for photon energies higher than 3.5 eV the polar Kerr rotation for Fe$_{0.5}$Pt$_{0.5}$ exceeds that of pure Fe metal. It is thought that the additional contribution to the polar Kerr rotation is due to transitions from spin-polarized 5d electrons in the case of Pt and from spin-polarized 4d electrons in the case of Pd, the effect being stronger in the Pt alloys owing to the spin–orbit coupling for Pt being stronger than that for Pd (Buschow 1988).

Iron-based compounds. The polar Kerr rotation spectra of various ferromagnetic Fe intermetallics were studied by Buschow *et al* (1983). It was reported that the changes of the Kerr rotation spectrum on alloying Fe with metals such as Al, Ga, and Sn affect primarily the high-energy side of the spectrum. The influence of s and p metals on the magnetooptical properties of Fe does not seem to be very different from the effect of s and p metals on the spectrum of Co (Buschow *et al* 1983).

For intermetallic compounds of Fe with Pd and Pt, as for the Co compounds, there is a substantial enhancement of the Kerr intensity at photon energies near the high-energy Kerr effect maximum. This enhancement was attributed to transitions involving spin-polarized 4d or 5d electrons. The values of Φ_K at the wavelengths 633 and 830 nm, and corresponding data on the magnetic and crystallographic properties of the Fe-based compounds, are given in table 10.2 (Buschow 1988).

Fe$_7$Se$_8$. The optical and magnetooptical properties of a single crystal of the metallic ferrimagnet Fe$_7$Se$_8$ have been measured by Sato *et al* (1987).

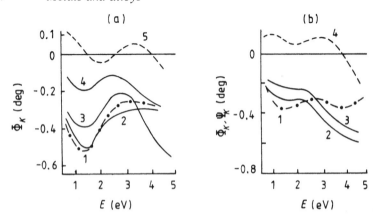

Figure 10.1. (a) The spectral dependence of the polar Kerr rotation for Fe (chain line 1), $Fe_{0.85}Pt_{0.15}$ (full line 2), $Fe_{0.5}Pt_{0.5}$ (full line 3), and $Fe_{0.5}Pd_{0.5}$ (full line 4); and the ellipticity (dashed line 5) for $Fe_{0.5}Pt_{0.5}$. (b) The spectral dependence of the polar Kerr rotation for Co (chain line 1), $Co_{0.5}Pt_{0.5}$ (full line 2), and $Co_{0.25}Pt_{0.75}$ (full line 3); and the ellipticity (dashed line 4) for $Co_{0.25}Pt_{0.75}$ (Buschow 1988).

Table 10.2. The Kerr rotations for several intermetallic compounds of Fe (Buschow 1988).

Alloy or compound	σ_r (A m² kg⁻¹)	Φ_K (deg) $\lambda = 633$ nm	Φ_K (deg) $\lambda = 830$ nm
Fe_3Al	156	−0.39	−0.48
Fe_7Ga_3	147	−0.45	−0.56
Fe_3C	100	−0.21	−0.24
Fe_3Si	138	−0.32	−0.36
Fe_2B	166	−0.45	−0.47
FeB	84	−0.26	−0.22
Fe_3Ge	133	−0.41	−0.44
Fe_3Sn	126	−0.31	−0.38
Fe_2Sc	52	−0.17	−0.17
$Fe_{74}Ti_{26}$	74	−0.10	−0.12
$Fe_{69}Zr_{31}$	91	−0.24	−0.26
Fe_2Hf	45	−0.32	−0.35
$Fe_{77}Nb_{23}$	36	−0.04	−0.06
$Fe_{75}Pd_{25}$	58	−0.30	−0.25
$Fe_{50}Pd_{50}$	96	−0.15	−0.19
Fe_3Pt	107	−0.51	−0.58
$Fe_{55}Pt_{45}$	45	−0.36	−0.44
$FePt$	42	−0.32	−0.39

This material crystallizes in a hexagonal NiAs structure with a superstructure associated with the ordering of vacant Fe sites. Fe_7Se_8 is one of the small number of chalcogenide crystals which show a net magnetic moment at room temperature. The magnetic properties of the compound were explained using an ionic model in which divalent iron and vacancy-induced trivalent iron couple antiferromagnetically. The easy axis of magnetization is in the c-plane at room temperature, and rotates toward the c-axis when the temperature is lowered below 220 K.

The polar Kerr rotation spectrum of Fe_7Se_8 at room temperature was measured in the spectral range 0.5 to 3.6 eV. A well-defined peak of rotation can be seen at around 0.8 eV, but the peak value of the Kerr rotation is very small: not more then $0.03°$. The Kerr rotation crosses zero near 1.5 eV, and in the spectral range 1.5 to 3.5 eV the value of the rotation is near $-0.01°$.

10.1.4 Cobalt-based alloys and compounds

Cobalt-based alloys. The Kerr effect in Co–Ga, Co–Si, Co–Sn, Co–Cr, Co–Mo, Co–W, Co–Rh, Co–Ir, and Co–Pt alloys has been investigated by Buschow *et al* (1983). For most of the Co-based alloys, compound formation already takes place at fairly high Co concentration, making the range of solid solubility of Co rather limited. Spectra for an extended Co concentration range were only reported for Co_xPt_{1-x} and Co_xPd_{1-x}. Results for the former series are shown in figure 10.1(b).

It is seen that increasing the Co concentration leads only to a moderate shift in energy of the first Φ_K minimum. There is a considerable increase in intensity of the second Φ_K minimum relative to the first minimum. It is assumed by Buschow *et al* (1983) that this intensity increase in the 4 eV region is due to transitions involving the spin-polarized 5d electrons of Pt. A similar behaviour was observed for alloys of Co with Pd by Buschow *et al* (1983), and with Rh or Ir by van Engen (1983), although the effect was strongest for the Pt alloys of comparable Co content.

This was explained in terms of a stronger spin–orbit coupling of the 4d electrons of the Pt compound compared to those of the 4d or 5d electrons of the other elements (Buschow 1988). The results of the investigations of the magnetooptical properties of Co alloys are summarized in table 10.3, where, in addition to the values for the polar Kerr rotation, the room temperature magnetizations of the various samples are also specified.

Cobalt-based compounds. Magnetooptical measurements on a fairly large number of ferromagnetic Co intermetallics have shown that, in general, only the Co-rich compounds are ferromagnetic at room temperature (Buschow *et al* 1983). For Co–Ga compounds with CsCl structure it was found that the Kerr rotation spectra change form and decrease in value when the Co content decreases. The most prominent difference is the disappearance of the high-energy Kerr effect maximum in the spectra of the CsCl-type compounds, while

Table 10.3. The Kerr rotations for several Co alloys (Buschow 1988).

Alloy	σ_r (A m^2 kg^{-1})	Φ_K (deg) $\lambda = 633$ nm	Φ_K (deg) $\lambda = 830$ nm
Co	156	−0.30	−0.36
$Co_{92}Ga_8$	133	−0.31	−0.37
$Co_{90}Si_{10}$	126	−0.31	−0.34
$Co_{90}Sn_{10}$	115	−0.29	−0.32
$Co_{94}Cr_6$	125	−0.26	−0.31
$Co_{80}Cr_{20}$	40	−0.10	−0.12
$Co_{90}Mo_{10}$	94	−0.19	−0.23
$Co_{90}W_{10}$	82	−0.16	−0.23
$Co_{85}Rh_{15}$	134	−0.30	−0.33
$Co_{85}Ir_{15}$	88	−0.24	−0.26
$Co_{90}Pt_{10}$	123	−0.37	−0.39
$Co_{80}Pt_{20}$	91	−0.29	−0.30
$Co_{70}Pt_{30}$	76	−0.40	−0.43
$Co_{30}Pt_{70}$	29	−0.25	−0.25

the low-energy Kerr effect maximum is seen to shift to higher energies with increasing Ga content (Buschow 1988).

A different concentrational behaviour was observed when Co was alloyed with early transition metals such as Ti, Zr, Hf, or Nb. It is found for Co–Hf compounds that the low-energy Kerr effect maximum is only relatively weak. The location of this latter maximum tends to shift to lower energies with increasing Hf concentration. Magnetooptical, magnetic, and structural data for a large variety of different intermetallic Co compounds are listed in table 10.4 (Buschow 1988).

Co–Pt. The changes in the polar Kerr rotation spectra observed by Buschow *et al* (1983) when Co formed intermetallic compounds with late transition metals such as Pt and Pd are about the same as those observed for the series of Co_xPt_{1-x} solid solutions, and shown already in figure 10.1(b). Here the prominent feature is an enhancement of the Kerr intensity at energies near the second Kerr effect maximum.

Magnetooptical investigations of disordered fcc-phase CoPt and the tetragonal compound CoPt have also been made by Treves *et al* (1975) using crystalline films prepared by sputtering. At photon energies between 2 and 4.6 eV the Faraday rotation was found to be approximately constant, but it showed a slight increase on going to energies up to 1.6 eV. It was also found that the conversion of the fcc phase to the tetragonal phase caused by annealing leads to a slight decrease (20%) of the saturation Faraday rotation.

Brandle *et al* (1992) have studied the polar Kerr rotation and ellipticity

Table 10.4. The Kerr rotations for several intermetallic compounds of Co (Buschow 1988).

Alloy or compound	σ_r (A m² kg⁻¹)	Φ_K (deg) $\lambda = 633$ nm	Φ_K (deg) $\lambda = 830$ nm
Co_3B	82	−0.25	−0.24
Co_2Mg	55	−0.33	−0.34
$CoZn$	39	−0.12	−0.12
$Co_{70}Ga_{30}$	62	−0.20	−0.24
Co_3Si	47	−0.13	−0.14
Co_7Ge_3	23	−0.07	−0.07
Co_2Sn	17	−0.05	−0.05
$Co_{17}Y_2$	131	−0.24	−0.29
$Co_{13}La$	130	−0.26	−0.36
$Co_{17}Gd_2$	—	−0.28	−0.35
$Co_{11}Zr_2$	72	−0.15	−0.17
Co_7Hf	74	−0.20	−0.23
Co_9Mo_2	35	−0.09	−0.11
$CoPd$	78	−0.19	−0.17
$CoPt$	30	−0.33	−0.36

of evaporated $Co_{1-x}Pt_x$ alloys in the spectral range 0.8–5 eV. The UV peak shows a maximum Kerr rotation of −0.66° at 4.2 eV for $Co_{47}Pt_{53}$. The authors claim that this magnetooptical peculiarity is due to the spin-polarized Pt 5d level, which shows a density-of-states peak near 4 eV.

10.1.5   Nickel-based alloys and compounds

Nickel-based alloys. The polar and equatorial Kerr effects for Ni alloyed with Al and Pd were investigated by Voloshinskaya and Fedorov (1973) and Voloshinskaya and Bolotin (1974). The group of carriers responsible for the magnetooptical effects for Ni in the spectral range 0.1–0.3 eV was identified as belonging to the sixth and fifth zones of the Brillouin zone, comprising 3d electron states with the minority-spin direction.

The Kerr rotation spectra of various Ni alloys in the energy range 0.5–4.5 eV were studied by Buschow *et al* (1983). It was found that the main features of the spectra of the alloys in which Ni is combined with Cu or s or p metals remained the same in relation to the spectrum of pure Ni metal, although one may notice a substantial reduction in intensity for the alloys with Al, In, or Sn.

It is noteworthy that there is almost no shift in the location of the low-energy minimum of Φ_K near 1.5 eV and the location of the sign reversal of Φ_K near 0.9 eV. However, marked changes occur on the high-energy side of the spectra, for the alloys with Al, In, or Sn in particular. It is seen that there is a pronounced shift of the high-energy minimum near 3 eV towards lower energy, and the same applies to the location of the second sign reversal of Φ_K. In all cases the values of the polar Kerr rotation for Ni-based alloys were lower than for pure Ni metal (Buschow 1988).

In these cases, too, the overall features of the spectrum have remained the same, small changes becoming visible mainly on the high-energy side of the spectra. In contrast to the change observed on alloying Ni with s or p metals, the effect of addition of Pt and Pd is to shift the location of the second sign reversal to higher rather than to lower energies (Buschow 1988).

Attempts were made by Buschow *et al* (1983) to interpret the change observed upon alloying of Ni in terms of the inter-band model and hybridization of the 3d electron states of Ni with the states of the valence electrons of the component added. Buschow *et al* noted that the energy and intensity shifts observed on the high-energy side in the Pt and Pd alloys are most probably due to transitions involving the 5d and 4d electron states of Pt and Pd, respectively. The intensity increase in the high-energy range is largest for the Pt alloys, which would correspond to the comparatively large spin–orbit coupling in Pt.

10.1.6 Manganese compounds

MnBi. The magnetooptical properties of the compound MnBi have been extensively studied by many authors, in connection with the possibility of using such material in the form of thin films in magnetooptical disk memory systems (Chen *et al* 1973). The Kerr rotation spectra, the Faraday rotation spectra, and the corresponding optical absorption coefficient are presented in figure 10.2.

A drawback of MnBi in memory applications is the occurrence of a first-order phase transition, resulting in a slow change of the crystal structure from a low-temperature to a quenched high-temperature phase, a relatively high Curie temperature, and a high grain-noise level. The Kerr rotation angle of the quenched phase is about half of that of the low-temperature phase.

Attempts to improve the thermomagnetic and magnetooptical characteristics of MnBi were made involving substitution of Ti for Mn. The compositional dependences of the Faraday and Kerr rotations for these systems were studied by Egashira *et al* (1977). Jin *et al* (1995) have shown that the doping of the elements Al, Ce, Pr, Nd, and Sm into MnBi films could improve the properties of these films. For MnBi films doped with the elements Cu, Ni, Rh, and Pd, the hcp structure of the MnBi film was changed, and the Kerr rotation angle was reduced. The doping of MnBi films with Ge enhanced their Kerr rotation angle ($\Phi_K = 2.1°$ at 633 nm). MnBi films doped with In or Sn lose their original NiAs-type hcp structure, and correspondingly the Kerr rotation angle is decreased.

Mn–Cu–Bi. The magnetooptical properties of the ternary system Mn–Cu–Bi with $T_c = 180$ °C—chiefly the magnetic compound $Mn_3Cu_4Bi_4$—have

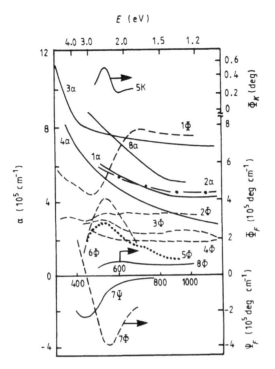

Figure 10.2. The spectral dependence of the specific magnetooptical Faraday rotation (Φ_F) and ellipticity (Ψ_F), Kerr rotation (Φ_K), and absorption coefficient (α) for normal (curve 1) and quenched (curve 2) MnBi (Chen *et al* 1973), MnSb (curve 3) (Sawatzky and Street 1971), Mn_5Ge_3 (curve 4) (Sawatzky 1971), $Mn_{0.33}(Cu_{0.8}Ni_{0.2})Bi_{0.34}$ (curve 5), Mn–Cu–Bi (curve 6) (Katsui 1976a), Mn–Cu–Bi (curve 7) (Shibukawa *et al* 1976), and Mn–Ga–Ge (curve 8) (Sawatzky and Street 1973).

been studied because it is a promising material as regards magnetooptical disk memory application (Katsui 1976a). Thin films of composition $Mn_1Cu_1Bi_1$ were studied by Shibukawa *et al* (1976). The Faraday rotation for $Mn_1Cu_1Bi_1$ is about 4×10^5 deg cm^{-1} at the wavelength 514 nm and $\Phi_F = 2.5 \times 10^5$ deg cm^{-1} at 633 nm. The dispersion of the specific Faraday rotation, and the ellipticity are presented in figure 10.2.

The polar Kerr rotation of Mn–Cu–Bi measured for nearly normal incidence at 633 nm is about $0.2°$ (Katsui 1976a). The Faraday rotation and ellipticity in the spectral range 450 to 700 nm were measured by Shibukawa *et al* (1976) (figure 10.2, curve 8Φ). In the measurements of the Faraday rotation and ellipticity, a film 35 nm thick was used. The sign of the Faraday rotation obtained by Shibukawa *et al* was opposite to that obtained in the work of Katsui (1976a).

The wavelength dependences of the specific Faraday and Kerr rotations for thin films of $Mn_{0.33}(Cu_{0.80}Ni_{0.20})Bi_{0.34}$ are shown in figure 10.2 (Katsui 1976a).

It should be noted that the specific Faraday rotation and Kerr rotation for these films are smaller than for MnBi in which no substitutions have been made. For the composition $Mn(Cu_{1-x}Dy_x)_{0.24}Bi_{0.86}$ the specific Faraday rotation decreased with increasing Dy content (Katsui 1976b). The polar Kerr rotation measured for nearly normal incidence at 633 nm is about 0.2°.

MnSb. The specific Faraday rotation and the corresponding optical absorption of stoichiometric and off-stoichiometric Mn–Sb compounds of NiAs structure are shown in figure 10.2 (Sawatzky and Street 1971). The room temperature absorption coefficient α and the specific Faraday rotation Φ_F are only slightly inferior to the corresponding values for MnBi in the wavelength region of interest.

The room temperature Kerr rotation spectrum of MnSb was studied by Buschow *et al* (1983). It was found that Φ_K does not vary much in the 1–3.5 eV range. This energy dependence of Φ_K agrees roughly with the corresponding energy dependence of Φ_F shown in figure 10.2.

The effect of substrate type on the polar Kerr rotation of rf-sputtered MnSb films was studied by Carey *et al* (1990). It was found that for the composition Mn–Sb at 633 nm, $\Phi_K = 0.16°$ when the film was sputtered on a glass substrate, $\Phi_K = -0.19°$ when the film was sputtered on a manganese-coated glass substrate, and $\Phi_K = 0.14°$ when the film was sputtered on an antimony-coated glass substrate.

Mn_5Ge_3. The Faraday rotation for Mn_5Ge_3 was studied by Sawatzky (1971) in the 1–4 eV range. As can be seen in figure 10.2, the specific Faraday rotation is nearly constant below 2 eV but increases strongly at higher photon energies.

MnAs. There is growing interest in transition metal compounds with NiAs-type crystal structure, as their behaviour has begun to be interpreted as exhibiting itinerant-electron magnetism. For example, recent energy band calculations for MnAs with NiAs structure suggest the appearance of a narrow d band with a high density of states in the vicinity of the Fermi surface, which explains the ferromagnetism in this material.

A Faraday rotation of $\Phi_F = 7.8 \times 10^4$ deg cm^{-1}, and an optical absorption coefficient of $\alpha = 4.5 \times 10^5$ cm^{-1} were found for polycrystalline stoichiometric MnAs at the wavelength 800 nm (Stoffel and Schneider 1970). The dispersion of the Kerr rotation shows a change in sign at 720 nm, the maximum Kerr rotation of +0.06° is found at 900 nm, and the maximum Kerr ellipticity of −0.15° is found at 720 nm. This intermetallic compound undergoes a first-order transition from a ferromagnetic state to the paramagnetic phase at $T = 40$ °C. Some magnetic and magnetooptical parameters for Mn-containing compounds are given in table 10.5.

Films of composition $Mn_{27}Al_{61}Cu_{12}$ have shown in the spectral range 500–1000 nm Kerr rotations of $\Phi_K = 0.1°$ without any spectral dispersion (Shen *et al* 1990). The films studied present a mixture of the ferromagnetic phase of MnAl and a highly paramagnetic amorphous phase. The structure of the

Table 10.5. Magnetooptical properties of several Mn-containing compounds (see Chen *et al* (1973), and references cited therein).

Material	$4\pi M_s$ (G)	T_c (°C)	$2\Phi_F/\alpha$ (deg)	λ (nm)
MnBi (normal)	7200	360	3.6	633
MnBi (htp)	5500	180	1.4	633
MnSb	9600	300	0.8	550
Mn_5Ge_3	12 400	37	0.8	550
MnAlGe	3600	245	0.54	633
MnGaGe	4170	185	0.28	633
MnAs	7900	45	0.2	600

magnetic phase is CsCl-type bcc, with the formula (Mn, Cu)Al.

Magnetooptical properties of several binary Mn intermetallics not mentioned above, such as Mn–Ga, Y_3Mn_{23}, Gd_6Mn_{23}, $MnNi_3$, $MnPt_3$, and Mn_2Sb, were also studied. In all of these cases the values of Φ_K reported are rather low and remain well below 0.1° at room temperature and at photon energies of practical interest (Buschow 1988).

MnPt$_3$. Kato *et al* (1995a, b) have investigated the magnetic and magnetooptical properties of ordered $MnPt_3$ alloy films. The $MnPt_3$ alloy films were prepared by annealing Mn/Pt multilayers at a temperature of 800 °C for one hour. It is found that the $MnPt_3$ alloy exhibits a very large Kerr rotation, reaching its maximum value of $-1.18°$ at the wavelength $\lambda = 1$ μm.

10.1.7 Chromium compounds

CrTe and Cr$_3$Te$_4$. The specific Faraday rotations and Kerr rotations for the compounds CrTe and Cr_3Te_4 were reported to be rather low (Atkinson 1977). For the intermetallic compound CrTe with the Curie temperature $T_c = 334$ K, the Faraday rotation and absorption decreased from 5×10^4 deg cm^{-1} at 550 nm to 4×10^4 deg cm^{-1} at 1000 nm, and from 2×10^5 cm^{-1} at 500 nm to 6×10^4 cm^{-1} at 2500 nm, respectively. The figure of merit F for CrTe at 900 nm was 0.8°. The CrTe films used for the magnetooptical investigations were prepared by sequential deposition of layers of Cr–Te–Cr of thickness 87 nm on cleaved mica substrates.

10.2 Rare-earth–transition metal compounds

Re–Co compounds. The magnetooptical properties of thin films of the compounds Nd_2Co_7 and $NdCo_5$, and of Co metal were investigated in the energy range from 1.8 to 3.1 eV by Stoffel (1968). The main rule was found to be that increasing the rare-earth content leads to a decrease of the longitudinal Kerr effect. Kerr rotation spectra of various Co–Y intermetallics, such as Co, Y_2Co_{17}, YCo_5, Y_3Co_7, and YCo_3, have shown a gradual decrease of the value of the rotation with decreasing Co content, because in these materials the magnetic and magnetooptical properties are determined by the Co atoms, Y being non-magnetic and having a relatively small spin–orbit splitting. In the spectral range 1.5 to 4.0 eV the polar Kerr rotations for YCo_3, Y_2Co_7, and YCo_5 were near $-0.1°$, and for Y_2Co_{17} the polar Kerr rotation in the spectral range 0.5 to 4.5 eV was between -0.35 and $-0.5°$ with extrema at 1.25 eV and at 3.5 eV (Buschow 1988).

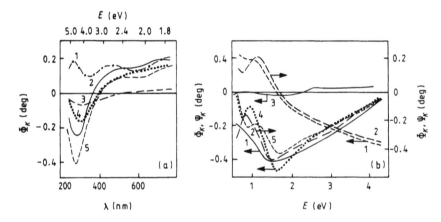

Figure 10.3. (a) The spectral dependence of the polar Kerr rotation at room temperature obtained for $GdFe_2$ (curve 1), $DyFe_2$ (curve 2), $ErFe_2$ (curve 3), $HoFe_2$ (curve 4), and $TbFe_2$ (curve 5) (Katayama and Hasegawa 1982). (b) The spectral dependence of the polar Kerr rotation (full lines) and ellipticity (dashed lines) obtained for several Heusler alloys: Co_2FeAl (curve 1); Co_2FeGa (curve 2); Cu_2MnAl (curve 3) (Buschow 1988); Co_2FeGe (curve 4); and Co_2FeSi (curve 5) (van Engen 1983).

ReFe$_2$. The spectral dependences of the polar Kerr rotation spectra of the cubic Laves phase compounds $GdFe_2$, $ErFe_2$, $HoFe_2$, $DyFe_2$, and $TbFe_2$ are shown in figure 10.3(a) (Katayama and Hasegawa 1982). The prominent feature in each spectral dependence is the strong negative peak near 300 nm (4 eV). Discussion of the magnetooptical properties of such compounds is usually based on a model of two magnetized sublattices formed of Fe and rare-earth (Re) atoms, which give the contributions $\Phi_K(Fe)$ and $\Phi_K(Re)$ to the Kerr rotation, respectively.

For $ReFe_2$ compounds the Re-sublattice magnetization at room temperature exceeds the Fe-sublattice magnetization. In an external magnetic field, the Re-sublattice magnetization aligns parallel to the applied magnetic field. Owing to the antiferromagnetic Re–Fe coupling between Re and Fe sublattices, the Fe-sublattice magnetization will be aligned antiparallel to the applied magnetic field. For this reason, the Fe sublattice gives a positive contribution to the polar Kerr rotation in all spectral ranges. The large negative peaks near 300 nm are usually attributed to the transition between 4f and 5d states of rare-earth atoms. The absolute Φ_K value corresponding to these peaks decreases with increasing number of 4f electrons, while the peak position tends to move to longer wavelengths.

GdCo$_5$. In the Kerr rotation spectrum of a single crystal of GdCo$_5$ at room temperature, a large negative Φ_K peak near 300 nm was observed, too (Katayama and Hasegawa 1982).

10.3 Pseudo-binary compounds

For some intermetallic compounds, AB_n, it is possible to gradually change the physical properties by substituting a third component C for either A or B without changing the crystal structure. In such pseudo-binary compounds, $A_{1-x}C_xB_n$ or $A(B_{1-x}C_x)_n$ the A and C (or B and C) atoms share the same crystallographic site in a random manner. Data on the Kerr rotation values for several pseudo-binary intermetallic compounds, including the room temperature magnetization, are given in table 10.6 (Buschow 1988).

Table 10.6. The Kerr rotations for several pseudo-binary intermetallic compounds (Buschow 1988).

Alloy or compound	σ_r (A m^2 kg^{-1})	Φ_K (deg) $\lambda = 633$ nm	Φ_K (deg) $\lambda = 830$ nm
GdFeAl	—	0.00	0.01
GdFe$_{1.2}$Al$_{0.8}$	—	0.04	0.06
Zr$_{0.8}$Mo$_{0.2}$Fe$_2$	19	−0.01	−0.1
Zr$_{0.75}$Nb$_{0.25}$Fe$_2$	58	−0.15	−0.16
ZrFe$_{1.76}$V$_{0.24}$	40	−0.05	−0.09
LaCo$_{11}$Si$_2$	77	−0.18	−0.24
Fe$_{2.4}$V$_{0.6}$Ge	43	−0.21	−0.20
Fe$_{2.7}$V$_{0.3}$Ge	97	−0.36	−0.35
Al$_5$Mn$_3$Ni$_2$	45.2	−0.00	−0.01
Al$_5$Mn$_3$Co$_2$	43.8	−0.01	−0.02
Al$_5$Mn$_3$Cu$_2$	70.2	0.00	0.01
Al$_5$Mn$_3$Fe$_7$	—	−0.01	−0.01

10.4 Intermetallic ternary compounds

Before discussing the magnetooptical properties of ternary compounds it is necessary to say a few words about the differences between ternary and pseudo-binary compounds. Ternary compounds differ from pseudo-binary compounds in that the different crystallographic sites are occupied by different types of atom. One might say that the physical properties of pseudo-binaries can be described as weighted means of the properties of the parent binary compounds. This does not extend to ternary compounds. Here one may expect novel properties not found for binary compounds. A useful example of this is that of the anomalous magnetooptical properties of the compound PtMnSb. Very often also, the crystal structure and the formula composition have no analogue in the corresponding binary system (Buschow 1988).

10.4.1   Heusler alloys

Heusler alloys are cubic ternary intermetallic compounds of general formula X_2YZ. They crystallize in the $L2_1$ structure type, which can be represented as being constituted of four crystallographic positions named a, b, c, and d. The positions a and c are occupied by the same type of atom (X). The interest in Heusler alloys has been focused mainly on the unusual result that some of these materials in these crystallographic phases are strongly ferromagnetic yet consist of a combination of elements which are considered to be non-magnetic (Buschow 1988).

Co_2FeAl, Co_2FeGa, and Cu_2MnAl. The magnetooptical properties of the compounds X_2YAl and X_2YGa were studied by Buschow and van Engen (1984). Kerr rotation spectra for the compounds Co_2FeAl, Co_2FeGa, and Cu_2MnAl are shown in figure 10.3(b). For both Co-based materials, Φ_K has a strong minimum near 1.5 eV, the value of the rotation corresponding to this minimum being almost as large as that for the pure Co–Fe alloys.

Results obtained on several Al- and Ga-based Heusler alloys are listed in table 10.7 (Buschow 1988). For Heusler alloys based on Sn, rather low values of the polar Kerr rotation in the energy range 0.5–4.5 eV were obtained, the only exceptions being a few alloys of the type Co_2YSn with Y = Hf, Zr, or Ti.

Co_2TiSn, Co_2ZrSn, and Co_2HfSn. The interpretation of these Co_2YSn alloys is based on the presence of narrow hybridized 3d, 4d, or 5d bands for Y = Ti, Zr, and Hf, respectively (Buschow 1988). The initial states associated with the magnetooptical transitions responsible for the spectra shown were taken by Buschow to be the 4d electron states of the Co atoms, whereas the final states were taken to be the d states of Y = Ti, Zr, and Hf atoms in Co_2YSn. We note that, in view of the nature of the initial and final states of the magnetooptical transitions, the corresponding spectra may be regarded as charge-transfer spectra. The substantial difference in intensity of the magnetooptical spectra for the compounds investigated was ascribed to the large differences in spin–orbit splitting of the hybridized d-band states of the Y component (Buschow 1988).

Co_2FeGe, Co_2FeSi, and Co_2FeGa. The spectral dependences of the polar Kerr rotations for the Heusler alloys Co_2FeGe and Co_2FeSi are shown in figure

10.3(b) (van Engen 1983, Buschow 1988). These spectra have in common with the Kerr rotation spectra for the pure transition metal alloy Co_2Fe that there is a deep minimum near 1.5 eV. The new feature in the spectra of the Heusler alloys is the additional Φ_K peak at photon energies below 1 eV. This trend towards forming a second peak seems also to be present for Co_2FeGa, although here the peak occurs at still lower photon energies (see figure 10.3(b)) (Buschow 1988).

$Ni_{3-x}Mn_xSn$. Buschow *et al* (1984) investigated the magnetooptical properties of the series $Ni_{3-x}Mn_xSn$. It is found that the intensity of the Kerr rotation first increases in this compound when the Mn content is increased, reaches a maximum when the Mn content $x = 1.3$–1.5, and then decreases again. The spectral dependences of the Kerr rotation and ellipticity for this alloy are rather complicated, but the extremum values $\Phi_K = -0.08°$ at 3.4 eV and $\Psi_K = 0.05^c$ at 2.8 eV are rather low, even for $x = 1.5$. Results for several other Heusler alloys based on Si, Ge, In, Sb, or Pb are listed in table 10.7 (Buschow 1988).

Table 10.7. Kerr rotations for various Heusler alloys (see Buschow *et al* (1983) and references cited therein, and Buschow (1988)).

Alloy or compound	T_c (K)	σ_r (A m^2 kg^{-1})	Φ_K (deg) $\lambda = 633$ nm	Φ_K (deg) $\lambda = 830$ nm
Fe_2NiAl	—	117	−0.25	−0.30
Fe_2MnAl	—	52	−0.12	−0.14
Co_2FeAl	—	138	−0.37	−0.42
Fe_2NiGa	—	55.0	−0.15	−0.15
Fe_2CoGa	—	120.9	−0.41	−0.47
Fe_2CrGa	—	42.5	−0.18	−0.20
Co_2FeGa	—	116.8	−0.41	−0.49
Co_2MnGa	694	87.5	−0.14	−0.15
Co_2FeSi	980	139.8	−0.33	−0.34
Co_2FeGe	—	124.2	−0.43	−0.51
Fe_2CoGe	—	118.2	−0.35	−0.39
Co_2FeIn	—	—	−0.44	−0.58
Co_2MnSb	—	59	−0.01	−0.06

10.4.2   PtMnSb, and Mn-based compounds

Closely related as regards structure to the Heusler alloys, PtMnSb, NiMnSb, and some related compounds crystallize in the *C*1*b* structure, which is face-centred cubic with space group *F*43*m* and which also—like the structure of Heusler alloys—contains four crystallographic positions a, b, c, and d. It differs from the normal Heusler structure only in the occupation of the c sites. In the normal

Heusler alloys X_2YZ, both the a and c sites are occupied by X atoms. In the $C1b$ structure, only the a site is occupied by X atoms, the c sites remaining empty. The ordinary $L2_1$ Heusler alloys have inversion symmetry, but the $C1b$ Heusler alloys do not, and one of the X sites in the ordinary $L2_1$ structure is vacant for the $C1b$ structure. The formula for the composition of these compounds is XYZ. The saturation magnetic moment of PtMnSb is about 4 μ_B.

PtMnSb. The PtMnSb alloy exhibits exceptionally strong magnetooptical properties at room temperature as compared with those of other compounds. The spectral dependence of the Faraday rotation, polar Kerr rotation, and ellipticity for PtMnSb are shown in figure 10.4. It is seen that near 1.72 eV the absolute value of Φ_K reaches a value of 1.27° (curve 1K) (Matsubara *et al* 1987, Buschow 1988). In fact, the value of the Kerr rotation is very sensitive to the composition, and the procedure for preparing the samples. For example, for thin film, Inukai *et al* (1987) reported a value for the rotation of about 2° at 633 nm.

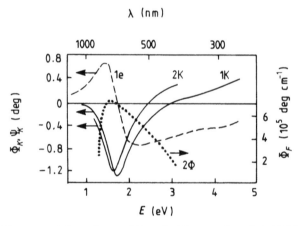

Figure 10.4. The spectral dependence of the polar Kerr rotation (Φ_K) (full line), ellipticity (Ψ_K) (dashed line), and Faraday rotation (Φ_F) (dotted line) of PtMnSb. Curve 1 shows results given by Buschow (1988), and curve 2 shows results given by Matsubara *et al* (1987).

The magnetic and magnetooptical (Kerr rotation) properties of Pt–Mn–Sb ternary films, and films in which various elements have been substituted, with the compositions $(Pt_{0.7}A_{0.3})MnSb$ (A = Fe, Co, Ni, Pb, Cu, and Au, substituted for Pt) and $Pt(Mn_{0.7}B_{0.3})Sb$ (B = Cr, Fe, and Co, substituted for Mn), deposited on glass substrates at room temperature by electron beam evaporation were reported by Inukai *et al* (1987). Single-phase films were obtained with the composition 26–33 at.% Pt, 30–44 at.% Mn, and 28–40 at.% Sb. The Kerr rotation angles obtained—up to 2.0° at 633 nm wavelength for the Mn-rich 31Pt–38Mn–31Sb films—were larger than those for bulk PtMnSb. In fact, they are the highest values reported thus far for room temperature Kerr rotation.

The magnetic properties, Kerr rotation, and ellipticity for Pt-substituted

MnSb films with film thicknesses of about 100 nm obtained by rf sputtering for the concentration ranges $44 < c_{Mn} < 55$, $37 < c_{Sb} < 44$, and $4 < c_{Pt} < 15$ at.% were investigated by Takahashi *et al* (1994). It was found that $Pt_6Mn_{50}Sb_{44}$ films annealed at 300 °C showed giant Kerr rotation angles of about 0.9° at the short wavelength of 500 nm. The results of structural x-ray and NMR analyses revealed that the giant Kerr rotation observed in these films is closely related to that for the compound MnSb including Pt atoms in the NiAs structure.

The M_s-values for these films also exceed the bulk value. Takahashi *et al* have observed a nearly linear relationship between Φ_K and M_s. For all of the films in which various substitutions had been made, the value of the Kerr rotation decreased. For example, for the $(Pt_{0.7}A_{0.3})MnSb$ films, the values of Φ_K were 0.5–1.1°. For the Co-, Ni-, or Cu-substituted $(Pt_{0.7}A_{0.3})MnSb$ films, the Kerr rotation peaks shifted to longer wavelengths than those for the stoichiometric PtMnSb. This suggests that the energy gaps in the minority-spin directions for the Co-, Ni-, or Cu-substituted films are narrower than those for PtMnSb films (Inukai *et al* 1987).

A description of the nature of the high level of magnetooptical activity discovered for PtMnSb based on band-structure calculations for several $C1_b$-type compounds was given by de Groot *et al* (1984). It was found that several of these ferromagnetic compounds have very unusual band structures. The electronic structure of the majority-spin electrons can be characterized as that of a normal ferromagnetic metal. The associated band structure of the minority-spin electron does not resemble that of a metal at all, but has the characteristics of a semiconductor. The minority-spin band structure has a semiconductor gap straddling the Fermi level, whereas the majority-spin band structure has normal metallic intersections. Since the density of states at the Fermi level is zero for the minority-spin direction, the peculiar band structure of these materials implies that the conduction electrons are fully spin polarized.

Another possibility for explaining the large Kerr effect for PtMnSb was presented by Feil and Haas (1987). These authors argued that a pronounced peak in the Kerr spectrum does not necessarily correspond to a particular magnetooptically active electron transition. They pointed out the fact that in metallic magnetic materials, the plasma resonance of the free charge carriers induces a resonance-shaped magnetooptical Kerr effect spectrum. A more detailed description of the proposed model is given in chapter 5, which is devoted to the theoretical consideration of magnetooptical phenomena in metals.

In order to clarify the nature of the giant Kerr rotation for PtMnSb, Wang *et al* (1994) provided, within the local spin-density approximation of density functional theory, calculations of both the diagonal and the off-diagonal conductivities of the compounds PtMnSb and NiMnSb. The similar electronic structures of these two materials yield similar diagonal conductivities, while the spin–orbit coupling for Pt being larger than that for Ni results in the off-diagonal conductivity for PtMnSb being larger than that for NiMnSb. The value of the Kerr rotation for PtMnSb is about four times the value of that for NiMnSb.

The results of the calculation of the spin density of states—in accordance with the results of de Groot *et al* (1984)—show that, for the minority-spin electrons, a semiconductor-like gap opens up at the Fermi level. Another observation is that although the majority-spin electrons are metallic, the density of states at the Fermi level is very low for compounds with the structure discussed above. At the same time, the theory does not give the correct plasma resonant edge for PtMnSb, and, as a result, the large Kerr rotation for PtMnSb was not reproduced by calculations, although the increase in magnetooptical activity on going from NiMnSb to PtMnSb was confirmed. It should be noted here that the low majority-spin density of states at the Fermi level, and the variations of the density of states related to the small composition changes for different samples will result in a wide spread of the experimental results.

Some progress was obtained in the theoretical description of the magnetooptical properties of PtMnSb and related compounds. On the basis of a density functional theory approach, Kulatov *et al* (1995) have reported successful reproduction of the huge Kerr effect for PtMnSb, and obtaining the correct positions of the main features for the related PdMnSb and NiMnSb compounds. Also, Oppeneer *et al* (1995) reported that calculations for Kerr spectra for the manganese–platinum compounds $MnPt_3$ and PtMnSb made using density functional band-structure theory give a good description of the huge Kerr rotations found experimentally for these compounds.

PdMnSb, NiMnSb and PtMnSn. The magnetooptical properties of the four compounds PtMnSb, PdMnSb, NiMnSb, and PtMnSn were compared by Buschow (1988). Kirillova *et al* (1995) reported frequency dependence measurements of the real and imaginary parts of the complex dielectric constant for the compounds XMnY, where X = Ni, Pd, Pt, or Cu, and Y = Sb or Sn, for the spectral range 0.05 to 8 eV. For all of the compounds, in addition to the main absorption band located in the spectral range 1 to 7 eV, a low-energy (from 0.3 to 0.7 eV) inter-band absorption is revealed.

MnAlGe and MnGaGe. The ternary compounds MnAlGe and MnGaGe crystallize in the tetragonal Mn_2Sb-type structure. The optical absorption coefficient α (uncorrected for reflection), and the Faraday rotation Φ_F as a function of wavelength for fully crystallized MnGaGe film at room temperature are shown in figure 10.2 (Sawatzky and Street 1973). Stoichiometric MnGaGe has a Curie temperature of 185 °C, which can be lowered considerably by varying the composition.

10.4.3 $Cr_{23}C_6$-type compounds

The polar Kerr rotation spectra and magnetic properties of numerous compounds crystallized in the cubic $Cr_{23}C_6$-type structure were investigated by Buschow *et al* (1983). The Kerr rotation spectra for $Co_{20}Al_3B_6$, belonging to this group, exhibit a peak value of $-0.15°$ near 1.4 eV. The Kerr rotations for several cubic $Cr_{23}C_6$-type compounds are given in table 10.8 (Buschow 1988).

Table 10.8. Kerr rotations for several cubic $Cr_{23}C_6$-type compounds (Buschow 1988).

Alloy or compound	T_c (K)	σ_r (A m² kg⁻¹)	Φ_K (deg) $\lambda = 633$ nm	Φ_K (deg) $\lambda = 830$ nm
$Co_{20}Al_3B_6$	409	39.5	−0.12	−0.13
$Co_{20}Ga_3B_6$	—	40.3	−0.15	−0.14
$Co_{21}Sn_2B_6$	—	50.5	−0.19	−0.19
$Co_{21}Ge_2B_6$	467	33.8	−0.11	−0.12
$Co_{21}V_2B_6$	—	57.4	−0.17	−0.14
$Co_{21}Nb_2B_6$	491	50.1	−0.15	−0.16
$Co_{21}Cr_2B_6$	—	63.6	−0.18	−0.19
$Co_{21}Mo_2B_6$	545	53.7	−0.15	−0.16
$Co_{21}W_2B_6$	538	52.6	−0.16	−0.16
$Co_{20}Ti_3B_6$	478	46.2	−0.11	−0.10
$Co_{21}Zr_2B_6$	—	62.5	−0.20	−0.21
$Co_{21}Hf_2B_6$	—	56.8	−0.20	−0.21
$Co_{21}Ta_2B_6$	—	54.6	−0.17	−0.21
$Co_{20}In_3B_6$	—	—	−0.16	−0.16
$Fe_{21}Mo_2C_6$	492	98.2	−0.16	−0.21
$Fe_{21}W_2C_6$	1006	97.4	−0.34	−0.21
$Fe_{23}C_3B_3$	742	178.5	−0.37	−0.43

10.4.4   Miscellaneous ternary compounds

PtFeSb.   Replacement of Mn by Fe in PtMnSb results in different crystal structure which is still cubic but involves atomic positions that are less symmetric than those in PtMnSb itself.   It was found that materials produced in this way have considerably lower Kerr effect intensities than their Mn counterparts (Buschow 1988).

Re₂Fe₁₄B. The Kerr rotation for $Re_2Fe_{14}B$ compounds (Re = Lu, Ce, Gd, La, Nd), $La_2Co_{14}B$, and $Gd_2Co_{14}B$, and the ellipticity for $Gd_2Fe_{14}B$ and $Gd_2Co_{14}B$, obtained at room temperature in a magnetic field 11.6 kOe, are given in figure 10.5 (van Engelen and Buschow 1986).

FeNiGe, FeNiSn, FeCoSn, and FeCoGe. The peak values that the polar Kerr rotation attains are −0.04° at 1.4 eV for FeNiGe, −0.07° at 1.2 eV for FeNiSn, −0.12° at 1.3 eV for FeCoSn, and −0.15° at 1.7 eV for FeCoGe compounds crystallized in the hexagonal Ni_2In-type structure (Buschow and van Engen 1983, van Engen 1983). The spectra show roughly the same features as those for the pure transition metals, though the intensity is considerably lower.

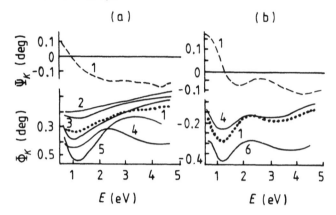

Figure 10.5. Kerr rotation (full and dotted lines) and ellipticity (dashed lines) spectra for $Re_2Fe_{14}B$ (a) and $Re_2Co_{14}B$ (b), taken at room temperature in a field of 11.6 kOe. Re = Gd (curve 1), Lu (curve 2), Ce (curve 3), La (curve 4), Fe (curve 5), and Co (curve 6) (van Engelen and Buschow 1986, Buschow 1988).

10.5   Cerium and thulium low-temperature compounds

10.5.1   Cerium-containing compounds

Kerr rotation and ellipticity measurements in the 0.5–5 eV energy range, and optical reflectivity data covering the photon energies from 0.03 to 12 eV were made by Reim *et al* (1986) on single crystals of CeSb, $CeSb_{0.75}Te_{0.25}$, and CeTe. The most striking feature of the results is the size of the Kerr rotation, Φ_K, associated with the fully spin-polarized $4f^1$ state of CeSb, which is more than 14° at $T = 2$ K and $H = 50$ kOe. Among all known magnetooptical materials, CeSb exhibits the highest value for the Kerr rotation. Figure 10.6 displays the Kerr rotations and ellipticities of CeSb and the pseudo-binary $CeSb_{1-x}Te_x$ compounds for $x = 0$, 0.25, and 1 (Reim *et al* 1986, Schoenes 1987). There is a maximum in the Kerr rotation which shifts from appearing at 2 eV for CeTe, to appear at 1 eV for $CeSb_{0.75}Te_{0.25}$, and then at below 0.5 eV for CeSb. At the same time, the peak value increases from 3.2° to 14°. From the shape of the magnetooptical spectra, Reim *et al* and Schoenes deduce that a single optical transition is involved, the size of the effect indicating that at least one of the states involved is highly spin polarized.

For CeSb, the absorptive part of the off-diagonal conductivity has been found to show a pronounced peak at 0.7 eV, and a zero crossing near 0.5 eV. Near this zero crossing it was possible to resolve a peak in the diagonal element of the conductivity σ_{1xx}, which is perfectly resolved after subtraction of a Drude term for the free-electron configuration (Schoenes 1987). The integrated weight of this peak has been shown to correspond exactly to the computed value for the f–$d_{t_{2g}}$ transition, and the line shape for σ_{xy} is the same as that observed for the f–d$_{t_{2g}}$ transition in uranium monochalcogenides. For this reason, it is possible that the 4f^1–5d (f–d$_{t_{2g}}$) transition occurs near 0.5 eV (possibly at 0.4 eV) for CeSb.

It should be noted that for the cerium monochalcogenides, the f–d$_{t_{2g}}$ peak observed in σ_{1xx} near 2.5 eV demonstrates the shift of the f–d$_{t_{2g}}$ transition on going from CeTe to CeSb (Schoenes 1987). The centre of the p valence band is located at 6.0 eV for CeTe. While for the latter material the valence band is nearly filled, the top of the p band extends up to Fermi level for CeSb.

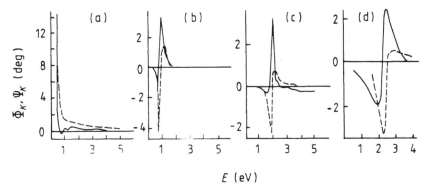

Figure 10.6. Kerr rotation (full lines) and ellipticity (dashed lines) spectra for CeSb (a), CeSb$_{0.75}$Te$_{0.25}$ (b), CeTe at $H = 50$ kOe (Reim *et al* 1986) (c), and TmSe at $H = 40$ kOe and $T = 2$ K (Reim *et al* 1984, Schoenes 1987) (d).

Antropov *et al* (1995) have reported calculations of the electronic structure of gadolinium metal and CeSb, made using a density functional method which explicitly includes the Coulomb parameter for the 4f electrons. It was found that the calculated density of states, total energies, Fermi surface, and magnetooptical properties are in better agreement with experiment than those obtained using the standard local density technique.

10.5.2 Thulium-containing compounds

TmS and TmSe. Relatively large values of the Kerr rotation were found by Reim *et al* (1984) in the magnetooptical spectra of the compounds TmS and TmSe. The Kerr rotation and ellipticity spectra at $T = 2$ K and $H = 40$ kOe for TmSe are shown in figure 10.6(d) (Schoenes 1987). The maximum Kerr ellipticity of 3.6° is attained at 2.35 eV. The occurrence of the strong Kerr peaks was attributed to an exchange splitting of the conduction electron plasma edge. The position of the plasma edge in this compound is near 2.3 eV.

10.6   Uranium-containing compounds

Many crystalline uranium compounds and intermetallic alloys are known to order magnetically, with some groups such as the rock-salt-structure mono-chalcogenides and the Th_3P_4-structure pnictides ordering ferromagnetically. The Curie temperatures T_c of these groups are below 200 K.

In contrast, the uranium monopnictides are antiferromagnets with complex spin structures. Néel temperatures as high as 300 K are attainable (for example, for UBi). Ferromagnetic uranium compounds have demonstrated interesting magnetooptical properties, including very large Kerr rotations in the near-infra-red region (Schoenes 1987).

Since the discovery of the large Kerr rotation and ellipticity for $USb_{0.8}Te_{0.2}$ at cryogenic temperatures, several attempts have been made to find uranium compounds possessing strong magnetooptical properties at room temperature. As has been shown, UCo_{5+x} (Brandle *et al* 1990), UMn_2Ge_2 and UMn_2Si_2 (van Engelen *et al* 1988), and $UFe_{10}Si_2$ (Brandle *et al* 1990) have rather high Curie temperatures. The common property of all of these compounds is the presence of magnetic transition elements such as Mn, Fe, and Co.

10.6.1   Uranium–transition metal materials

$UCo_{5.0+x}$. This material was found to be a uniaxial ferromagnet with a Curie temperature of $T_c = 360$ K and a large magnetic anisotropy ($H_k = 60$ kOe). The polar Kerr rotation and ellipticity were measured from 0.7 to 5 eV in an external magnetic field of 21.8 kOe for a polycrystalline sample of UCo_5 at room temperature. The maximum Kerr rotation, $\Phi_K = -0.25°$, was attained near 4 eV. Brandle *et al* have noted that the absence of extra neutron diffraction peaks and magnetooptical structures indicates that uranium does not carry a substantial magnetic moment in this compound. Instead, some of the uranium 5f electrons are probably transferred to Co 3d states, causing a decrease of the magnetic moment on Co, and a shift by 0.5 eV of the main maximum of σ_{2xy} to lower energy.

UMn_2Ge_2, UMn_2Si_2. These compounds order ferromagnetically at 390 and 377 K, respectively. However, as shown by neutron diffraction, the uranium sublattice acquires a substantial magnetization only below 150 and 90 K, respectively. As has been shown by van Engelen *et al* (1988), the absolute value of the Kerr rotation for UMn_2Si_2 does not exceed 0.25° above 1 eV at room temperature.

$UFe_{10}Si_2$. The compound $UFe_{10}Si_2$, which orders ferromagnetically at around 620 K, crystallizes in the tetragonal $ThMn_{12}$ structure. The maximum of the Kerr rotation, $\Phi_K = -0.55°$, is obtained at 0.75 eV; this value is comparable to the maximum Kerr rotation for pure Fe (Brandle *et al* 1990).

It should be noted that $UFe_{10}Si_2$ shows a Kerr effect which is as high as that for pure iron, while Curie temperatures of $T_c = 700$ K for Fe and $T_c = 620$

K for this compound were obtained. In the spectral range between 0.6 and 3 eV, the Kerr rotation is about 60% higher than for $NdFe_{10}V_2$, although the magnetizations of the two compounds are nearly equal (Brandle *et al* 1990). For this reason it is possible that uranium atoms play a substantial role in the magnetooptical properties of this compound at room temperature.

UCu_2P_2. This compound orders ferromagnetically at 216 K. The Kerr rotation, as has been shown by Schoenes and Brandle (1991), exceeds 3.5° near 0.8 eV at 10 K and for an applied magnetic field of 40 kOe. At 200 K and for an applied field of 40 kOe, the rotation amounts to 60% of its saturation value. It was found that the low-energy transition is of diamagnetic line shape, i.e. the absorptive component of the off-diagonal element of the conductivity tensor has a dispersive form, while the high-energy transition has a paramagnetic line shape, i.e. the off-diagonal element of the conductivity tensor has an absorptive form. This behaviour indicates that the low-energy transition is very probably a 5f–6d transition within the uranium atom.

UGa_2. This compound orders ferromagnetically at 140 K. The Kerr rotation angle attains 0.4° at 85 K and at a wavelength of 633 nm. When the wavelength is increased, the Kerr rotation slowly grows, and attains $\Phi_K = 0.6°$ at 1000 nm (Gambino and Ruf 1990).

UFe_2. UFe_2 orders ferromagnetically at 180 K. The Kerr rotation is less than 0.2° at 633 nm and $T = 85$ K (Gambino and Ruf 1990).

10.6.2 Uranium chalcogenides and pnictides

US, USe, UTe. Most uranium chalcogenides and pnictides are materials for which atomic coupling schemes are either inappropriate or useless, due to the delocalization of the 5f electrons. The ferromagnetic uranium monochalcogenides show magnetooptical Kerr rotation maxima of 3 to 4° (Schoenes 1987, Reim and Schoenes 1984). The largest contribution to the off-diagonal components of the conductivity arises from an f–d transition, but the d–f transition and free electrons also contribute. The f–d inter-band transition is identified as causing the minimum in the off-diagonal component of the conductivity at low energies.

The spectral dependences of the real and imaginary parts of the off-diagonal components of the conductivities for US, USe, and UTe were studied by Reim and Schoenes (1984). The enhancement of the off-diagonal conductivity of the uranium monochalcogenides is due to an increased overlap of the 5f and 6d wave functions of the uranium compounds compared to that of the 4f and 5d wave functions of the europium chalcogenides. The fact that this enhanced conductivity is five times larger than the off-diagonal conductivity of US results not in a larger, but—on the contrary—in a 2–3 times smaller Kerr rotation than that for EuO. For metallic US, the maximum Kerr rotation of 2.5° is observed at 1.8 eV (Schoenes 1987).

USbTe. Uranium compounds with superior Kerr signals can be engineered.

The technique starts with uranium monopnictides, which have local magnetic moments larger then those of the uranium monochalcogenides, but which in their pure form are not of much interest due to their antiferromagnetic order. By substituting tellurium for 20% of the antimony atoms in USb, one can change the antiferromagnetic to ordering ferromagnetic ordering. Figure 10.7(a) displays the Kerr rotation for $USb_{0.8}Te_{0.2}$ (Schoenes 1987). Comparing the data for the (100) plane with those for UTe, one finds an increase of the maximum Kerr rotation by a factor of two. The maximum Kerr rotation is further increased to 9° if the measurement is performed on a (111) plane. This single Kerr rotation value is the second largest ever reported. Because of the fact that the largest Kerr rotation occurs for CeSb at 2 K and 50 kOe, while $USb_{0.8}Te_{0.2}$ orders ferromagnetically at 204 K, the uranium pseudo-binaries look promising as regards potential applications.

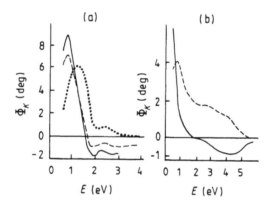

Figure 10.7. (a) The spectral dependences of the polar Kerr rotations for polished (111)- and (100)-oriented (full and dashed lines) and cleaved (100)-oriented (dotted line) $USb_{0.8}Te_{0.2}$ samples at low temperature. (b) The spectral dependences the polar Kerr rotation (full line) and ellipticity (dashed line) of U_3As_4 (Schoenes 1987).

U_3As_4. Figure 10.7(b) (Schoenes 1987) displays the polar Kerr effect spectra for U_3As_4, which is a uranium pnictide crystallizing in the Th_3P_4 structure. It orders ferromagnetically at 198 K. Its Kerr rotation spectrum is dominated by a strong increase to low energies, reaching 6° at the low-energy limit of the measurements at 0.5 eV.

10.6.3 Amorphous films of U_xSb_{1-x} and TbFeU

Amorphous films offer advantages as regards use as magnetooptical recording media, including the reduction of grain noise associated with the random orientation of anisotropic grains in polycrystalline films. Amorphous films of U_xSb_{1-x} and TbFeU exhibit a large degree of magnetization in moderate fields, even at temperatures over 100 K. These materials appear to be single-ion-anisotropy-dominated amorphous ferromagnets.

Amorphous films with the compositions $U_x Sb_{1-x}$ (for x between 0.2 and 0.8) prepared by magnetron sputtering show ferromagnetic properties, with Curie temperatures as high as 135 K (for $x = 0.45$). Large positive Faraday rotation values of over 2×10^6 deg cm^{-1}, and large negative Kerr rotation values of over 3° were reported. Still higher values of the Faraday rotation of $\Phi_F = 2.2 \times 10^6$ deg cm^{-1} and $\Phi_F = 2.1 \times 10^6$ deg cm^{-1} were obtained for films with the compositions $U_{0.48} Sb_{0.52}$ (McElfresh *et al* 1990) and $U_{0.53} Sb_{0.47}$ (McGuire *et al* 1990), respectively, at 10 K and $H = 40$ kOe. The values of the Kerr rotation for the spectral range 1.55–2.76 eV were between -3.0 and $-3.5°$, with the maximum, $\Phi_K = -3.5°$, being found at 1.8 eV. The large values of the Faraday rotation are correlated with a large Hall effect, having corresponding Hall angles up to $\theta_H = 17°$.

Polar Kerr rotation spectra in the spectral range down to 0.65 eV for magnetron-sputtered thin films of amorphous U–Sb with various U concentrations, x, have been investigated by Fumagalli *et al* (1992). For $x > 50$ at.%, the polar Kerr spectra are dominated by a broad negative peak near 1 eV. For $x < 50$ at.%, this peak is absent, and the negative Kerr rotation increases strongly towards low energies, with some indication of a peak at 0.7 eV. The maximum rotation measured is $-2.4°$ at 1 eV for $U_{51} Sb_{49}$ at a temperature of 10 K and in a field of 30 kOe.

The amorphous $U_x Sb_{1-x}$ system shows two compositional regions of characteristic energy dependences of the polar Kerr rotation: for $x > 50$, a rock-salt-structure-like local coordination of the U atoms leads to a broad negative peak near 1 eV; and for $x < 50$, the local coordination of the U atoms changes, presumably into a $Th_3 P_4$-structure-like coordination, as manifested by a strong increase of the negative Kerr rotation toward lower energies and the absence of a broad feature near 1 eV.

Amorphous films of U–Sb, and films with ten and twenty per cent of Co and Mn, respectively, added were obtained, and their magnetooptical properties were studied by McGuire *et al* (1990). In general, addition of Co and Mn leads to decreased magnetooptical activity.

TbFeU amorphous films. The influence of uranium addition on the Kerr rotation for amorphous TbFe films was studied by Dillon *et al* (1987). At room temperature, addition of four atomic per cent of uranium does not change the value of the Kerr rotation, but when the content of uranium was increased up to seven atomic per cent, the value of the Kerr rotation decreased due to the decrease of T_c.

10.7 Amorphous alloys

Amorphous alloys can be fabricated either by liquid quenching or by vapour quenching. In both cases, formation of the solid occurs at a rate high enough to suppress the ordered arrangement of the metal atoms usually associated with crystallization.

From the practical point of view, amorphous alloys have a number of advantages over crystalline alloys, including: (i) the possibility of varying an alloy composition over a wide range; (ii) the absence of grain boundaries, resulting in a low noise level in magnetooptical application; and (iii) the presence of a large degree of atomic disorder, resulting in a large and almost temperature-independent electrical resistivity as well as a low thermal conductivity. In this section, amorphous alloys made by alloying transition metals with non-magnetic elements will be considered. In these alloys, upon alloying, electrons are transferred from the less electronegative non-magnetic element to the 3d metal, where they fill up the 3d band and thus lower the 3d moment. The rare-earth–transition metal alloys used in magnetooptical disk memory will be discussed in the following subsection.

10.7.1 Amorphous alloys based on 3d transition metals

Amorphous Co–B and Co–Mg films. The Kerr rotation at 633 nm and 830 nm for amorphous $Co_x B_{1-x}$ alloys and $Co_x Mg_{1-x}$ films as a function of Co concentration were investigated by Buschow and van Engen (1981). It was found that for $Co_x B_{1-x}$ the Kerr rotation decreases much more rapidly with decreasing Co concentration than it does for $Co_x Mg_{1-x}$.

Amorphous Co–Si alloys. Amorphous $Co_x Si_{1-x}$ alloys were investigated by Afonso *et al* (1981) over the range of x between 0.55 and 0.77, by means of a study of the transverse Kerr effect. The maximum intensity change of the reflectivity, $\delta R/R = 6 \times 10^{-3}$, was obtained in the spectral range 3–4 eV for $x = 0.73$–0.77, with the angle of incidence of the light equal to 70°. Similar measurements, made for crystalline Co metal for the same angle of incidence of the light, were reported by Krinchik and Artem'ev (1968). Comparison of the spectra obtained by Afonso *et al* (1981) with the latter data shows that the Kerr intensities for amorphous alloys with relatively high Co concentrations are only slightly smaller than that for Co metal, although the spectra of the amorphous alloys show considerably less pronounced structure.

Amorphous Fe–Si films. The transverse Kerr effect for amorphous $Fe_x Si_{1-x}$ films, in the energy range 1.2–5 eV, was investigated by Afonso *et al* (1980), and the results were compared with the results of similar measurements made on crystallized samples. It is found that the Kerr intensities for the amorphous alloys are larger than those for the corresponding crystalline alloys.

Amorphous Fe–B alloys. The transverse Kerr effect was studied for amorphous $Fe_{0.80} B_{0.20}$, in the range 0.8–2.8 eV, by Ray and Tauc (1980). The alloy was prepared by liquid quenching. It was found that the magnitude of the Kerr effect is reduced for amorphous alloys by a factor of about five as compared with that for pure Fe metal.

Amorphous $Fe_x B_{1-x}$ alloys were also studied by Buschow and van Engen

(1981), and it was found that the concentrational dependence of the Kerr rotation Φ_K, measured at 633 nm, was as follows: Φ_K increased from nearly zero at $x = 0.4$, up to $0.57°$ (for light incident on the substrate) at $x = 0.8$, and up to $0.41°$ (for light incident on the film) at $x = 0.7$, with some reduction of the rotation for x close to one.

Amorphous Fe–B–F films.   Strong magnetooptical properties were obtained for Fe–B–F amorphous films prepared via a high-deposition-rate sputtering by Sugawara *et al* (1987). These films are ferromagnetic and rather transparent.   It is interesting to note that the magnetic and magnetooptical properties of Fe–B–F films are quite different from those of FeF_2 and FeF_3, which are antiferromagnetic and paramagnetic compounds, respectively.

The values of the Faraday rotation were between 3×10^4 deg cm^{-1} and 1×10^5 deg cm^{-1} at 633 nm. The absorption coefficient decreased monotonically from 8×10^4 to 3×10^4 cm^{-1} in the wavelength region 300–800 nm. The Curie temperature of these films is highly dependent upon the B content. For instance, the T_c-values of amorphous Fe–B–F films containing about 10 at.% of B are almost constant at around 523 K for rather different Fe contents.

Resulting rotation angles up to 0.6–$0.7°$, at wavelengths in the range 700–800 nm and for $H = 9.8$ kOe, were obtained for films with the compositions $Fe_{44}B_4F_{52}$ (thickness 670 nm) and $Fe_{49}B_5F_{46}$ (thickness 760 nm). At the same time, the reflectivities of these films, in the spectral range for which the high rotation angles were obtained, were reduced by up to 4–7%. The figure of merit of the films investigated has a value of about 0.1 deg dB^{-1}.

10.7.2   Amorphous rare-earth–transition metal films

RE–Co films. Magnetooptical polar Kerr rotation spectra of amorphous Gd–Co, Tb–Co, Y–Co, Ce–Co, and Nd–Co were investigated by Choe *et al* (1987). It has been found that the Kerr rotation of RE–Co is mainly due to the Co constituent in the long-wavelength region, whereas at short wavelengths, the presence of Nd led to an increase in the absolute value of the Kerr rotation for RE–Co amorphous alloy films, which attained its maximum at about 300 nm, and the presence of Tb decreased the absolute value of the Kerr rotation for RE–Co amorphous alloy films. It should be noted here that the Nd–Fe composition also shows a significant increase in the absolute value of the Kerr rotation at short wavelength: it attains the value $\Phi_K = 0.48°$ at 380 nm (Kanaizuka *et al* 1987).

GdTbFeCo films. The amorphous alloys $(GdTb)_{1-x}(FeCo)_x$ have attracted considerable attention, as a result of their possible application in magnetooptical recording. The basic magnetic and magnetooptical properties of these alloys can be tailored well via selection of appropriate preparation conditions and compositions. A careful investigation of the magnetooptical properties of $RE_{1-x}T_x$ (RE = Gd, Tb, GdTb; T = Fe, Co, FeCo) was carried out by Hansen *et al* (1987). Amorphous films were prepared by codeposition from three electron-gun sources in a high-vacuum system onto rotating glass substrates.

For the amorphous alloys discussed, the spectral dependences of the Faraday and Kerr rotations show almost no structure, due to the absence of a well-defined band structure. Figure 10.8(a) shows the wavelength dependences of the Kerr rotations for some compositions (Hansen *et al* 1987). Except a broad minimum in the infra-red region related to magnetooptically active transitions of Co and Fe, the measured spectra reveal no characteristic features. The small differences near 350 nm for Tb-containing alloys may be related to the Tb giving rise to a very strong magnetooptically active transition in crystalline alloys near 270 nm (Katayama and Hasegawa 1982).

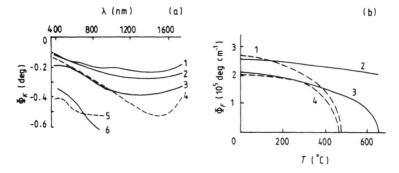

Figure 10.8. (a) The spectral dependences of the polar Kerr rotations ($T = 295$ K) obtained for amorphous $Gd_{0.33}Co_{0.67}$ (curve 1), $Gd_{0.31}Co_{0.69}$ (curve 2), $Gd_{0.44}Fe_{0.56}$ (curve 3), and $Gd_{0.17}Tb_{0.10}Fe_{0.73}$ (curve 4) (Hansen *et al* 1987). The corresponding spectral dependences for Co (curve 5) and Fe (curve 6) (van Engen 1983) are also shown for comparison. (b) The temperature dependences of the Faraday rotations ($\lambda = 633$ nm) obtained for $Gd_{0.24}Fe_{0.76}$ (curve 1), $Gd_{0.21}Co_{0.79}$ (curve 2), $Gd_{0.31}Co_{0.69}$ (curve 3), and $Gd_{0.44}Fe_{0.56}$ (curve 4) (Hansen *et al* 1987).

The cobalt-containing alloys, as compared to iron-containing alloys, show smaller rotations in the visible region, in accordance with the results observed for the pure metals. It should be noted here that the increase in the Kerr rotation for alloys with increasing Co content at room temperature is not based on a larger magnetooptical activity of Co, but is associated with the significantly different magnetic behaviours of amorphous Co and Fe alloys (Hansen *et al* 1987). Magnetization and Curie temperature data reveal a much stronger effect of the amorphous structure on the magnetic ordering of Fe-based alloys, as compared with Co-based alloys, which is reflected in the very low Curie temperature of amorphous iron of about 200 K, contrasting with the high T_c for Co-rich alloys.

This difference in behaviour strongly affects the concentration dependences of the Faraday and Kerr rotations, as shown in figure 10.9 for $\lambda = 633$ nm (Hansen *et al* 1987). The data for zero values of the Faraday and Kerr rotations correspond to the compositions with $T_c = 295$ K. From these plots it is obvious that the magnetooptical effects for the Fe-based alloys are basically controlled by the Curie temperatures on the Fe-rich and the RE-rich side.

The Co-based alloys exhibit a room temperature T_c-value shifted to higher x; this shift is related to the fact that the Co atoms carry no magnetic moment for $x < 0.5$. The T_c-values for Co-rich alloys are far above room temperature, causing considerably stronger magnetooptical effects as compared to those for Fe-based alloys of corresponding compositions (Hansen *et al* 1987). TbFe and GdFe alloys with $0.6 < x < 0.9$ are thus expected to show almost constant values of the Faraday and Kerr rotations.

$RE_{1-x}(Fe_{1-y}Co_y)_x$ alloys with $0.6 < x < 0.9$ and increasing Co content are expected to show first an increase in the Faraday and Kerr rotations due to the rise of T_c. This is confirmed by Φ_K and Φ_F data at 633 nm and $T = 295\ K$, as shown in figure 10.9. The temperature dependence of Φ_F at 633 nm is shown in figure 10.8(b) for some iron- and cobalt-based alloys incorporating gadolinium (Hansen *et al* 1987). The curves demonstrate again the difference between the influences of Fe and Co on the value of T_c, and thus on the Faraday rotation.

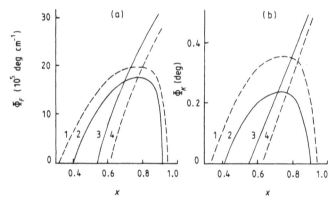

Figure 10.9. The concentration dependences of the Faraday rotation (a) and polar Kerr rotation (b) ($\lambda = 633$ nm, $T = 295$ K) obtained for $Gd_{1-x}Fe_x$ (curve 1), $Tb_{1-x}Fe_x$ (curve 2), $Gd_{1-x}Co_x$ (curve 3), and $Tb_{1-x}Co_x$ (curve 4) (Hansen *et al* 1987).

As has been shown by Hansen *et al* (1987), the temperature dependences of Φ_F and Φ_K can be well described in terms of the magnetizations of the rare-earth and transition metal sublattices, inferred from the fits of the mean-field theory to the measured saturation magnetizations. The sublattice Faraday rotations turn out to be of comparable magnitude at low temperatures, while in the high-temperature range, the transition metal contribution dominates.

Chapter 11

Semimagnetic semiconductors

The intense research effort into semimagnetic semiconductors (SMS), which constitute a new class of materials combining the properties of ordinary and magnetic semiconductors, originated in the studies of manganese-doped crystals (Komarov *et al* 1977), and $Cd_{1-x}Mn_xTe$ solid solutions (Gaj *et al* 1978), which reported giant spin splitting in the band states of electrons, holes, and excitons, as well as a giant Faraday effect caused by the exchange interaction between the band carriers and the localized magnetic moment of Mn^{2+} ions (Nikitin and Savchuk 1990). A relatively small external magnetic field (of $H = 30$ kOe) induced in these materials a large spin splitting of the exciton line (up to tens of meV), and significantly enhanced the Faraday rotation in the appropriate spectral range.

The fundamental qualitative difference between SMS and ordinary semiconductors (for example, between $Cd_{1-x}Mn_xTe$ and $Cd_{1-x}Zn_xTe$) is that an external magnetic field induces a significant exchange interaction between magnetic ions and charge carriers in the bands of SMS. In the absence of an external field, however, the electronic processes and modifications of the band structure as functions of x are quite similar in these two classes of crystal.

On the other hand, SMS also can be viewed as an intermediate class of materials, lying between magnetic and non-magnetic semiconductors. By varying the magnetic component concentration together with the external parameters (temperature, magnetic field), one can induce a transition from one type of semiconductor to the other, and extract the properties introduced by the appearance of the localized magnetic moment.

There have been predictions that exchange interactions between electrons and magnetic ions in SMS should lead to large light-induced spin polarizations. Indeed, magnetization by intense, circularly polarized radiation (this phenomenon is also known as the inverse Faraday effect) has been observed for $Cd_{1-x}Mn_xTe$ and $Hg_{1-x}Mn_xTe$ SMS materials. However, the magnitude of this effect turned out to be very small.

11.1 Manganese-doped semiconductors

$Cd_{1-x}Mn_x Te$. The sharp enhancement of the Faraday rotation was first reported by Komarov *et al* (1977), who studied CdTe crystals doped with Mn^{2+} ions up to a concentration of 8.2×10^{18} cm^{-3}. Investigations of the dispersion of the Verdet constant $V = \Phi_F / Hh$ measured for samples of thickness 0.2 mm with different Mn concentrations at $T = 1.7$ K in a magnetic field of $H = 3.65$ kOe revealed the non-linear dependence of the Verdet constant on the concentration of Mn ions, and the sharp growth the Verdet constant near 1.5 eV.

The sharp rise in value of the Verdet constant coincides with the absorption edge of the crystal, or, more precisely, with the long-wavelength tail of the main exciton band. It was found that the Faraday rotation increased non-linearly with the doping level. Komarov *et al* investigated the nature of the observed Faraday rotation by pump exciting the samples at microwave frequencies, and observed a sharp decrease of the Faraday rotation at fields corresponding to the electron paramagnetic resonance of Mn^{2+} ions in cadmium telluride.

This made it possible to relate the observed Faraday effect enhancement for CdTe(Mn) uniquely to the spin polarization of Mn^{2+} impurity ions. Direct spectroscopic observation of large exciton band splittings in magnetic fields indicates that the Faraday effect is due to the splitting of the exciton band. The magnitude of this splitting, like the extent of Faraday rotation enhancement, is determined not by the external magnetic field itself, but rather by the field-induced magnetization of the paramagnetic Mn^{2+}-ion subsystem in CdTe.

The giant Faraday rotation effect for the solid solutions $Cd_{1-x}Mn_x Te$ with Mn concentration $x < 0.5$ was first reported by Gaj *et al* (1978). Their results indicate (figure 11.1(a)) that the Verdet constant of $Cd_{1-x}Mn_x Te$ is at least an order of magnitude large than that of pure CdTe crystals. The actual value of the constant varies strongly with temperature and manganese content, while the spectral dependence of the Verdet constant deviates from the spectrum of the inter-band Faraday effect observed for non-magnetic semiconductors.

It was found that at room temperature, as the concentration of manganese increases and we go from 'pure' CdTe to $Cd_{1-x}Mn_x Te$ solid solution, not only does the Faraday rotation angle increase, but also it changes sign. Similarly complicated Faraday rotation dispersion is observed as a function of temperature when the sample composition is fixed.

Many advances in the epitaxial growth of thin SMS films have been achieved in recent years (Imamura *et al* 1994). Investigations of the Faraday effect in thin films have markedly increased our understanding of SMS processes—above all because the shift towards shorter wavelength expanded the accessible spectral range. Koyanagi *et al* (1987) obtained $Cd_{1-x}Mn_x Te$ ($0 < x < 0.76$) films by simultaneous evaporation of CdTe and MnTe onto sapphire and glass substrates. As can be seen in figure 11.1(b), the Faraday rotation for thin films is dispersive, with sharp extrema, and rotation angle sign reversal.

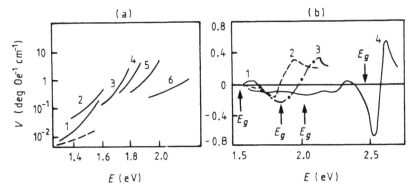

Figure 11.1. (a) Faraday rotation spectra of $Cd_{1-x}Mn_xTe$ crystals at $T = 77$ K for various compositions: $x = 0.02$ (curve 1); 0.05 (curve 2); 0.15 (curve 3); 0.2 (curve 4); 0.3 (curve 5); and 0.5 (curve 6) (Gaj *et al* 1978). Dashed line: the Faraday rotation for CdTe. (b) The spectral dependence of the Faraday rotations for $Cd_{1-x}Mn_xTe$ films of various compositions at $T = 300$ K: $x = 0$ (curve 1); 0.26 (curve 2); 0.36 (curve 3); and 0.76 (curve 4) (Koyanagi *et al* 1987).

The films investigated were strongly dichroic in a magnetic field; the absorption edge shifts towards longer wavelengths for right-polarized light and towards shorter wavelengths for left-polarized light. As the Mn content increases, the curves shift towards shorter wavelengths, and the extremal values of the Verdet constant increase significantly. Analogous changes in the Faraday rotation curves are observed if Cd–Mn–Te films are cooled.

A detailed theoretical study of the Faraday effect for $Cd_{1-x}Mn_xTe$ epitaxial films was reported by Koyanagi *et al* (1988), who took into account the splittings and oscillator strengths of the exciton transitions to second order in perturbation theory.

Effects of an electric field on the Faraday rotation for $Cd_{1-x}Mn_xTe$ films, prepared by ionized cluster beams, have been investigated by Koyanagi *et al* (1989). A marked change in the Faraday rotation was observed near the band-gap energy under exposure to an electric field perpendicular to the plane of the film. The maximum change in the Verdet constant obtained for $Cd_{0.49}Mn_{0.51}Te$ film was of the order of 0.22 deg cm^{-1} Oe^{-1} at an electric field of 174 kV cm^{-1}. It was found that this effect could be attributed to the effects of the electric field on the optical absorption by excitons.

$Zn_{1-x}Mn_xTe$. The Faraday rotation spectra of ZnTe crystal at $T = 300$ °C are characterized by a positive Verdet constant which changed from 0.004 deg cm^{-1} Oe^{-1} at 1.5 eV to 0.03 deg cm^{-1} Oe^{-1} at 2.2 eV. For $Zn_{1-x}Mn_xTe$, the Faraday rotation behaviours as the concentration and temperature are varied are analogous to those of $Cd_{1-x}Mn_xTe$, but the sign reversal of the Faraday rotation angle occurs at a higher Mn content, $x = 0.05$, for the full spectral range from 1.5 to 2.1 eV. For this composition, the Verdet constant is nearly zero between 1.4 and 2.0 eV, and then the absolute value of the Verdet constant increases up to -0.02 deg cm^{-1} Oe^{-1} at 2.1 eV (Bartholomew *et al* 1986).

Hg$_{1-x}$Mn$_x$Te. Dillon *et al* (1990) investigated the Faraday rotation of Hg$_{1-x}$Mn$_x$Te, using two samples prepared by solid-state recrystallization, with energy gaps of interest as regards isolator application: at 1300 and 1550 nm. The Verdet constant measured for the sample with $x = 0.36$ is shown in figure 11.2 as a function of the photon energy. Plots of the room temperature absorption coefficient for three samples with $x = 0.26$ (curve 1), 0.31 (curve 2), and 0.36 (curve 3) are shown in inset of figure 11.2. Sample 1 was grown by the slow vertical Bridgman method, used traditionally for the preparation of Hg$_{1-x}$Mn$_x$Te with x less then 0.2.

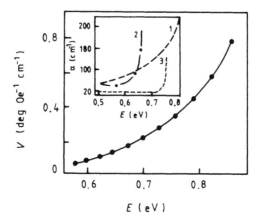

Figure 11.2. In the main figure, the measured spectral dependence of the Verdet constant for Hg$_{0.64}$Mn$_{0.36}$Te at 295 K is shown. The inset shows the absorption curves measured at 295 K, with no correction for reflection (Dillon *et al* 1990).

Cd$_{1-x}$Mn$_x$S. Gubarev (1981) studied manganese-doped cadmium sulphide crystals, and observed a several-fold enhancement of the Faraday rotation as compared with the value for the undoped crystal when the concentration of the Mn ions reached 2.2×10^{19} cm^{-1}. It should be noted that at helium temperatures, the absorption edge spectrum of hexagonal cadmium sulphide crystals is dominated by A-, B-, and C-exciton transitions, and consequently the Faraday rotation spectrum is largely determined by the spin splitting of the lowest-lying A-exciton resonance. The above-described measurements on doped semiconductors demonstrated that the magnitude of the Verdet constant depends strongly on the magnetic ion concentration.

GaAs(Mn). Sharp growth of the Faraday rotation near the fundamental absorption edge has been observed for GaAs doped with Mn up to $c_{Mn} = 5 \times 10^{18}$ cm^{-3} (Hennel *et al* 1965). Despite this, the magnitude of the Faraday rotation for GaAs(Mn) was much smaller than that for CdTe(Mn). The sign of the Faraday rotation for GaAs(Mn) is the same as for undoped GaAs. The spectral dependence of the Faraday effect for GaAs(Mn) is temperature invariant, shifting to follow the temperature-induced change in the band-gap. Furthermore, the magnetic field dependence of the Faraday rotation for this material is nearly linear with field. Evidently the value of the Faraday rotation for GaAs(Mn) is determined by the s–d exchange interactions between $Mn^{3+}(3d^4)$ ions and band carriers, as well as by the van Vleck paramagnetism of the manganese ions. Indirect confirmation of this picture is provided by the absence of electron paramagnetic resonance in GaAs(Mn) crystals (Hennel *et al* 1965).

$CdP_2(Mn)$. Abramishvili *et al* (1987) investigated the Faraday rotation for Mn-doped CdP_2 crystals. The doping the crystal with Mn leads to a rather small change in the spectral dependence of the Faraday rotation, and does not alter its sign. At the same time, the field dependence of the rotation becomes markedly non-linear at high magnetic fields, and can be approximated by the Brillouin function for a spin-5/2 system.

This behaviour corresponds to the magnetization of the Mn^{2+}-ion spin system, suggesting that the effect of Mn^{2+} ions on the magnetooptical characteristics of CdP_2 is not related to the carrier–impurity exchange interaction affecting the energy structure and electronic transitions of the matrix, but rather to the direct contribution of electronic transitions within Mn^{2+} ionic shells (Abramishvili *et al* 1987). The spectral range investigated in this work is close to the intra-centre $^6A_1(^6S) \rightarrow {}^4T_1(^4G)$ electronic transition in Mn^{2+}. The low symmetry of the CdP_2 crystal makes these transitions more probable.

There are two features that help to explain the weakness of the carrier–impurity exchange interaction in Mn-doped CdP_2: the relatively small values of the exchange integrals in comparison with those for the $A^{II}B^{VI}(Mn)$ system, and the strong carrier scattering by the Mn^{2+} impurity ionic potentials in CdP_2 (Nikitin and Savchuk 1990).

$Pb_{1-x}Mn_xI_2$. Solid solutions of the PbI_2–MnI_2 system provide another class of SMS that exhibit giant enhancement of the inter-band Faraday effect due to the magnetic component. This enhancement has been attributed to the exchange interaction between photoexcited carriers and localized Mn^{2+} ionic spins (Abramishvili *et al* 1987).

The binary PbI_2 and MnI_2 compounds have layered crystal structures (D_{3d} symmetry group), characterized by hexagonal close packing of the I^- anions. The Pb^{2+} and Mn^{2+} cations are located in the octahedral spaces between adjacent iodide layers. The similarity of the PbI_2 and MnI_2 crystal lattices, and the closeness of the lattice parameters promote the formation of PbI_2–MnI_2 solid solutions (Nikitin and Savchuk 1990).

The spectral dependence of the Faraday rotation for several compositions of $Pb_{1-x}Mn_xI_2$ was studied by Abramishvili *et al* (1987). It was found that the Faraday rotation for $Pb_{1-x}Mn_xI_2$ is opposite in sign and markedly larger in value than that for the binary PbI_2 compound. It is interesting to note that $Pb_{1-x}Mn_xI_2$ exhibits non-linear dependence of the Faraday rotation on the value of the applied magnetic field, which can be explained by the difference between the field dependences of the Brillouin function and the linear Zeeman contributions.

Analogously to the case for the $A_{1-x}^{II} Mn_x B^{VI}$ crystals, the Faraday rotation for $Pb_{1-x} Mn_x I_2$ can be adequately described by the 'excitonic' contribution, as long as detuning is small. For larger values of the detuning, the inter-band contribution should be taken into account. For $x = 0$ and $x = 0.01$, there is good agreement, whereas for $x = 0.03$, there is a discrepancy. This is due to the inhomogeneous broadening, and when the latter is taken into account the discrepancy disappears.

Summarizing the experimental results on the magnetooptical properties of various SMS materials, Nikitin and Savchuk (1990) have noted the main effects: (i) the giant Faraday effect, caused by exchange interactions of excitons and carriers with magnetic ions, is most pronounced for $A^{II} B^{VI}(Mn)$ and $A_{1-x}^{II} Mn_x B^{VI}$ semiconductors; (ii) among other SMS, only $Pb_{1-x} Mn_x I_2$ solid solutions exhibit Faraday behaviour analogous to that for the $A_{1-x}^{II} Mn_x B^{VI}$ system, despite a markedly smaller sum of exchange constants that characterizes the carrier–ion exchange interactions; and (iii) for some SMS, the complicated character of the Faraday rotation spectra, involving concentration- and temperature-dependent sign reversal of the rotation angle, is due to the competition of three different mechanisms (exciton, inter-band, and intra-centre transitions) contributing to the Faraday effect.

11.2   The Faraday effect induced by the free carrier

Faraday rotation measurements on $Hg_{1-x} Mn_x Te$ solid solutions were carried out on samples with large values of x, i.e., composition with $E_g > 0.5$ eV. Recently, Yuen *et al* (1987) have reported measurements on free-carrier-spin-induced Faraday rotation for narrow-gap $Hg_{1-x} Mn_x Te$. Experimental results point to a number of differences between the spin-induced Faraday rotation for $Hg_{0.78} Cd_{0.22} Te$ and for $Hg_{0.89} Mn_{0.11} Te$ samples.

The Verdet constant of $Hg_{1-x} Mn_x Te$ is somewhat larger, probably because of a higher g-factor. As the laser intensity increases, so does the Faraday rotation saturation level for $Hg_{1-x} Cd_x Te$, but in the linear region, the rotation is not sensitive to the laser intensity.

For $Hg_{1-x} Mn_x Te$, on the other hand, the linear Faraday rotation decreases at higher laser intensity, while the saturation level increases, even though the linear slope is smaller. These discrepancies have been explained by spin thermal effects. For $Hg_{1-x} Cd_x Te$, the laser beam increases the electron translational energy, but it cannot alter the frozen electron spin position, since spin-flip-scattering events are rare. For SMS, in contrast, the exchange interaction of carriers and Mn^{2+} ions promotes the rapid transfer of translational energy to the spin system, eventually raising the spin temperature. Complete spin alignment is also impeded by the laser-induced depolarization, leading to the increase of the Faraday rotation saturation field for $Hg_{1-x}Mn_xTe$ at high laser intensities (Nikitin and Savchuk 1990).

11.3 Magnetooptics of quantum objects

Detailed analysis of the giant excitonic Faraday rotation for CdTe/CdMnTe quantum wells in the excitonic resonance condition was performed by Buss *et al* (1995). The value found for the excitonic Verdet constant, 7×10^2 deg Oe^{-1} cm^{-1}, is about two orders of magnitude larger than that measured for inter-band transitions in the bulk or superlattices of semimagnetic semiconductors, making this effect attractive as regards use in the study of magnetooptical properties of single quantum wells. Yanata and Oka (1994) have investigated the photoluminescence of $Cd_{1-x}Mn_xSe$ quantum dots embedded in quartz glass matrices, via selective-excitation spectroscopy. Photovoltaic magnetospectroscopy was used in the study of the 2D electron gas in narrow-gap HgCdMnTe by Dudziak *et al* (1995). A strong circular dichroism was found, which agrees with the phenomenological description of the optical properties of the 2D electron systems in the quantized Hall regime.

11.4 Light-induced magnetization

Van der Ziel *et al* (1965) discovered that intense, circularly polarized radiation can induce magnetization in some materials. This light-induced magnetization is usually called the inverse Faraday effect. It has been observed for $CaF_2(Eu^{2+})$ crystals, diamagnetic glasses, and liquids (Zapasskii and Feofilov 1975).

Among semimagnetic materials, light-induced magnetization has been observed for $Cd_{1-x}Mn_xTe$ (Ryabchenko *et al* 1982) and for $Hg_{1-x}Mn_xTe$ (Krenn *et al* 1985). Since these semiconductors exhibit strong exchange interaction between carriers and magnetic ions, leading to reciprocal spin orientation, stronger magnetization was expected after irradiation with polarized light. The first, indirect indications of induced magnetization in SMS were obtained in magnetization measurements on laser-irradiated $Cd_{0.95}Mn_{0.05}Te$ (Ryabchenko *et al* 1982). The observed dependence of the excitonic luminescence lines on the pump intensity was explained in terms of changes in $\langle S_z \rangle$ caused by exchange scattering of photoexcited carriers by localized Mn^{2+} ionic spin moments.

Subsequently, light-induced magnetization for $Hg_{1-x}Mn_xTe$ was directly observed by Krenn *et al* (1985). The experiments employed a laser beam of fixed intensity, whose polarization was modulated from linear to circular. The measurements were made for two compositions of $Hg_{1-x}Mn_xTe$ ($x = 0.07$ and 0.12), and for non-magnetic $Hg_{1-x}Cd_xTe$ ($x = 0.23$), and also for InSb semiconductors with approximately the same band-gap E_g. The signal was recorded only for the SMS $Hg_{0.88}Mn_{0.12}Te$ irradiated with circularly polarized light.

The absence of light-induced magnetization for $Hg_{1-x}Cd_xTe$ and InSb confirms that the inverse Faraday effect for SMS cannot be attributed to spin polarization of electrons and holes. Furthermore, the effect cannot be explained by invoking the direct light-induced polarization of Mn^{2+} ions, for then it would appear for the zero-gap compound $Hg_{0.93}Mn_{0.07}Te$ as well. Hence, as in earlier experiments, the dominant mechanism leading to the appearance of magnetization is the alignment of paramagnetic ions by polarized carriers. In turn, the alignment of localized Mn^{2+} ionic spin moments occurs via spin-flip scattering of polarized free carriers (Nikitin and Savchuk 1990).

11.5 Applications of the Faraday effect in SMS

Materials with large Verdet constants have always attracted considerable interest, because of their possible applications in magnetooptic devices. Table 11.1 lists the Verdet constants of the major crystalline and glass-like materials active in the visible and near infra-red (Nikitin and Savchuk 1990). Clearly the Verdet constant of $Cd_{1-x}Mn_xTe$ is the highest, outdistancing those of other materials by more than an order of magnitude.

Table 11.1. Verdet constants of various materials (633 nm, $T = 300$ K).

Crystals	V (rad T^{-1} m^{-1})	Glasses	V (rad T^{-1} m^{-1})
$Cd_{1-x}Mn_xTe$	2000	FR-123	−71.0
EuF_2	−262	FR-5	−72.0
$Tb_3Al_5O_{12}$	−180	$Pr(PO_3)_3$	−39.6
$LiTbF_4$	−128	FR-7	−34.9
ZnSe	118	FR-4	−30.5
CeF_3	−114	SF-59	28.5
Bi_4GeO_{12}	29.8	SiO_2	4.0
LaF_3	3.5	SF-N64	1.5

Turner *et al* (1983) analysed the feasibility of employing $Cd_{1-x}Mn_xTe$ as the active medium for a compact Faraday valve (an optical isolator). The required thickness of an optical isolator made of crystalline $Cd_{0.55}Mn_{0.45}Te$ is only 1.31 mm at a wavelength of 633 nm in an applied field of 3 kOe; the figure of merit at this wavelength attains 464.6 deg dB^{-1} (Walecki and Twardowski 1989). Analysis of the spectral dependencies of various Cd–Mn–Te compositions indicates that, in principle, these crystals can also be employed in the near infra-red, particularly at the wavelength 1060 nm.

Chapter 12

Bilayer, multilayer, superlattice, and granular structures

Magnetic multilayers have attracted much attention due to their novel properties, which make them suitable for a variety of applications. One interesting phenomenon is the perpendicular magnetic anisotropy that has been found in Co/non-magnetic metal multilayers where the metal can be Pt, Pd, Au, or Ir. The large perpendicular magnetic anisotropy and excellent magnetooptical properties shown in these multilayers make them potential candidates for use in future magnetooptical recording media. With a view to enhancing Φ_K, several types of multilayered film have been proposed and evaluated: a simple structure characterized by multiple reflections in a dielectric layer coated on a magnetic layer; a four-layer structure consisting of a thin magnetic layers sandwiched between two dielectric layers and a reflecting layer; bilayer structures composed of non-magnetic and magnetic layers; and bilayer structures composed of two magnetic layers and compositionally modulated structures.

In these multilayer structures, GdTbFe/Ag bilayer films are unique in the respect that their Kerr rotations are enhanced with almost no change in the reflectivity, and TbFe/SiO$_2$ compositionally modulated films are remarkable in that their Kerr rotations are enormously enhanced to 25° at the sacrifice of reflectivity.

A new class of phase-separated magnetic materials in which the two phases couple antiferromagnetically across the phase boundary has been discovered by Gambino *et al* (1994). A prototypical example is cobalt containing ($\sim$10 nm) diameter precipitate particles of EuS with NaCl structure. These materials show many of the properties of ferrimagnets, such as compensation points, so the term 'macroscopic ferrimagnet' was adopted to distinguish them from ferrimagnets in which the antiferromagnetic exchange couples individual atoms. In such a material, the exchange between the Co and EuS also orders the EuS well above its normal Curie point of 16 K.

12.1 Bilayer structures

12.1.1 Rare-earth–transition metal bilayers

Amorphous rare-earth–transition metal compounds are now widely used as high-density magnetooptical recording media in magnetooptical disk memory systems. Because magnetic recording media must meet some contradictory demands, serious efforts have been made to improve the magnetic, optical, and magnetooptical properties of such materials. One of the interesting possibilities for obtaining high-quality recording media for magnetooptical memory devices is related to the development of multilayer films in which the different layers meet the contradictory demands that are impossible to meet in a monolayer structure.

Several double-layer films, such as Gd–Fe/Tb–Fe, Gd–Co/Gd–Co, Tb–Fe/Tb–Fe, Tb–Fe–Co/Tb–Fe–Co, and Pr, Ce, Nd–Fe/Tb–Fe–Co, have already been investigated. Their magnetization behaviours have been explained in terms of the exchange interaction between two layers. Combination of one magnetic layer with improved magnetic properties and a second layer, also with improved magnetooptical properties, produces a magnetooptical medium which combines the improved properties of both layers.

12.1.1.1 Tb–Fe–Co/Tb–Fe–Co

Let us consider the properties of such double-layer films, looking, for example, at a bilayer which consists of two Tb–Fe–Co layers with the transitional metal dominant in the first layer, and the rare earth dominant in the second layer. Such bilayers have superior properties for magnetooptical memory use (Fujii *et al* 1987). The first layer has a Curie temperature of about 250 °C, and the second is characterized by a compensation temperature of 105 °C and a Curie temperature of about 180 °C. It was found that the Kerr rotation at 830 nm increased with the increase of the first-layer thickness from 0.31° for zero thickness, and approached the intrinsic Kerr rotation value of the first layer, 0.49°, when its thickness approached 35 nm. It was also found that the temperature dependencies of the switching properties change variously with the thicknesses of the two layers, and are rather complicated.

12.1.1.2 Ce, Pr, Nd–Fe/Tb–Fe–Co bilayers

Kanaizuka *et al* (1987) demonstrated the growth of the Kerr rotation angle when $Tb_{20}Fe_{72}Co_8$ amorphous films were coated with light rare-earth–iron alloys ($Ce_{40}Fe_{60}$, $Pr_{52}Fe_{48}$, and $Nd_{52}Fe_{48}$), in which the amorphous film contained more transition metal than in the compensation composition. The value of the Kerr rotation goes up to $-0.34°$ at 780 nm for a $Pr_{52}Fe_{48}$ coating about 6 nm thick on a Tb–Fe–Co amorphous film, while it is $-0.28°$ for the uncoated film. On the other hand, a significant enlargement of the rotation angle is not observed in the case of a terbium-rich underlayer.

12.1.2   PtMnSb/TbFe bilayers

It is known that PtMnSb films have an easy axis of magnetization in the film plane, because of their low anisotropy energies. In order to overcome this, a PtMnSb/TbFe double-layered structure rf sputtered onto a glass substrate was made by Matsubara *et al* (1987). The maximum value of the Kerr rotation obtained by them was about 1.6°. The position of the peak in the Kerr rotation spectrum shifts from 1.6 eV for PtMnSb to 2.4 eV for the PtMnSb/TbFe double-layered film, while the size of the peak slightly decreases. Matsubara *et al* stated that this shift arises from the sum of the polar Kerr rotation and the Faraday rotation for the PtMnSb layer. The value of the coercive force for the double-layered films was of the order of 600 Oe, and was five times as large as that for PtMnSb monolayered films (Matsubara *et al* 1987).

12.1.3   Magnetic/non-magnetic bilayers

Fe/Cu. Katayama *et al* (1988) showed that the Kerr rotation of a thin iron film on top of bulk copper is enhanced near the plasma energy of the copper.

GdTbFeCo/Ag and GdTbFeCo/Cu. Reim and Weller (1988) demonstrated that the Kerr rotation for rare-earth–transition metal films can be strongly enhanced using bilayer thin-film structures, where the magnetooptical material is deposited onto a non-magnetic metallic film with low optical constants n and k. The optical reflectivity of such a bilayer remains almost the same as that of the bulk magnetooptical material, which in turn leads to an enhancement of $R\Phi_K$ by more than a factor of two over a rather wide spectral range.

The Kerr rotation spectrum of a rf-sputtered 20 nm $Tb_{22}(Fe_{80}Co_{20})_{78}$ film on top of copper is shown in figure 12.1(a). The Kerr rotation spectrum of a 100 nm $Tb_{25}(Fe_{80}Co_{20})_{75}$ film is also shown, for comparison. The films were covered with AlN protective layers, 15 and 20 nm thick, respectively. A dielectric layer of such thickness does not affect Φ_K up to photon energies of 3.5 eV. The Kerr rotation of the bilayer film is found to be mainly enhanced at photon energies between 1.5 and 2.3 eV, where the index of refraction of copper is very low ($n \simeq 0.3$) and the index of absorption has moderate values of k ($2.6 < k < 5$). At 2 eV, the reflectivity of a 20 nm layer of TbFeCo on top of copper is only reduced by about 10% compared to that for bulk TbFeCo, which leads to an enhancement by about a factor of 1.8 of $R\Phi_K$.

Figure 12.1(b) shows polar Kerr spectra of 11 and 23 nm thick films of $Gd_{11}Tb_{11}(Fe_{80}Co_{20})_{78}$ on top of bulk silver, prepared by ion beam sputtering. The spectrum of a film of $Tb_{25}(Fe_{80}Co_{20})_{75}$, 100 nm thick, is shown for comparison. At 3 eV (413 nm), the Kerr angle is enhanced by more than a factor of two for both films on top of silver, while the reflectivity remains

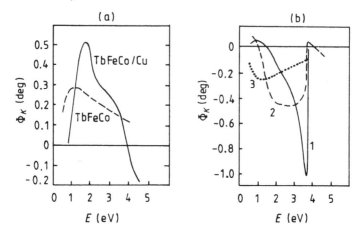

Figure 12.1. Room temperature polar Kerr rotation spectra of the bilayer film $Tb_{25}(Fe_{80}Co_{20})_{75}$(20 nm)/Cu, and bulk $Tb_{25}(Fe_{80}Co_{20})_{75}$, covered with 15 and 20 nm of AlN, respectively (a); and of $Gd_{11}Tb_{11}(Fe_{80}Co_{20})_{78}$/Ag bilayers with magnetic layer thicknesses of 11 nm (curve 1) and 23 nm (curve 2), respectively, and of a $Tb_{22}(Fe_{80}Co_{20})_{78}$ film 100 nm thick (curve 3) (b) (Reim and Weller 1988).

almost unchanged, at about 40%.

Near the plasma edge of silver at 3.75 eV, Φ_K for the film of thickness 11 nm reached a value of 1.06°—that is, an increase in the Kerr rotation by a factor of six is observed—but the reflectivity is reduced by a factor of three. The energy at which optimum enhancement of the figure of merit occurs can be tuned by varying the thickness of the top layer.

Nd–Fe/Cu and Nd–Fe/Al. Zhou *et al* (1990) studied the magnetooptical properties of Nd–Fe films, and of Nd–Fe/Cu and Nd–Fe/Al bilayers on glass substrates, using vacuum evaporation in the spectral range 400–700 nm in a magnetic field of 7 kOe. The thickness of the Nd–Fe layer was changed from 5 nm to more than 30 nm, and the underlying Cu and Al layers became thicker than 100 nm. For Nd–Fe/Cu bilayers with Nd contents less than about 37 at.% and with various thicknesses, it was found that the value of Φ_K was larger than that for a single thick film, and that a peak of rotation appears. The peak position is located consistently near the absorption edge of Cu, at about 560 nm. At the same time, its reflectivity shows an abrupt drop at the same wavelength.

Figure 12.2 shows Kerr rotation spectra of thick Nd–Fe films, and of Nd–Fe/Cu and Nd–Fe/Al bilayers with different magnetic layer thicknesses. The main difference between the Kerr rotation spectra of the films with Cu and Al underlayers is that there is a rotation peak near 560 nm for bilayers with copper underlayers, and the bilayers with aluminium underlayers are characterized by monotonic growth of the rotation with decreasing wavelength.

TbFeCo/Al. The bilayer films composed of amorphous TbFeCo and pure Al layers deposited sequentially on a glass slide substrate without plasma

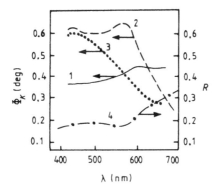

Figure 12.2. Kerr rotation spectra of a single thick $Nd_{32}Fe_{68}$ film (curve 1), and $Nd_{32}Fe_{68}/Cu$ (curve 2) and $Nd_{32}Fe_{68}/Al$ bilayer films (curve 3) each with a magnetic layer thickness of 14.5 nm, together with the reflectivity spectrum of copper (curve 4) (Zhou *et al* 1990).

exposure, using the facing-targets sputtering system, were studied by Song *et al* (1993). The amorphous TbFeCo layers have thicknesses from 5 to 300 nm. The Al layers, each 100 nm thick, were deposited on the magnetic layers.

The film with the large thickness of the magnetooptical layers, 300 nm, showed a monotonic increase of the Kerr rotation with increase of wavelength. On the other hand, the film with the small thickness of the magnetooptical layers, 8.8 nm, showed a monotonic decrease of the Kerr rotation with increase of wavelength. The film with the 14 nm magnetooptical layers revealed a broad peak of Φ_K, of value $\simeq 0.36°$, at a wavelength of about 450 nm.

The figure of merit F for the film that had a TbFeCo layer 100 nm thick increased monotonically with the increase of the wavelength, from 0.16° at 400 nm up to 0.38° at 900 nm; but for the film with a 14 nm TbFeCo layer, F decreased monotonically with the increase of the wavelength, from 0.25° at 400–500 nm down to 0.16° at 700–800 nm. Also, the Tb–Fe–Co(5 nm)/Al(5 nm) bilayer had a magnetic anisotropy energy K_u as large as 2×10^6 erg cm^{-3} and exhibited a Kerr rotation angle Φ_K as large as 0.58° at 830 nm.

Tb–Fe–Co/Ta, Tb–Fe–Co/C. Song and Naoe (1994) studied the magnetic and magnetooptical characteristics of bilayer films composed of Tb–Fe–Co layers with overlayers of Ta, C, Co–Cr, and Ni–Fe, obtained by using a facing-target sputtering apparatus. It was found that the Tb–Fe–Co/Ta bilayer showed a Kerr rotation angle as large as 0.85° ($\lambda = 633$ nm) when the thickness of the Tb–Fe–Co layer was about 30 nm, compared with 0.43° when the thickness was 100 nm. The Tb–Fe–Co/C bilayer exhibited a slight increase in the rotation, from 0.42° for the magnetic layer thickness 60 nm to 0.52° for the magnetic layer thickness 10 nm, at the same wavelength. The Tb–Fe–Co/Co–Cr bilayer also exhibited growth of the rotation angle, from 0.43° for the magnetic layer thickness 100 nm up to 0.6° for the magnetic layer thickness 30 nm.

12.1.4 Dielectric/magnetic film bilayers

SiO layer/Dy–Co compositionally modulated films. A general expression for
the polar Kerr effect for a bilayered configuration in which the optical constants
of two media differ was presented, and the optical and magnetooptical properties
of SiO-coated Dy/Co compositionally modulated alloys were investigated
experimentally by Chen *et al* (1990). The calculated values were found to
compare well to the experimental results. The uncoated films were characterized
by a smooth spectral dependence of the rotation near 0.2°, while the SiO-coated
film revealed a sharp rotation peak near 2 eV, with a rotation value near 2°,
in good accordance with the calculated results. The reflectivity of the structure
appeared to be less than one per cent at this wavelength.

ZnS(O, N)/TbFeCo bilayers. In order to obtain a large Kerr rotation in
the short-wavelength region where magnetooptical memory can be raised to
high density, ZnS(O, N) films with an absorption edge in the short-wavelength
region were employed as coating layers on amorphous TbFeCo layers (Kudo *et
al* 1990). The films of ZnS(O) contain 7 to 8% oxygen.

For a ZnS(O, N)/$Tb_{24}Fe_{66}Co_{10}$ bilayer structure, an enhanced Kerr rotation
angle of $\Phi_K = -17.3°$ at a short wavelength of 373.5 nm was detected.
The peak wavelength of the enhanced Kerr rotation is situated between the
high-transmittance region and the fundamental absorption region in the optical
transmission spectrum of the ZnS(O, N) layer. It was found that the increase
of the N_2 flow rates of a mixture of sputtering Ar and N_2 gases moves the
threshold of the fundamental absorption in the ZnS(O, N) layer towards shorter
wavelengths.

As examples, Kerr rotation spectra of ZnS(O, N)/TbFeCo bilayer films with
the ZnS(O, N) layer 20.2 nm thick, for different N_2 flow rates, are shown in
figure 12.3 (Kudo *et al* 1990). It should be noted that the position, form, and
amplitude of the curve are sensitive to the thickness of the coating layer and the
flow rates of the N_2. Kudo *et al* assumed that the strong enhancement of Φ_K is
induced by a singular phenomenon in the optical constant spectra of the ZnS(O,
N) layer on the TbFeCo layer.

12.2 Trilayer and quadrilayer structures

NdDyTbFeCo/NdTbFeCo/NdDyTbFeCo. In order to increase the Kerr rotation
angle in rare-earth–transition metal recording media at short wavelengths,
Miyazawa *et al* (1993) proposed using a multilayered structure containing three
magnetic layers, NdDyTbFeCo/NdTbFeCo/NdDyTbFeCo, which are exchange
coupled. The central film has a large Kerr rotation angle at short wavelength.
The films on either side have large coercivities. The total thickness of the
magnetic layer was less than 30 nm. For the structure investigated, the layers
were deposited on a polycarbonate substrate, by magnetron sputtering, in the
following order: a dielectric layer, AlSiN (50 nm); the first magnetic layer,

$Nd_6Dy_{16}Tb_5Fe_{57}Co_{16}$ (8 nm); the second magnetic layer, $Nd_{23}Tb_7Fe_{33}Co_{37}$ (4 nm); the third magnetic layer, $Nd_6Dy_{16}Tb_5Fe_{57}Co_{16}$ (8 nm); another dielectric layer, AlSiN (20 nm); and the reflecting layer, AlTi (100 nm). The Kerr rotation angles of the three magnetic film layers appear to be 0.1° higher than those of single-layer recording media, with the composition of the first layer and similar magnetic properties resulting in a 2 dB higher signal-to-noise ratio. The reflectivity at 532 nm was 12% for both samples.

Al/Co–Cr/Al trilayers. Hirata and Naoe (1992) studied the magnetic and magnetooptical properties of a trilayer Al(0.7 nm)/Co–Cr(150 nm)/Al(0.7 nm) structure, and measured a Kerr rotation of about 0.41° at 400 nm.

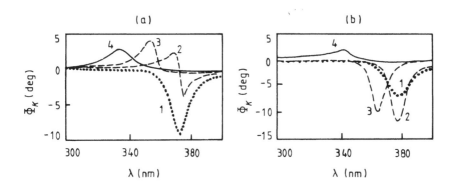

Figure 12.3. Kerr rotation spectra of ZnS(O, N)/TbFeCo bilayer films, with the ZnS(O, N) layer 20.2 nm thick (a), and with the ZnS(O, N) layer 21.6 nm thick (b), for different N_2 flow rates (0 (curve 1), 0.3 (curve 2), 3.0 (curve 3), and 20.0 cm^2 min^{-1} (curve 4) (Kudo *et al* 1990).

Co–Pt/SiO$_2$/Al trilayers. Atkinson *et al* (1994a) performed a calculation for the trilayer phase-optimized structure having zero Kerr ellipticity, and made an experimental investigation of the proposed structure. The Co–Pt multilayer consisted of a Pt(1.1 nm) layer and six Co(0.4 nm)/Pt(1.1 nm) layers, giving a total thickness of 10.1 nm. In the experiment, the maximum rotation attained was 0.9° at 520 nm, at zero ellipticity, and a reflectance of about 4% was attained, while the calculation predicts a 1.5° rotation for the optimal thickness, 9.6 nm.

CrO$_2$/Co multilayers on Cu(100). Chen *et al* (1995) have fabricated CrO_2/Co multilayers, each consisting of four layers of Co and one layer of CrO_2 on Cu(100), via a new technique utilizing the thermodynamic sink of CrO_2. The multilayers exhibit an easy magnetization axis in the plane, and the rotation angle increases rapidly with the film thickness. Four CrO_2/Co repeats are sufficient to yield a Kerr rotation angle of 0.09°. The coercivity decreases from 105 Oe for four clean layers of Co to 80 Oe for four clean CrO_2/Co repeats. They discussed the effect of carbon doping, excess Cr, and other modifications on the magnetic properties.

12.3 Compositionally modulated structures

12.3.1 Transition metal/transition metal multilayers

Daalderop *et al* (1992) predicted the existence of a perpendicular magnetic anisotropy in Co/Ni multilayer thin films; the prediction was confirmed for samples that were electron beam evaporated and sputtered. It was found that the perpendicular magnetic anisotropy is due to the positive interface anisotropy, and is strongly dependent on the choice of buffer layer. To obtain perpendicular magnetic anisotropy, use of a metallic underlayer was necessary, since direct deposition on glass substrates gave strong in-plane anisotropy.   Zhang *et al* (1994) investigated the magnetic and magnetooptical properties of Co/Ni multilayers deposited on Ag and Au buffer layers. The samples with thick Au buffer layers show perpendicular magnetic anisotropy, but those with Ag buffer layers do not. For Au buffer layers, the anisotropy constant changed dramatically with variation of the thickness of the buffer layer. The perpendicular magnetic anisotropy arises from the interface anisotropy, with $K_i = 0.23$ erg cm^{-2} for these sputtered samples.

Measurements of the polar Kerr rotation as a function of wavelength and layer thickness for the multilayers revealed that Φ_K for Co(0.2 nm)/Ni(0.8 nm) × 16 multilayers without any buffer layer changed from $-0.1°$ at 300 nm to 0.3° at 800 nm, crossing the zero level of rotation at 350 nm. The same multilayers on a Au buffer showed an increase of Φ_K from 0.1° at 300 nm to 0.2° at 360 nm, and to 0.23° at 530 nm, and then a slow decrease down to 0.1° at 800 nm. The magnetooptical figure of merit $((\Phi_K^2 + \Psi_K^2)R)^{1/2}$ of the films investigated is larger than 0.1° over the entire spectral region, and the peak gets as high as 0.22° at about 550 nm. Note that the figure of merit for Co/Ni multilayers on Au buffer layers is larger than that for heavy rare-earth–transition metal alloys at short wavelengths. The figures of merit for Co/Ni samples are larger than those for Co/Pt and Co/Pd for wavelengths in the range 500–600 nm, and are smaller than those for Co/Pt but larger than those for Co/Pd for wavelengths in the range 300–400 nm. The spectral dependence for multilayers on Ag buffers gives rise to a substantial increase in Φ_K, from 0.07° at 700 nm to a peak value of 0.44° at 350 nm, and a decrease to 0.03° at 800 nm.

12.3.2 Rare-earth/transition metal multilayers

Tb–Co/Tb–Co compositionally modulated films. Honda *et al* (1987) studied the magnetic and magnetooptical properties of compositionally modulated Tb–Co/Tb–Co films in which the terbium content was modulated between 14.4 and 20.1 at.%. The period of modulation varied from 0.8 to 300 nm. The short-period-modulation films, with layer thicknesses of 0.4 nm, showed the highest perpendicular anisotropy energy of 2.5×10^6 erg cm^{-3}, and the highest Kerr effect of 0.31° at 780 nm.

Tb/Fe–Co multilayers. Tb/Fe–Co–Ti multilayered films, 20–80 nm thick, were prepared, and their magnetooptical and recording characteristics were investigated by Toki *et al* (1990). The Kerr rotation at 830 nm varied between 0.37 and 0.48° for SiN-protected films, depending on the composition and modulation length.

Co/Ni and Tb/Fe multilayers on noble-metal layers. Hilfiker *et al* (1994) studied the magnetooptical properties of Co/Ni and Tb/Co multilayers deposited onto Au_xAg_{1-x} alloy buffer layers which were sputtered onto a glass substrate. It was found that the maxima in the Kerr rotation, appearing in the mid-visible region, shift towards the blue as the x-value is decreased from one toward zero.

12.3.3   Transition metal/non-magnetic metal multilayers

Ultra-thin Co/Pt and Co/Pd multilayered films (superlattices) stand out as suitable materials for use in high-density magnetooptical recording using a shorter-wavelength laser. The multilayers with thin Co layers, one to three atoms thick, exhibited a Kerr loop that was perfectly square, and an enhancement of the Kerr rotation, when the total film thickness become less than several hundred ångströms. These ultra-thin films have a high Kerr rotation—up to 0.37–0.45° at 400–800 nm (Hashimoto *et al* 1989), and even 0.6° at 280 nm (Sugimoto *et al* 1989).

These films also have a strong perpendicular magnetic anisotropy—up to 3×10^6 erg cm^{-3} even for very low film thicknesses. The figure of merit of the ultra-thin Co/Pt films is superior to that of the TbFeCo system even at 780 nm, and becomes even larger at shorter wavelength. Very importantly, the ultra-thin multilayers have very high corrosion resistance, and the artificially made layered structure is thermally stable up to 400 °C.

Co/Pt multilayers. Co/Pt multilayers with strong magnetooptical properties and perpendicular magnetic anisotropy were prepared by two-source dc magnetron sputtering from Co and Pt targets of 100 mm diameter (Hashimoto *et al* 1989), or by the rf-sputtering method with two targets (Sugimoto *et al* 1989) onto a glass substrate. It was found that perfect squareness of the Kerr rotation loop for Co/Pt films is attained only for the ultra-low-thickness region (Hashimoto *et al* 1989). Uba *et al* (1995) studied dc-sputtered Co/Pt multilayers, and concluded that the strong magnetooptical activity in the UV range is due to alloying effects at the sublayer interface.

Figure 12.4 (Hashimoto *et al* 1989) shows the wavelength dependence of the Kerr rotation angle Φ_K for Co(0.47 nm)/Pt(0.80 nm) multilayer films, with total film thicknesses of 15 nm and 10 nm, compared with those for the alloy films of the same composition. The Kerr rotations for $Tb_{22}Fe_{67}Co_{11}$ and $Tb_{24}Fe_{76}$ films are also given in figure 12.3 for comparison.

The film thickness of 15 nm is the optimized value for attaining a high magnetooptical Kerr rotation. It was found that when the total film thickness was decreased from 12 nm to 40 nm, there was no major change in the magnetooptical

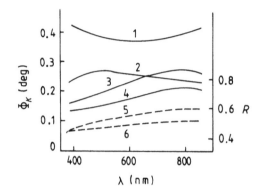

Figure 12.4. The Kerr rotations (solid lines) for Co(0.47 nm)/Pt(0.80 nm) (curve 1) and Co(0.42 nm)/Pd(0.85 nm) (curve 2) multilayer films, with total film thicknesses of 15 nm, and of the $Tb_{22}Fe_{67}Co_{11}$ (curve 3) and $Tb_{24}Fe_{76}$ (curve 4) alloy films, and also the reflectivity R (dashed lines) for TbFeCo (curve 5), and Co/Pt and Co/Pd (curve 6) films (Hashimoto *et al* 1989).

properties of the films; then when the film thickness was decreased from 40 nm to 15 nm, there was growth of the Kerr effect, to almost double the value at 40 nm. When the film thickness became less then 15 nm, there was a decrease in the Kerr rotation.

The thickness at which Φ_K reached its maximum value did not depend upon the periodicity, and Φ_K enhancement occurs in the transparent region of film thickness; the same result was observed for the alloy films. It is related to the optical and magnetooptical parameters of the prepared films, and can be explained in terms of the optical interference effect, with multiple reflection (Hashimoto *et al* 1989).

The Kerr rotation spectra of the thick (30–100 nm) multilayers were almost the same as those of the alloy films with the same compositions, and the Kerr rotation increased remarkably with shortening wavelength. The changes of the reflectivities R of ultra-thin (15 nm) Co/Pt films and TbFeCo films are shown in figure 12.4. The reflectivities of the opaque films that were 100 nm thick attained 0.6–0.7 for the wavelength range of 400–850 nm (Hashimoto *et al* 1989).

The film thickness at which a rectangular Kerr loop was attained depended upon the modulation periodicity, and decreased with increasing average Co content of the films. It was found also that the coercivity corresponded completely to the nucleation field. The aging properties of Co/Pt films at 85 °C and 75% relative humidity were examined: Co(0.48 nm)/Pt(1.08 nm) films with thickness 15 nm, without protective layers, were studied. The value of H_c and the remanent value of Φ_K were hardly changed after an aging test lasting 300 h. The film did not lose its perpendicular magnetic anisotropy during the aging test. The magnetooptical properties did not change at all for the coated films.

Brandle *et al* (1992) reported a maximum Kerr rotation of $-0.66°$ for the alloy $Co_{0.5}Pt_{0.5}$, compared with $-0.49°$ for the corresponding multilayers. They related the difference in rotation to the type of exchange polarization mechanism, which is 3D for alloys and 2D for multilayers, and a lower room temperature magnetization due to the reduced Curie temperature for the multilayers.

Višňovský *et al* (1994) studied the polar and longitudinal Kerr effects for Co/Pt multilayers and alloys in the spectral range 1.5–5.2 eV. For the alloys, the Kerr rotation reached a value of 0.66° at 4.3 eV for the composition $Co_{59}Pt_{41}$; for the multilayers, the measured values of the Kerr rotation and ellipticity attained 0.47° at 4.2 eV and 0.4° at 5.2 eV, respectively. The longitudinal Kerr rotation for the multilayers attained $-0.1°$ at 5 eV, and for the alloys it was double that.

Krishnan *et al* (1990a, b) have reported the growth of the Faraday rotation for Co/Pt multilayers with decreasing Co layer thickness. Typical optical and magnetooptical parameters of 77 × [Co(0.3 nm)/Pt(1.0 nm)] multilayer films measured at three wavelengths through the 1.2 mm thick glass substrate are given in table 12.1 (Lin and Do 1990).

Table 12.1. The optical and magnetooptical properties of Co/Pt multilayered film (Lin and Do 1990).

λ (nm):	488	633	820
R	0.54	0.60	0.64
Φ_K (deg)	0.29	0.22	0.20
$R^{1/2}\Phi_K$ (deg)	0.21	0.17	0.16

Angelakeris *et al* (1995) studied the Kerr rotation and ellipticity at room temperature, between 1.5 and 5.2 eV, for Co/Pt, Co/(CoPd), and Pt/(CoPt) multilayers grown on Kapton, Si, and glass in ultra-high vacuum. The magnetooptical properties were found to depend on the modulation layer thicknesses. The results were reproduced theoretically, in conjunction with the alloying modulations.

Co–Bi/Pt and Co/Pt–Bi multilayers. Suzuki *et al* (1994) investigated the magnetic and magnetooptical properties of $Co_{1-x}Bi_x$/Pt and $Co/Pt_{1-x}Bi_x$ multilayers, made by rf sputtering in Ar onto silicon (111) substrates. The bismuth content, x, changed from 0 to 0.58 and from 0 to 0.11 for the first and second compositions, respectively. The thicknesses of Co or Co–Bi and Pt or Pt–Bi were 0.4 and 1.7 nm, respectively. The total number of layers was 25, and Pt seeding layers about 100 nm thick were sputtered onto the substrates.

It was found that the magnetization, Curie temperature, and perpendicular magnetic anisotropy at 300 K decrease with the Bi content x in $Co_{1-x}Bi_x$/Pt multilayers, whereas the magnetization and Curie temperature increase but the anisotropy decreases with x in $Co/Pt_{1-x}Bi_x$. The maximum Kerr rotation increases with x up to $x = 0.12$ (from 0.28 to 0.31° at 4 eV), then decreases with

x for Co–Bi/Pt. On the other hand, the rotation decreases with x in Co/Pt–Bi. The photon energy for the maximum rotation decreases with x for Co/Pt–Bi, but it remains unchanged for Co–Bi/Pt.

Co/Pd multilayers. Figure 12.4 shows the wavelength dependences of the Kerr rotations Φ_K for Co(0.42 nm)/Pd(0.85 nm) films with total film thicknesses of 15 nm and 100 nm, compared with those for alloy films. The Kerr rotation for the ultra-thin Co/Pd films (15 nm) has a small peak at 500 nm, and decreases at shorter wavelength, in contrast with that for the ultra-thin Co/Pt film, and for thick alloy Co/Pd films. The change of the reflectivity R for Co/Pd film with wavelength is shown in figure 12.4 (Hashimoto *et al* 1989).

Fe/Pt multilayers. Fe/Pt multilayers having perpendicular magnetic anisotropy were prepared by the rf-sputtering method with two targets (Sugimoto *et al* 1989). It was found that compositionally modulated Fe/Pt films turn into perpendicularly magnetized films when the Fe layer becomes thinner than about 0.5 nm.

Fe–Co/Pt multilayers. Iwata *et al* (1992) studied the magnetic and magnetooptical properties of Fe–Co/Pt multilayers. The surface anisotropy, and the volume anisotropy, are found to depend both on the Pt layer thickness and on the Fe–Co layer thickness. The intrinsic perpendicular magnetic anisotropy was found to be nearly same as that for Co/Pt multilayers for Fe–Co layer thicknesses above 0.4 nm. However, the effective perpendicular anisotropy for Fe–Co/Pt was considerably smaller than that for Co/Pt films, because of the larger saturation magnetization of the Fe–Co/Pt films. The value of the Kerr rotation angle for Fe–Co/Pt was lower than that for Co/Pt films in the spectral range 250–800 nm, in spite of the larger saturation magnetization of Fe–Co/Pt.

Ni/Pt multilayers. Krishnan *et al* (1995) investigated the polar Kerr effect spectra for Ni/Pt multilayers at 80 K, and found a strong increase in the spectral amplitudes, which exceeds that predicted by the model based on electromagnetic theory. An enhancement of the Kerr effect is observed at photon energies above 3 eV, similar to that for Co/Pt and Fe/Pt.

Fe/Ag multilayers. The first spectroscopic studies of magnetooptical polar Kerr rotation of iron/noble-metal bilayer and multilayer systems were performed by Katayama *et al* (1986a, b). Katayama *et al* discovered anomalies in the Kerr spectra near the region of noble-metal plasma resonance. In order to correctly interpret the experimental results, the virtual-optical-constant method based on the theory of electromagnetic waves in layered structures was used. The inter-layer coupling between Fe–Ag and Co–Cu layers in Fe/Ag and Co/Cu multilayers was investigated by Jin *et al* (1994).

Fe/Ag multilayers offer the interesting possibility of keeping the interface as sharp as possible, because these two metals do not form an alloy. The results of measurements of the polar Kerr rotation and ellipticity for the multilayer (figure 12.5) revealed the existence of sharp peaks near 4 eV in the rotation and ellipticity spectra (Krishnan *et al* 1993). The number of Fe layers in this sample was 20. The first and last layers were of Ag.

The measurements were carried out in the photon energy range 1.5–3.3 eV at room temperature, in a magnetic field of 4.5 kOe. This is about 25% of the field required to saturate a thin plate of pure bulk iron in the normal direction. For this reason the real values of the Kerr rotation and ellipticity of the samples investigated in the state of magnetic saturation must be about 0.6°.

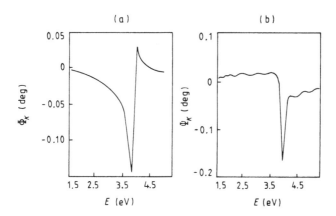

Figure 12.5. The polar Kerr rotation (a) and the ellipticity (b) for the multilayer Fe(0.6 nm)/Ag(6 nm) (Krishnan *et al* 1993).

Anomalous line shapes of both the Kerr rotation and the ellipticity in the region of the plasma resonance in silver, near 3.9 eV, are observed. The effect was particularly strong for the sample in which the thicknesses of the individual iron layers were of the order of three atomic monolayers. The calculated spectra for the complex Kerr effect show good agreement with the experimental results for thick samples, and less satisfactory agreement for the thinner samples.

The theory still correctly reproduced the shape of the observed spectra, but the amplitudes of the computed Kerr rotation and ellipticity were smaller by a factor of two than those obtained from the experiment if a correction factor of four for the saturation is used. One of the possible explanations for the differences between the experimental curves and the calculated ones is an increase in the specific Faraday rotation by a factor of two for Fe layer thicknesses close to 3 nm (Krishnan and Tessier 1990), and a reduction of the saturation magnetization field for a thickness of about three Fe monolayers.

The theory of electromagnetic interactions in magnetic multilayers (Višňovský 1986, 1991a, b) provided rather a good explanation when the factors discussed above were taken into account.

Fe/Au/Fe superlattices. Katayama *et al* (1994) revealed an intensity oscillation of the polar Kerr rotation related to the interlayer thickness d_{Au} for Fe(0.5 nm)/Au(d)/Fe(0.5 nm) (100) and Fe(0.6 nm)/Au(d)/Fe(0.6 nm) (100) sandwiched films. The periods of the oscillations are about 7–8 and 5–6 monolayers of Au, respectively. These phenomena were considered by

Katayama *et al* to be closely related to a formation of spin-polarized quantum-well states of the minority Δ_1 band in the Au layer sandwiched between two Fe barrier layers.

Fe/Cu multilayers. Katayama *et al* (1986a, b) reported on the anomalous polar Kerr effect for Fe/Cu, Co/Cu, Fe/Au, and Co/Au compositionally modulated films. Let us discuss the magnetooptical properties of the compositionally modulated Fe/Cu films investigated by Katayama *et al* (1986a, b, 1988). The composition modulation length, defined as the thickness of one period of Fe plus a Cu layer, was in the experiment varied from about 1 nm to more than 40 nm. It was found that the optimal Fe/Cu ratio is near 0.62. In the spectral range 250–800 nm, the values of the Kerr rotation are low (of the order of 0.2°), and the spectrum is relatively simple for a compositionally modulated Fe(0.5 nm)/Cu(0.9 nm) film. This spectrum is similar to that of the sputtered solid solution of Fe and Cu metals. However, a new peak of Φ_K begins to appear at about 560 nm, corresponding to the absorption edge of Cu metal, for Fe(1.5 nm)/Cu(2.5 nm) and Fe(3 nm)/Cu(5 nm). At the same time, the values of the rotation are lower than that for pure Fe—excluding the peak value of the rotation for the latter composition, which for the compositionally modulated film exceeded that for Fe by two minutes of arc.

Moreover, it is found that the peak position moves towards the longer-wavelength side with increasing Fe/Cu layer thickness ratio. Here, the absorption edge is near the wavelength below which the reflectivity decreases abruptly. It was also found that the Kerr rotation for compositionally modulated films in which the uppermost surface is of Fe is larger than that for compositionally modulated films in which the uppermost surface is of Cu. The value of Φ_K in the spectrum for the wavelength region from 560 to 800 nm has a tendency to increase with increasing L. A rotation value of 0.7° at 600 nm was attained for the compositionally modulated Fe(15 nm)/Cu(24.5 nm) films in which the upper surface was of Fe. It is interesting to note that for films in which the upper surface was of Cu, the rotation angle firstly increases with the growth of the composition modulation length up to 6 nm, and then decreases; while for films in which the upper surface was of Fe, the rotation angle continuously grows with increasing composition modulation length.

Fe/Zr multilayers. The magnetooptical and optical properties of Fe/Zr multilayer films with bilayer periods in the range 1.6–25 nm, and Fe/Zr thickness ratios of 1:1, 1:2, and 1:4, in the spectral range 1 to 4 eV, were studied by Kudryavtsev *et al* (1995). For Fe/Zr multilayers with a nominal thickness of the Fe layer of 5 nm, the Kerr spectra can be satisfactorily simulated using an effective-medium approximation, whereas a large discrepancy between the experiment and the model is observed for multilayers with a nominal thickness of the Fe layer <2.4 nm, which suggests a transformation to an amorphous structure.

Co/Au multilayers. Gambino and Ruf (1990) studied Co/Au multilayers prepared by electron beam evaporation. These multilayers have perpendicular magnetic anisotropy in the as-deposited condition, and the Faraday rotation of these square-loop multilayers is about 9×10^5 deg cm^{-1} for Co or 1×10^5 deg cm^{-1} for the total thickness, at a wavelength of 633 nm. These values indicate an enhancement of the Faraday rotation for Co at this wavelength by a factor of two. It was found that the rotation per unit thickness of Co increases with the thickness of the Au layer, i.e., with the increasing volume fraction of gold.

Faraday rotation, at a wavelength of 633 nm, of about 9×10^5 deg cm^{-1} for Co or 1×10^5 deg cm^{-1} for the total thickness of Co/Au multilayers, with a perpendicular hysteresis loop squareness ratio of 1, was observed by Gambino and Ruf (1990). These values indicate an enhancement of the Faraday rotation for Co (Φ_F(Co) $= 5.9 \times 10^5$ deg cm^{-1} at 633 nm) at this wavelength by about a factor of two.

It was found that the Faraday rotation per unit thickness of Co increased with the Au layer thickness, from 7×10^5 deg cm^{-1} for a layer thickness of Au of 0.5–1 nm up to 9–10×10^5 deg cm^{-1} for a layer thickness of Au of 2–3 nm, i.e. with increasing volume fraction of gold. Gambino and Ruf proposed that this may be related to a plasma edge enhancement of the magnetooptical effects. Some effect of the plasma edge of gold on the Faraday rotation was also observed in the magnetooptical spectra. The Faraday rotation as a function of photon energy shows an overall trend like that of elemental cobalt, with a decrease in rotation with photon energy; however, the decrease is less pronounced for the multilayered films. Also, for the layered films there is an increase in rotation, indicating a small peak in the vicinity of 2.5 eV where n and k for Au change abruptly. The magnitude of this upturn is sample dependent. The samples investigated were prepared by electron beam evaporation.

Atkinson *et al* (1994a) and Višňovský *et al* (1994) investigated Co-rich Co(0001)/Au(111) multilayers grown by molecular beam epitaxy on GaAs(111) substrates. The polar Kerr rotations and ellipticities were measured at photon energies between 1.5 and 5.2 eV. The maximum Kerr rotation attained was $-0.34°$ at 3.1 eV, for Co(1.5 nm)/Au(1 nm) multilayers.

Co/Cu multilayers. Co/Cu multilayers made by rf sputtering were studied by Xu *et al* (1995). It was found that the magnetization increases with decreasing thickness of the Cu layer; this was attributed to the joint effect of inter-layer coupling and 2D magnetism. Spin polarization of Cu related to inter-layer coupling was indicated by the experimental data. Various features of the experimental and calculated magnetooptical spectra further suggested that the Cu is spin polarized, and gives an additional contribution to the magnetooptical activity.

Co–Cr/Al multilayers. $Co_{81}Cr_{19}$/Al multilayer films were prepared on glass slide substrates at room temperature, by alternately sputtering pairs of $Co_{81}Cr_{19}$ and Al targets. The thickness of the $Co_{81}Cr_{19}$ layer was in the range 5–17 nm, while the thickness of the Al layer was 0.7 or 1.4 nm. The number of periods making up the total thickness of the Co–Cr/Al multilayers was in the range 9–30 (Hirata and Naoe 1992).

When the thickness of the aluminium layer was 0.7 nm, the perpendicular magnetic anisotropy was very large. On the other hand, when this thickness was 1.4 nm, in-plane magnetic anisotropy was observed. Films with thicknesses of the Co–Cr and Al layers of 13.8 and 0.7 nm, respectively, had a Kerr rotation of 0.21° and a reflectivity of 0.7. The Kerr rotation angle for the multilayer with the thickness of the Co–Cr layer equal to 15 nm was about 0.45° at 400 nm. The measurement of the Kerr rotation for the Co–Cr/Al multilayers was made on a sample whose top layer was of Al. The Kerr rotation and reflectance were measured at wavelengths of 780 and 830 nm. It should be noted here that, according Buschow *et al* (1983), the Kerr rotation at this wavelength for the alloy $Co_{80}Cr_{20}$ is −0.12°, and for the alloy $Co_{75}Cr_{25}$ it is −0.05°.

Tb–Fe–Co/Ta multilayers. Song and Naoe (1994) investigated multilayers of $Tb_{20}(Fe_{90}Co_{10})_{80}$/Ta deposited on 'plasma-free' substrates by using a facing-target sputtering apparatus. The enhancement of the Kerr rotation angle was increased with decreasing thickness of the paramagnetic Ta layer.

12.3.4   Co/U–As multilayers

Fumagalli *et al* (1993) studied the possibility of exchange coupling in sputtered multilayer films composed of Co and U–As layers. They investigated Co/U–As multilayers with a fixed U/As ratio of 1.5, with Co and U–As film thicknesses ranging from 0.2 to 8 nm. For $h_{Co} \geq 1$ nm, the films are ferromagnetic at 300 K. Multilayer films with thinner Co layers do not show magnetic ordering at room temperature, because of the inter-diffusion of U and As into the Co. Films with $h_{Co} = 2$ nm and with varying U–As layer thicknesses $h_{U-As} = 0.2$–8 nm provided magnetooptical evidence of an induced magnetic moment in U at room temperature, arising through exchange coupling to Co layers. The number of Co/U–As bilayers is varied from 14 to 170, giving a total film thickness of 140 nm. The topmost layer is always Co; this is in order to prevent corrosion of the underlying U–As layer.

For the composition Co(2 nm)/U–As(8 nm) at 10 K and 3 T, the polar Kerr rotation reached −1.1° at 0.7 eV, and in the spectral range 0.7–3 eV the rotation changed from −1.1° to 0.4°. The Kerr rotation and ellipticity spectra at room temperature for films with $h_{U-As} = 2$ nm are completely different to the corresponding Co spectra. The Co peak at 3.8 eV has disappeared, and the peak at 1.3 eV is shifted to lower energies. A comparison with the corresponding UAs spectra reveals some resemblance to the spectra of the binary UAs compounds at low temperature. It is concluded that the disappearance of the Co peak at 3.8 eV is due to the optical constant effect, and that the large room temperature rotation in the multilayered films with thick U–As layers is due to an exchange-induced moment in the U, reaching 20% of the low-temperature moment of binary U–As amorphous thin films.

12.3.5 Magnetic/dielectric multilayers

TbFe/SiO multilayers. Compositionally modulated films composed of thin layers of magnetic and dielectric material are rather interesting, because the magnetic and magnetooptical properties of a given material will be controlled by the compositionally modulated structure. In such a structure, the incident light is transmitted thorough several semitransparent magnetic layers, so multiple reflection of the light by several layers is expected. As the thickness of each layer is small compared to the wavelength of light, the refractive index averaged over a whole structure is considered as an effective refractive index of the film. One advantage of a compositionally modulated film is that the effective index of a given magnetic material can be tuned to an optimum value for a given choice of structure, with a single material and a simple structure of the coating layer. As a result, the Kerr rotation of compositionally modulated films can be enhanced for an arbitrary wavelength of light.

Sato *et al* (1988) studied magnetooptical properties of compositionally modulated TbFe/SiO films. The thicknesses of the TbFe layers are chosen to be below 10 nm; the number of layers was 10. The thicknesses of the SiO layers were varied from 5 to 20 nm. The thicknesses of the protective layers were in the range from 50 to 75 nm. The glass substrate was also coated with a SiO layer. The Kerr rotation was measured on the film surface side, in an applied field of 4 kOe normal to the film surface, over the wavelength range from 400 to 800 nm. It was found that the TbFe/SiO structure exhibits uniaxial magnetic anisotropy perpendicular to the film with a TbFe layer thickness of 5.2 nm. The experimental results have shown that a Kerr rotation value of 25° is obtained at 650 nm for the film with a protective layer of thickness 65 nm, TbFe layers of thickness 10 nm, and a SiO layer of thickness 15 nm. At the same time, the enhancement of the Kerr rotation by means of multiple reflection is inevitably accompanied by a reduction of the reflectivity, to below 0.1%. Sato *et al* compared the enhancements of the rotations in three films: (film 1) a SiO-coated TbFe film; (film 2) an uncoated, compositionally modulated TbFe/SiO film; and (film 3) a SiO-coated, compositionally modulated TbFe/SiO film. The total thickness of TbFe in the three films was fixed at 41 nm. The value of the rotation is multiplied by 2.5 in film 1, by 1.9 in film 2, and by 55 in film 3, as compared with the bulk value for TbFe. It was found that the value of the rotation increases with the total thickness of the magnetic layers, until the total thickness of the TbFe layers reaches about 50 nm. These results indicate that the penetration depth of light is about 50 nm for a TbFe alloy.

$(BiDy)_3(FeAl)_5O_{12}$/Fe, Co, Dy multilayers. Shen *et al* (1995) investigated $(BiDy)_3(FeAl)_5O_{12}$/Fe, Co, Dy multilayer thin films prepared by magnetron sputtering and subsequent processing by multipulse rapid thermal annealing. It was found that these multilayer garnet samples have very square hysteresis loops, large coercivities, and nanoscale-size grains. Large effective rotation angles were found in the blue wavelength region, which make these films attractive candidates for use in short-wavelength magnetooptical recording media.

12.3.6   Cd/Mn/Te superlattices

Koyanagi *et al* (1987) grew $Cd_{0.9}Mn_{0.1}Te–Cd_{0.5}Mn_{0.5}Te$ multilayers on a sapphire (0001) substrate by the ionized-cluster-beam deposition technique, and investigated their magnetooptical properties. In the Faraday rotation spectra of the multilayered structures, two band-edge dispersions, corresponding to the well and the barrier layer, were observed. As the well width decreases, the dispersion peak for the well layer shifts towards higher photon energies, and becomes higher, while the dispersion for the barrier layer remains unchanged. This can be explained by a quantum-size effect in the superlattice. Quantum-well structures, with their unique magnetic properties, not only have the potential for use in fabricating new optical devices, but also facilitate the investigation of the exchange interaction in two-dimensional structures.

The measurements of the Faraday rotation spectra were performed under a magnetic field of 5 kOe, at room temperature. Figure 12.6 shows the Faraday rotation spectrum for $Cd_{0.9}Mn_{0.1}Te–Cd_{0.5}Mn_{0.5}Te$ multilayer film, with $h_{bar} = 4$ nm and $h_{well} = 9$ nm. Two dispersions, corresponding to the well layer ($Cd_{0.9}Mn_{0.1}Te$) and the barrier layer ($Cd_{0.5}Mn_{0.5}Te$), are observed at 1.7 eV and 2.3 eV, respectively. Only the dispersion spectrum corresponding to the well layer was varied, according to the well width. For the multilayered structure, the lower-energy peak height is larger than that for the monolayered film, and increases monotonically with decreasing well width. In the experiment, a doubling of the Verdet constant was obtained. This result suggests that the oscillator strength of the excitons is enhanced, owing to the confinement of the electrons in the well layer. The inset in figure 12.6 shows the energy band diagram for the $Cd_{0.9}Mn_{0.1}Te–Cd_{0.5}Mn_{0.5}Te$ quantum-well structure.

12.4   Composite magnetooptical materials

12.4.1   Transition metal/dielectric granular thin films

Anodized aluminium films having electrodeposited micropores of ferromagnetic metals, such as Fe, Ni, and Co, have attracted interest recently, considered as magnetooptical films (Abe and Gomi 1987). Composite alumina films exhibit a prominent magnetic anisotropy perpendicular to the film plane, due to the shape anisotropy of the metal rods formed in the micropores.

The values obtained agreed with a theoretical calculation based on the 'effective dielectric tensor' (Abe and Gomi 1987). By electronically depositing particular metals it is possible, in principle, to adjust the magnetooptical effect in the composite fills to a desired value, as predicted on the basis of the 'effective dielectric tensor' (Abe and Gomi 1987). A composite alumina film containing

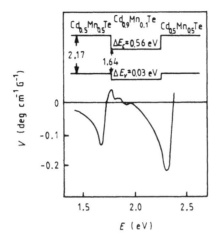

Figure 12.6. The Faraday rotation for a Cd–Mn–Te/Cd–Mn–Te multilayer, and the energy band diagram for the $Cd_{0.9}Mn_{0.1}Te$–$Cd_{0.5}Mn_{0.5}Te$ quantum-well structure (inset) (Koyanagi *et al* 1987).

deposited micropores of Fe was described by Tsuya *et al* (1986).

Co–Al$_2$O$_3$ granular thin films.   Composite alumina films with Co deposited electronically in the micropores exhibited a fairly large Faraday rotation and ellipticity of 0.41° and 1.14 deg μm^{-1}, respectively, at 633 nm (Abe and Gomi 1987). The micropore structures are much smaller in size than the optical wavelength (e.g. $D_c = 30$ nm and $D_p = 12$ nm). Such composite films behave like continuous magnetooptical media.

Ni–SiO$_2$ granular thin films.   Another example of a composite structure is the case of granular ferromagnetic particles in a SiO_2 insulator matrix. Thin films of dispersed systems have been prepared by cosputtering a metal and an insulator, under controlled conditions. The films of nickel particles, with diameters in the range 1–10 nm, in a SiO_2 matrix, have been the subject of an extended study, which included the measurement of both magnetic and magnetooptical properties (Abe and Gomi 1987). Effective optical and magnetooptical parameters for granular cobalt films were calculated for different volume fractions of spherical cobalt particles, with radii in the range 1.5–6 nm (Abe and Gomi 1987). Films of dispersed particles, each with a radius of 4.5 nm, are shown to produce polar Kerr rotations approximately double those for continuous films of the same thickness.

Fe–Al$_2$O$_3$ granular thin films. Polar Kerr rotation measurements on Fe–Al$_2$O$_3$ granular thin films, composed of iron particles, with different average diameters ($50 < \emptyset < 10$ nm), dispersed in an amorphous Al_2O_3 matrix, were made by Doormann *et al* (1991). The measurements were performed at 4 and 300 K, over the wavelength interval 0.5–2.5 μm. Values of the Kerr rotation as high as 2.8° ($T = 4$ K) and 1.9° ($T = 300$ K) in the visible region ($\lambda = 500$ nm), and 0.8° (4 K) and 0.4° (300 K) in the infra-red region, are obtained when a magnetic field of 4 kOe is applied perpendicular to the surface of the sample.

Fe–SiO$_2$ granular thin films. Polar Kerr rotation spectra of Fe–SiO$_2$ granular films were measured at room temperature, over the wavelength region between 400 and 850 nm, by Jiang *et al* (1995a, b). A maximum Kerr rotation angle of 0.3° was obtained between 700–850 nm when a magnetic field of 10 kOe was applied perpendicular to the film plane. It was found that Φ_K increases with increase of the Fe volume fraction (f) for small contents of Fe, and peaks at $f = 0.4$, and then the rotation decreases as the Fe particle size increases. For samples with $f = 0.4$, the polar Kerr rotation increases from 0.2° at 400 nm up to 0.3° at 700–850 nm. In these films a large Faraday rotation (up to 10^5 deg cm^{-1}) was also observed (Jiang *et al* 1995a). The Faraday rotation increases with the Fe volume fraction, and peaks for f approximately equal to 0.55.

The ferrite particles–glass matrix granular composition. Petrovskii *et al* (1994) studied a composite system based on oxide glasses (matrix: K$_2$O–Al$_2$O$_3$–B$_2$O$_3$) containing magnetically ordered ferrite microparticles, 5–20 nm in size. The ferrite microparticles were formed from paramagnetic oxides such as Fe$_2$O$_3$, MnO, Co$_2$O$_3$, and Gd$_2$O$_3$, during the preparation of the glasses. The glasses have a high magnetic susceptibility and magnetic anisotropy for very low concentrations of the paramagnetic element oxides when they remain transparent in the visible and near-IR regions. The Faraday effect for these glasses is an order of magnitude stronger than that for paramagnetic glasses containing the same elements in the same concentrations. As for magnetically ordered materials, the field dependence of the Faraday effect is non-linear, and exhibits both hysteresis and magnetic saturation.

12.4.2 Macroscopic ferrimagnetics—EuS precipitates in a cobalt matrix

A new class of two-phased materials, called macroscopic ferrimagnetics, which displays many properties associated with ferrimagnetism, including a magnetic compensation point, was studied by Gambino and co-workers (Gambino *et al* 1994, Gambino and Fumagalli 1994). A typical material system of this type, which shows this phenomenon, is cobalt containing ($\sim$10 nm) precipitates of EuS.

The antiferromagnetic exchange in such systems is relatively weak, so the remanent magnetooptical properties indicate a ferrimagnetic state; but at fields above 2 to 3 T, alignment of both the Co and the EuS with the field can occur. Because of the antiferromagnetic exchange between the Co and the EuS, the net magnetization of the composite is low in the vicinity of the compensation point. Thin films of Co–EuS that have been electron beam evaporated have a perpendicular easy-axis growth-induced anisotropy. The high anisotropy field produces square hysteresis loops and high coercivity, as for conventional ferrimagnets. A film with 70 mol% Co and 30 mol% EuS has a compensation point above 16 K, showing that the Eu^{2+} ion has a large fraction of its moment at this temperature.

The diffraction peaks observed for the films of 10–60 mol% EuS in Co could be indexed like EuS: cubic with a lattice constant of 5.97 Å, and the Co matrix is probably amorphous. The diffraction broadening indicates that the EuS particles are about 100 Å in diameter. The strong Co/EuS exchange increases the ordering temperature of EuS to at least 60 K. Gambino and Fumagalli estimate the interfacial exchange energy to be about 10 erg cm^{-2}.

The composition $Co_{70}(EuS)_{30}$ is dominated by EuS up to the compensation point, at about 19 K, and then it is dominated by Co. In contrast, the $Co_{80}(EuS)_{20}$ composition is dominated by Co at all temperatures up to the Curie point. It was found that the wavelength dependence of a $Co_{90}(EuS)_{10}$ sample is very similar to the published Kerr spectra for cobalt. The remanent-state spectra for $Co_{80}(EuS)_{20}$ appear to be very similar to those for $Co_{70}(EuS)_{30}$ samples, but with opposite sign; the remanent-state spectrum for a $Co_{70}(EuS)_{30}$ sample bears a strong resemblance to the Kerr spectrum for EuSe, with negative peaks at about 2.2 and 4.2 eV. For the composites investigated, the peak rotation values near 3 eV were about 0.6–0.7° for the remanent states, and for $T = 5.5$ K.

PART 3

APPLIED MAGNETOOPTICS

Chapter 13

Thin-film magnetooptical devices

13.1 Magnetooptical modulators

Magnetooptical modulators are devices that control the intensity of optical radiation in optical communication and data-processing systems via the applied magnetic field. A schematic diagram of a magnetooptical modulator is presented in figure 13.1. The magnetooptical element, which is enclosed inside the solenoid coil that creates the modulating magnetic field, is placed between the polarizer and the analyser. When no field is applied, the magnetooptical element does not affect the light that is being transmitted, and its intensity after passing through the polarizer–magnetooptical element–analyser system is described by the Malus law:

$$I = I_0 \cos^2 \beta \tag{13.1}$$

where β is the angle between the transmission axes of the polarizer and the analyser, and I_0 is the intensity of the incident radiation.

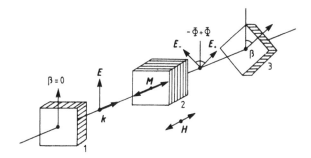

Figure 13.1. A schematic diagram of a magnetooptical modulator. Components: 1: the polarizer; 2: the magnetooptical (Faraday) cell; 3: the analyser.

Real systems have losses that are caused by reflection from the elements, and by insertion losses in the elements. In thin-film dichroic polarizers, the total losses on reflection and absorption are 5–10%; in a magnetooptical element made of iron garnet, losses in the visible spectral region can reach several dozen per cent.

After transmission through a sample having parallel surfaces, the light intensity is described by the expression

$$I = I_0(1 - R)^2 \exp(-\alpha h)[1 - R^2 \exp(-2\alpha h)]^{-1}$$

where R is the reflection coefficient, α is the absorption coefficient of the material, and h is the sample thickness. The expression in the square brackets takes into account multiple reflections.

If between the polarizer and the analyser there is a magnetooptical element that rotates the plane of polarization of the radiation by the angle Φ, instead of formula (13.1) we should write

$$I = I_0 \cos^2(\beta \pm \Phi). \tag{13.2}$$

As real polarizers do not fully suppress the light in the crossing position, formula (13.2) should contain a coefficient p which takes into account the finite transmission in the polarizer–magnetooptical element–polarizer system. Then

$$I = cI_0 \exp(-\alpha h)[(1 - p) \cos^2(\beta \pm \Phi) + p].$$

Here the coefficient c takes into account radiation losses in the system.

This device can be used in two modes—as a modulator and as an optical switch (Scott and Lacklison 1976). In the first case the angle $\beta = 45°$, and in the second mode $\beta - \Phi = 90°$. When the current flows through the modulation coil, the magnetooptical element magnetizes, and the plane of polarization of the light rotates due to the Faraday effect, by the angle

$$\Phi = \Phi_F(M/M_s)h \cos \gamma$$

where Φ_F is the specific Faraday rotation of the ferromagnet, M and M_s are the magnetic moment per unit volume of the sample and the saturation magnetization, respectively, and γ is the angle between the magnetization and the light propagation directions.

Analysis of the expression (13.2) shows that the derivative $dI/d\beta \sim I_0 \sin 2\beta$ reaches its maximum at $\beta = \pi/4$. In this case the intensity of the light that has passed through the modulator changes by

$$\Delta I = cI_0 \exp(-\alpha h) \sin 2\Phi. \tag{13.3}$$

If the device operates as an optical switch, the β-angle is chosen to be equal to $-\Phi_F h$, i.e., when the magnetooptical element is magnetized in one direction, the intensity of the light after passing through the element is zero; when the magnetization direction changes, the intensity of the light is

$$\Delta I = cI_0 \exp(-\alpha h) \sin^2 2\Phi.$$

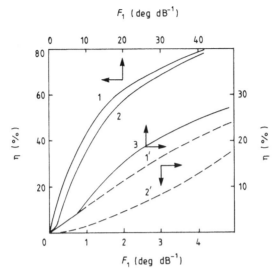

Figure 13.2. The optical efficiencies of the magnetooptical modulator (1, 1′), the switch (2, 2′), and the deflector (3) (Scott and Lacklison 1976).

The main characteristic of a magnetooptical device is its optical efficiency η, which is defined as the ratio of the signal intensity change when the cell switches to the intensity of the incident radiation I_0. The efficiency of a magnetooptical device depends on the Faraday rotation and on the absorption coefficient of the material. From (13.3) we can determine an optimal thickness h_{opt} that provides the maximum efficiency of the magnetooptical modulator:

$$h_{opt} = (2\Phi_F)^{-1} \tan^{-1}(2\Phi_F/\alpha).$$

For the magnetooptical switch

$$h_{opt} = (2\Phi_F)^{-1} \tan^{-1}(4\Phi_F/\alpha).$$

The corresponding expressions for the efficiencies of the magnetooptical modulator and the switch take the following forms:

$$\Delta I/I_0 = \exp[-(F_1)^{-1} \tan^{-1} F_1] \sin(\tan^{-1} F_1)$$
$$\Delta I/I_0 = \exp[-(F_1)^{-1} \tan^{-1} 2F_1] \sin^2(\tan^{-1} 2F_1)$$

where $F_1 = 2\Phi_F/\alpha$ is the magnetooptical figure of merit of the material. Figure 13.2 shows the calculated dependencies of the optical efficiencies of the magnetooptical modulator, the switch, and the deflector on the figure of merit of the material (Scott and Lacklison 1976).

Optical radiation can be modulated by means of transparent ferromagnetics using the following physical processes: the shift of the domain wall, the rotation of the magnetization vector of a saturated sample in an external magnetic field, and the precession of the magnetization of a sample that is inserted in a microwave resonator. The main advantages of magnetooptical modulators over mechanical choppers lie in the long-term reliability, small size and low power consumption. Magnetooptical modulators, like any other modulators, are characterized by the modulation coefficient, the linearity of the transfer function, the quality factor, the power efficiency, and the frequency band of the modulation.

When optical radiation is modulated by the shift of the domain wall, the frequency dependence of the extinction ratio is defined by the equation of motion for the domain wall:

$$m x + \beta \dot{x} + k x = 2 M_s H \cos \omega t$$

where $m = (2 \pi \gamma^2 \Delta)^{-1}$ is the mass of the domain wall, β is the loss coefficient, which characterizes the mobility of the domain wall, $k x$ is a restoring force, where

$$k = -P_0^{-1} 32 \pi M_s^2$$

and P_0 is the equilibrium period of the domain structure. The amplitude of the domain wall displacement is defined by the expression

$$x_0 = (16 \pi M_s)^{-1} \{ [1 - \omega_r^{-2} \omega^2]^2 + \omega_c^{-2} \omega^2 \}^{-1/2} P_0^2 H \tag{13.4}$$

where $\omega_r = (k/m)^{1/2}$ is the resonant frequency of the domain walls, and $\omega_r = k/\beta$ is the relaxation frequency.

In the more common case, where the magnetic field is applied to a sample containing stripe domain structure, as has been shown by Red'ko (1989) the coefficient k determining the restoring force can be written in the form

$$k = -32 \pi M^2 / (w_+ + w_-) - (32 M^2 / h) \sum_{n=1}^{\infty} (1/i) \cos(2 \pi i w_+ / (w_+ + w_-))$$
$$\times [1 - \exp(-2 \pi i h / (w_+ + w_-))]$$

where w_+ and w_- are the widths of the stripe domains oriented along the magnetic field direction and in the opposite direction, respectively.

The modulation percentage M is related to x_0 as follows:

$$M = P_0^{-1} 4 x_0 \times 100 \ (\%).$$

The epitaxial structures of iron garnets and orthoferrite plates, which are characterized by substantial uniaxial anisotropy with $H_a > 4 \pi M_s$, are often used in modulators of this kind.

There is a widely adopted view that magnetooptical modulators that employ the domain wall shift can operate only at low frequencies. In fact, as follows from (13.4), the frequency limit is determined by the m-, β-, and P_0-coefficients.

The lower the domain structure period, the smaller the shift of the domain structure needed for reaching the desired modulation depth. Increase of the displacement speed of the domain wall for a given applied field also leads to higher frequency limits. Early magnetooptical modulators of this type employed materials with domain wall displacement speeds less than 10 m s^{-1}. Obviously, for $v \approx 10 \text{ m s}^{-1}$ and $P_0 = 10 \text{ } \mu\text{m}$, the modulation frequency was restricted to several megahertz.

Modern successes in magnetooptical material technology have changed the situation. Epitaxial iron garnet films with orthorhombic anisotropy, which have been developed recently, have speeds of domain wall displacement up to $v = 1400 \text{ m s}^{-1}$ (Breed *et al* 1978/79). Therefore, for $P_0 = 10 \text{ } \mu\text{m}$, it is possible to provide modulation depths of several dozen per cent at frequencies of hundreds of megahertz and applied fields of dozens of oersteds. Thus we can conclude that the response of magnetooptical modulators based on the domain wall displacement is restricted by the technological difficulty of achieving a magnetic field strength of several dozen oersteds at a frequency of several hundred megahertz.

In addition to traditional films with stripe and maze-like domain structures, a special domain structure—the one that appears when the sample is placed in a magnetic field gradient—can also be used for magnetooptical modulators. An alternating magnetic field, applied perpendicular to the sample surface, produces a displacement of the domain wall, and therefore provides modulation of the light that passes through the element.

The modulation frequency can be substantially increased if the optical radiation is modulated due to the rotation of the magnetization vector of a saturated sample (LeCraw 1966). In this design, the magnetooptical element is magnetized up to the saturation point in a constant magnetic field; the magnetization direction is perpendicular to or makes some angle with the direction of propagation of the light. When a high-frequency signal is applied to the modulation coil, an alternating component of the projection of the magnetic moment on the direction of propagation of the light appears in the element. In practice, the modulation frequency of this modulator is limited by our ability to create a high-frequency magnetic field with a strength of several dozen oersteds.

A low-frequency magnetooptical modulator, for the infra-red spectral band, based on a cylindrical yttrium iron garnet crystal magnetooptical element 2 mm long, with a working aperture of 4 mm, was described by Scott and Lacklison (1976). The modulator, in which a coreless coil was used, provided 20% modulation in the wavelength interval from 3 to 6 μm when a magnetic field of strength 2 kOe was applied. The modulation frequency was limited to 60 kHz, because the bias field was created by a coil with high inductance. The applied field can be reduced if we use an yttrium iron garnet crystal in which gallium ions are substituted for some of the iron ions; in these crystals the saturation magnetization can be reduced to $4\pi M_s \sim 100–300 \text{ G}$.

13.2 Bistable optical switches

Faraday rotation can be used to realize optical bistability. A 1×2 optical switch was constructed by combining an $Y_3Fe_5O_{12}$ Faraday rotator with phase-matching films, a wire-wound magnet of semihard magnetic material, a polarizer, a polarization splitter, and semicylindrical lenses, as shown in figure 13.3(a) (Shirasaki *et al* 1981).

A Glan–Thompson prism was used as the polarizer, and a Wallaston prism as the polarization splitter. The light of a laser diode at 1.3 μm wavelength was introduced to the device by focusing the light beam onto the polished end of an $Y_3Fe_5O_{12}$ plate using a semicylindrical lens.

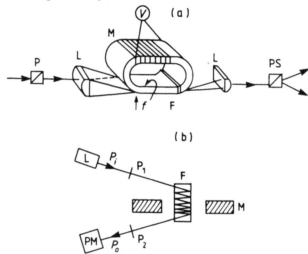

Figure 13.3. A schematic diagram of bistable optical switches based on an $Y_3Fe_5O_{12}$ thin-plate Faraday rotator with coatings for phase matching (a) (Shirasaki *et al* 1981), and a plate of FR-5 glass (b) (Umegaki *et al* 1981).

The phase-matching films used were three layers of SiO_2–Si:H–SiO_2, which were deposited on both surfaces of the $Y_3Fe_5O_{12}$ crystal plate. The phase-matching films are needed because when the light rays travelling through the thin crystal plate of $Y_3Fe_5O_{12}$ suffer total internal reflection at the crystal's upper and lower surfaces, the phase changes resulting from total internal reflection are different for TE and TM modes, and as a result the phase velocities of these modes are different. The switching was performed by applying an electric pulse to the coil; the switching time was 20 μs, and the extinction ratio of the switch was 24 dB.

The switching operation is as follows. When linearly polarized light travels through a Faraday rotator along the axis of magnetization, the plane of polarization rotates by 45°. The direction of rotation depends on the direction of rotation; if the direction of magnetization is reversed by inverting the direction of the applied magnetic field, the direction of rotation is reversed too. The optical paths of the two orthogonal polarizations are separated by the polarization splitter.

Another bistable set-up is shown schematically in figure 13.3(b) (Umegaki *et al* 1981). A polarizer, P_1, a Faraday cell, F, and an analyser, P_2 constitute the magnetooptical modulator. The Faraday cell is placed in a small electromagnet, M. The cell consists of a plane-parallel plate of FR-5 glass, of surface area 15×15 mm, and thickness 3 mm; its Verdet constant V is -0.24 min G^{-1} cm^{-1} at 633 nm. The two surfaces of the plate are coated to produce 100% reflection for the wavelength 633 nm; its entrances and exits are antireflection coated.

As expected, the transmittance $T(I, \beta)$, where I is the coil current and β is the relative orientation angle of the polarizer and analyser, changes non-periodically with respect to the coil current I. Owing to this feature, this modulator will find an interesting use of its optical bistability in stabilizing cw laser intensity. It should be noted that very high stability can be obtained with a small current in the magnetic coil.

13.3 Non-reciprocal devices

The development of optical fibres with extremely low losses ~ 0.2–2 dB km^{-1} stimulated basic studies and applied investigations in the field of optical communication lines and devices designed for operation in the near-infra-red spectral band. One of the most important of the problems that arise in the communication systems is the suppression of the reflected signals which appear in fibre-optic connectors and other contact elements of functional optical devices, and which badly affect the optical performance of the laser sources.

In a practical optical fibre communication system, the feedback light of the laser diode reflected at several points in the transmission path causes noise generation, and degradation of the transmission characteristics. Of particular interest are non-reciprocal magnetooptical elements (optical isolators), which transmit the signal propagating in an optical communication channel in one direction, and block the backward-propagating signal. Optical isolators are useful for stabilizing the oscillation of laser diodes.

A non-reciprocal optical device must satisfy the following requirements: low insertion losses; a high ratio of the transmission in one direction to the attenuation in the opposite direction; compactness; low power consumption (zero consumption preferable); and operation over the given temperature interval.

Early non-reciprocal devices employed yttrium iron garnet single-crystal plates. The drawbacks of these devices are the high cost of the single crystals, and the necessity of using a cobalt–samarium magnet to create the required bias field of more than 1.5 kOe. More promising are devices that utilize epitaxial iron garnet films. Three types of non-reciprocal device can be created on the basis of such films: with transverse, longitudinal, and waveguide configurations (figure 13.4).

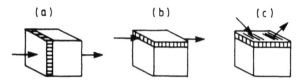

Figure 13.4. Transverse $L \gg \lambda$ (a), longitudinal $h \gg \lambda$ (b), and waveguide (c) geometries for non-reciprocal elements.

Magnetooptical transverse elements can process wide light beams, but demand high saturation fields; furthermore, the growth of epitaxial films that are 300–900 μm thick is a rather difficult technological task. Longitudinal geometry utilizes lower saturation fields, but the film thickness must be more than 100 μm to provide an effective input of the laser radiation through the film edge.

In the waveguide configuration, epitaxial films several micrometres thick are used. Devices of this kind require special radiation couplers, and, to provide non-reciprocal light propagation, a periodic variation of the material parameters of the film is necessary.

Figure 13.5(a) shows a schematic diagram of a non-reciprocal magneto-optical device that comprises two polarizers, which have their transmission axes positioned at 45°, and the Faraday magnetooptical element, which is placed between them, the element thickness being chosen such that the rotation of the plane of polarization is 45°. When light propagates from the left along the x-axis, the beam passes the polarizer (1), which has its transmission axis directed along, say, the z-axis, then through the Faraday element (2), where its plane of polarization rotates by +45°, and then through the polarizer (3), whose transmission axis coincides with direction of polarization of the light that has passed through the Faraday element. The light propagating in the opposite direction passes through the polarizer (3), and in the Faraday element its plane of polarization rotates a further +45°; therefore, the light polarization turns out to be perpendicular to the transmission axis of the polarizer (1). Thus the backward-propagating light is completely suppressed by the polarizer.

Figure 13.6 shows a cross-section of a non-reciprocal magnetooptical device, in which the Faraday element (1) has a disc of yttrium iron garnet, 1.8 mm thick and 8.6 mm in diameter, oriented in the [111] direction, i.e., along the light magnetization axis of the sample. The polished faces of the disc were antireflection coated, so that the losses on reflection were less than 0.1 dB, and the absorption losses in the magnetooptical material at the 1.153 μm wavelength were less than 0.9 dB. The magnetic system, which consisted of a permanent cobalt–samarium magnet (2), and a magnetic circuit (3) of soft iron, created the magnetization field (saturation of the disc required a magnetic field strength of 1.2 kOe). The Glan–Thompson prisms, with a 2 mm aperture, which were made of antireflection-coated calcite, worked as the polarizer (4). The whole

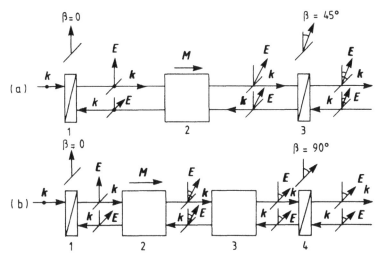

Figure 13.5. A schematic diagram of non-reciprocal devices based on a Faraday element (a), and on Faraday and optically active elements (b) (Iwamura *et al* 1978).

set-up was encased in an aluminium housing (5), 25 mm long and 5 mm in diameter. The magnetooptical element, together with the magnetization system and the prisms, were fixed by screws. The total losses of the entire set-up at the wavelength 1.153 μm were 1.3 dB in one direction, and 30 dB in the opposite direction (Iwamura *et al* 1978).

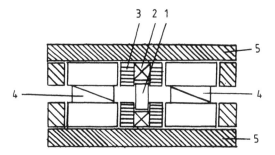

Figure 13.6. A cross-section of a non-reciprocal magnetooptical set-up. Key: 1: the Faraday element; 2: the permanent magnet; 3: the magnetic circuit; 4: polarization prisms; and 5: the housing (Iwamura *et al* 1978).

As the Faraday rotation of the yttrium iron garnet depends on the wavelength, the optical losses introduced by the non-reciprocal elements vary substantially through the spectrum. We should note that the Faraday rotation of yttrium iron garnet in the near-infra-red spectral region depends on the method of formation of the single crystal, and on the presence of dopants; hence, at the wavelength 1.153 μm, the specific Faraday rotation Φ_F can vary from 240 to

270 deg cm^{-1}, and at 1.3 μm, it can vary from 190 to 220 deg cm^{-1} (Nakano *et al* 1984). Dispersion characteristics of the above-described non-reciprocal element are shown in figure 13.7. Curve 1a characterizes losses in one direction, and curve 2a is for attenuation in the opposite direction.

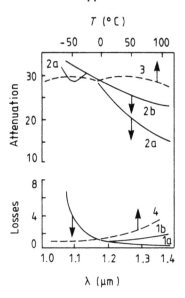

Figure 13.7. Spectral and temperature dependencies of the losses and attenuation of non-reciprocal devices (Iwamura *et al* 1978).

Non-reciprocal optical elements must provide effective discrimination of optical signals over a given temperature interval. Figure 13.8 shows the temperature dependence of the Faraday effect for yttrium iron garnet at the wavelength 1.153 μm. When the temperature changes over a wide range, the attenuation that is created by the non-reciprocal element for the signal propagating in the opposite direction decreases substantially, because of the variation of the Faraday rotation angle with temperature.

In order to create non-reciprocal devices that could work over a wide temperature range, Iwamura *et al* (1978) proposed another design for the non-reciprocal element; a schematic diagram is shown in figure 13.5(b). To expand the wavelength and the temperature interval, it utilizes a combination of the magnetooptical element (2), which is made of yttrium iron garnet single crystal, and an optically active element (3), which is made of crystalline quartz. Each element provides rotation of the plane of polarization of the light at the working wavelength by +45° when the light propagates in one direction, so the total rotation of the plane of polarization after the light had passed through both elements was 90°.

When the axes of the polarizers 1 and 4 are fixed at the angle 90°, the system transmits the light, without obstacle, in one direction. When the light propagates

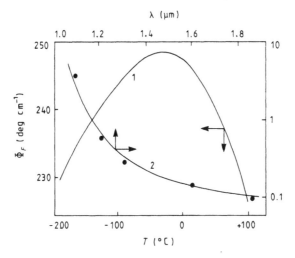

Figure 13.8. The temperature dependence of the specific Faraday rotation Φ_F, for $\lambda = 1.153$ μm (curve 1), and the spectral characteristic of the absorption coefficient α (curve 2) of an $Y_3Fe_5O_{12}$ single crystal (Nakano *et al* 1984).

in the opposite direction, the optically active element and the magnetooptical elements rotate the plane of polarization in opposite directions; therefore, the overall effect is that there is no rotation, and the light is blocked by polarizer 1.

The dispersion characteristics of the composite element are presented in figure 13.7. Curve 1b corresponds to the losses in one direction, and curve 2b to attenuation in the opposite direction. The total dispersive dependence of the whole device is determined by the dispersion of the Faraday rotation of the yttrium iron garnet, and by the optical activity of the *c*-cut quartz. The composite non-reciprocal element exhibits substantially better temperature characteristics. The dependencies of the losses in one direction, and the attenuation in the opposite direction are shown in figure 13.7 (curves 3 and 4).

A non-reciprocal magnetooptical device, which was constructed according to the schematic diagram shown in figure 13.5(b), and designed for an operating wavelength of 1.3 μm, provided attenuation of more than 32 dB in the temperature interval -20 to $+40$ °C, with losses in the forward direction of less than 1.8 dB, the loss at room temperature being 1.4 dB (Iwamura *et al* 1984). The device employed a magnetooptical element which was 2.1 mm thick and 3 mm in diameter. The dependence of the absorption coefficient of the yttrium iron garnet single crystal utilized on the wavelength is shown in figure 13.8 (curve 2) (Nakano *et al* 1984).

In recent years, bismuth-substituted iron garnet single crystals and epitaxial films have also been used in non-reciprocal devices. The characteristics of the non-reciprocal elements that employed $Bi_{1.2}Gd_{1.8}Fe_5O_{12}$ single crystal are summarized in table 13.1 (Tamaki and Tsushima 1987).

At wavelengths of 780, 1153, and 1300 nm, Rochon prisms were used, and at the wavelength of 1550 nm, a polarizing beam splitter was used. The above-described insertion losses, and the discrimination which can be attained using non-reciprocal magnetooptical elements, characterize the quality of the elements, and are measured in laboratory conditions. Connections with real optical fibres usually result in additional optical losses; a reduction of the attenuation is also observed. Below, we consider several different variants for connections of the non-reciprocal elements with fibres.

Table 13.1. Characteristics of non-reciprocal elements (Tamaki and Tsushima 1987).

Wavelength (μm):	0.78	1.15	1.3	1.55
Insertion losses (dB)	1.5	1	0.35	0.5
Attenuation (dB)	32	36	34	32
Diameter of beam (mm)	1.5	2	1.5	2
Dimensions (mm)	$7 \times 7 \times 6$	$7 \times 7 \times 6$	$7 \times 7 \times 8$	10×8
Weight (g)	1.6	1.6	1.7	1.9
Losses (dB)	1	0.25	0.025	0.04
Thickness (μm)	41	155	191	377
Saturation field (kOe)	1	1	1	1.1

Suzuki *et al* (1996) described an optical isolator intended for 1.3 μm semiconductor lasers. $La_{0.08}Sm_{2.15}Bi_{0.77}Fe_5O_{12}$ film was used in this non-reciprocal element; it has a high Faraday rotation stability against wavelength variation at around 1.3 μm. This stability results from the peaks in the Faraday rotation at 1.24 μm and 1.38 μm due to the Sm^{3+} ions. Suzuki *et al* calculated that isolators produced using this film could provide a 60 nm bandwidth (1.29–1.35 μm) in which isolation would be maintained at 35 dB or higher. This bandwidth is approximately three times wider than the bandwidth that can be obtained using $(TbBi)_3(FeAl)_5O_{12}$ film.

Single-mode fibres can be matched with non-reciprocal elements via graded-index lenses (GRIN-rod lenses) placed at the end faces of the input and output optical fibres. The lenses are usually glued to the fibres' ends with a glue that polymerizes under ultra-violet radiation. Antireflection-coated Glan–Thompson prisms (the reflection losses do not exceed 0.25% at the wavelength $\lambda = 1.3$ μm) are used as the input and output polarizers. The whole set-up usually has losses of 3.5 dB in one direction, and attenuation of 26 dB in the opposite direction at the operating wavelength $\lambda = 1.3$ μm (Green and Georgiou 1986).

Further development of the design of matching elements for the fibres and the non-reciprocal elements (the fibre-embedded isolator) circumvented the problem of the lenses having to couple light in and out of the fibres. The fibres were matched with the element via a special fibre that expands the fibre mode (a beam-expanding fibre). The characteristics were further improved when a new type of polarizer (Lamipol) was used instead of standard prism polarizers. The polarizers were multilayered thin-film structures of metal–dielectric type. As a result, at the wavelength $\lambda = 1.3$ μm, discrimination of about 36.9 dB was achieved, while the direct losses were 1.6 dB (Shiraishi *et al* 1989).

If the required optical discrimination of the laser diode with respect to the reflected signal is -60 dB, the recommendation is to use yttrium iron garnet plane–convex lenses for excitation of the fibre (Drogemuller 1989).

Usually, in a conventional laser diode module, the Faraday rotator has been set between a coupling lens and a polarizer, because the focusing lens has to be placed just in front of the laser diode to focus the divergent laser beam on a fibre. In such a construction, however, the light reflected from the window of the hermetic seal for the laser diode, the coupling lens, and the front surface of the $Y_3Fe_5O_{12}$ cannot be cancelled by the isolator.

Matsuda *et al* (1987) developed a new type of module, in which a thin Faraday rotator acts as the window of the hermetic seal. In this construction, the Faraday rotator is set just in front of the laser diode, and there are no reflective surfaces between the laser diode and the Faraday rotator, which is made of an iron garnet epitaxial film with a substantial amount of bismuth substituted in it—of composition $(BiLuGd)_3Fe_5O_{12}$. The isolation ratio of the optical isolator in the module was about -35 dB, while the coupling efficiency of the laser diode to the single-mode fibre was -5 dB.

An optical isolator for an orthogonally polarized two-frequency laser was developed by Umeda and Eguchi (1988). An orthogonally polarized two-mode laser, e.g. an axial or transverse Zeeman laser, was used as a light source for interferometric and polarimetric measurements. In these optical measurements there is a serious problem: the instability of the laser frequency induced by optical feedback from the measurement system to the laser source.

The principle of the optical isolator is shown schematically in figure 13.9. The incident light from the orthogonally polarized two-mode laser is split into two polarization components by the first polarization beam splitter PBS_1. One component is reflected at a mirror, M_1, and so becomes parallel to the other component. The polarization directions of the two components rotate by $+45°$ on passing through the Faraday rotator, FR, and rotate by an additional 45° at the following half-wave plate, HWP, of the fast-axis azimuth of 67.5°.

With the aid of two polarization devices, FR and HWP, the polarization direction of each component is rotated by 90°. Therefore, the polarization states of the beams emerging from the HWP are changed to those of the input beams before the beams are incident on the optical isolator. After the two emerging beams are combined by the second reflecting mirror, M_2, and a polarization beam splitter, PBS, an optical beam composed of two orthogonally polarized components enters the external optical system.

In the case of the backward direction, the light beams reflected from the external optical system are assumed to have the same polarizations as those of

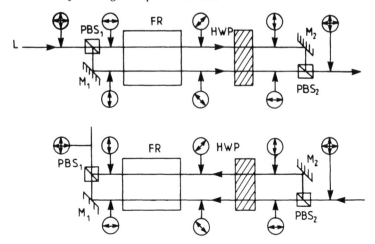

Figure 13.9. A schematic diagram of the optical isolator for an orthogonally polarized two-frequency laser. The arrows show the polarization states of the light beams passing through the optical isolator. Key: 1: the polarization beam splitter (PBS$_1$); 2: the Faraday rotator; 3: the half-wave plate (HWP); 4: a mirror (M$_1$); 5: another mirror (M$_2$); 6: another polarization beam splitter (PBS$_2$) (Umeda and Eguchi 1988).

the emerging beam in the forward direction. Although the polarization states of the two components are the same as in the case of the forward direction at the HWP, the light beams output from the FR have polarization states that are orthogonal to those of the light beams input in the forward direction. Therefore, a light beam reflected from the external optical system is redirected at a right angle to the incident direction of the optical isolator at PBS$_1$, i.e., the light source is isolated from the external optical system.

The Faraday rotator used in the experiment was an FR-5 glass rod, and had a Verdet constant of -0.242 min Oe^{-1} cm^{-1} for a wavelength of 633 nm. It was 30 mm in length and 10 mm in diameter. Umeda and Eguchi tripled the optical path length in the rod, with some margin, by coating a semicircular reflecting thin film on the front and back surfaces of the rotator rod. Isolation ratios of 23.5 and 26.9 dB for the two components were measured, and their insertion losses were 2.4 and 2.9 dB, respectively.

A Faraday rotator in which the light beam travels back and forth five times between two multilayered dielectric mirrors coated onto the glass rod, making the path length about 10 cm, was described by Kobayashi and Seki (1980).

13.4 Magnetooptical circulators

Magnetooptical circulators can be used to eliminate the influences of reflected rays upon an optical transmission system. The construction and characteristics of a polarization-independent magnetooptical circulator were discussed by

Kobayashi and Seki (1980), and Yan and Xiao (1989).

Optical circulators for the 800 nm band have been realized with all of the four ports lying in a single plane (the flat-port configuration) by inserting an optically active quartz rotator, 4 mm thick, between the Faraday rotator and Rochon prism in the optical isolator for the 800 nm band, without input/output lenses, as shown in figure 13.10(a). Insertion loses of 0.5–0.6 dB and isolation of 28–31 dB were obtained for an 830 nm overall size (24 × 24 × 33.5 mm). An FR-5 glass rod was used as the Faraday rotator element (Kobayashi and Seki 1980).

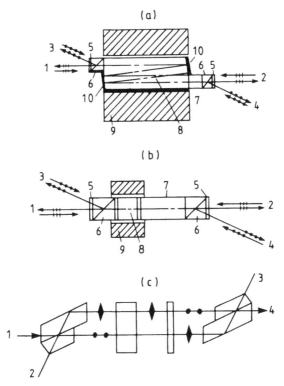

Figure 13.10. Configurations of optical circulators for 800 nm (a), 1300 nm (b) (Kobayashi and Seki 1980), and 1300 nm (c) bands (Yan and Xiao 1989). Key: 1, 2, 3, 4: light beams; 5: an antireflection-coated glass plate; 6: Rochon prisms; 7: the quartz rotator; 8: the Faraday rotator; 9: a permanent magnet; 10: a mirror.

Optimal circulators for the 1300 nm band have also been realized with a flat-port configuration, as shown in figure 13.10(b). A quartz plate, 10 mm long, was inserted between the $Y_3Fe_5O_{12}$ and the Rochon prism in the 1300 nm optical isolators, without SELFOC lenses. The insertion losses were 0.6–0.7 dB, and the isolation was 25–35 dB, both at 1280 nm. The overall size was $11 \times 10.5 \times 22$ mm (Kobayashi and Seki 1980).

A circulator which consists of a 45° $Y_3Fe_5O_{12}$ Faraday rotator followed by a half-wave plate oriented such that the set-up constitutes a 45° reciprocal rotator, and two polarization splitters on either side of each rotator, is shown in figure 13.10(c) (Yan and Xiao 1989). The $\lambda/2$ plate is of mica, and its fast axis is at 22.5°, out of the plane of the figure. For light passing through the two rotators in one direction, the rotations of non-reciprocal and reciprocal rotators add, and the resultant rotation is 90°. For light transmitted in the reverse direction, the two rotators impose rotations of opposite signs, and the resultant rotation is 0°.

Table 13.2. Insertion and isolation of the circulator (dB) (Yan and Xiao 1989).

Input Output:	Port 1	Port 2	Port 3	Port 4
Port 1	—	25	2.0	20
Port 2	25	—	21	2.1
Port 3	21	2.2	—	25
Port 4	2.1	20	25	—

The circulator has four ports. The non-reciprocal paths of the circulator are 1–3, 2–4, 3–2, and 4–1. A Faraday rotator consists of an yttrium iron garnet crystal and a Sm–Co permanent magnet. The size of the $Y_3Fe_5O_{12}$ block is 2.25 mm (length) × 4.5 mm (width) × 1.5 mm (thickness). The two polarized beams are transmitted along the longitudinal direction of the crystal; the distance apart of the two beams is 3.2 mm. The signal attenuations for 1300 nm light incident and emerging at each port are given in table 13.2.

13.5   Magnetooptical deflectors

Propagation of linearly polarized light through a sample that has a stripe or maze-like domain structure exhibits a number of peculiarities. In particular, because of the opposite rotations of the planes of polarization of light in neighbouring domains, passage of radiation through the sample is accompanied by diffraction of the light in the magnetic domains, i.e., the domain structure of a thin magnetic film can be considered as a phase diffraction grating (Haskal 1970). (A phase diffraction grating modulates the phase of a light wave, while in an amplitude diffraction grating, the amplitude of a light wave is modulated.)

For the first time, diffraction by a stripe domain structure was observed in single-crystal yttrium orthoferrite plates that were cut perpendicular to the optical axis of the crystal by Chetkin and Didosjan (1968). The variation of the period of the domain structure in an external magnetic field, and consequent changes of the diffraction angle, can be used in magnetooptic deflectors of optical radiation (Johansen *et al* 1971).

Consider a thin plate of a magnetic dielectric possessing uniaxial magnetic anisotropy. Stripe or maze-like domain structure is realized in such plates (figure 7.1), the magnetization direction in the domains being oriented perpendicular to the sample's surface. As the neighbouring domains are magnetized in opposite directions, the plane of polarization in them rotates in opposite directions:

$$\Phi_+ = (\lambda n)^{-1}\pi g h \qquad \Phi_- = -(\lambda n)^{-1}\pi g h. \qquad (13.5)$$

Let us assume first that the thickness h is chosen such that $\Phi_+ = +90°$, and $\Phi_- = -90°$. Then, for the wave at the input end that is polarized along the y-axis, the E-vector of the wave is parallel to the y-axis, and the output light will be polarized along the x-axis; but when the light passes from one domain to another, its phase will change by 180° (figure 13.11). From the optical point of view, the domain structure behaves like a phase diffraction grating.

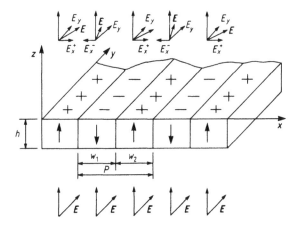

Figure 13.11. The propagation of linearly polarized light through a plate with stripe magnetic domain structure.

It is obvious that, in the symmetric phase diffraction grating considered here (for which $w_1 = w_2 = P_0/2$, where P_0 is the period of the domain structure, and w_1 and w_2 are the widths of neighbouring domains with opposite orientations of the magnetization), there will be no zero diffraction order, because, due to the interference of the beams that have passed through the neighbouring domains, the total intensity of the radiation along the z-axis is zero. Considering interference of the light beams at arbitrary incidence angle θ, where θ is an angle in the $x0z$-plane with respect the to z-axis, it is possible to derive the following expression, which determines the angles of diffraction from the stripe domain structure:

$$P \sin \theta = n\lambda \qquad (13.6)$$

where $n = \pm 1, \pm 2, \ldots$ is the diffraction order, and λ is the wavelength of light. For the symmetric phase grating which is shown in figure 13.11, even diffraction orders are absent, because the diffraction intensity is zero at $n = 0, \pm 2, \pm 4, \ldots$.

In most of the cases, the Faraday rotation turns out to be less than 90°; therefore at the output of the crystal, the light wave has both E_y- and E_x-components. The E_y-components all have the same phase; therefore this component of the light wave propagates along the z-axis, and bears no information about the domain structure, i.e., the radiation that is polarized along the y-axis is background light. Thus, when the light passes through the domain system in an arbitrary case, in the zero diffraction order, radiation is polarized along both the y- and x-axes (an asymmetric phase diffraction grating), and in the other diffraction orders, radiation is polarized only along the x-axis.

The radiation intensity in the zero diffraction order, which is polarized along the y-axis, is defined as (Haskal 1970)

$$I_{y.0} = I_0 \exp(-\alpha h) \cos^2(\Phi_F h)$$

and the light intensity in other diffraction orders, which is polarized along the x-axis, is defined by the formulae

$$I_{x.0} = I_0 (2w_1/P - 1) \exp(-\alpha h) \sin^2(\Phi_F h) \tag{13.7}$$

$$I_{x.\pm n} = I_0 \pi^{-2} n^{-2} 4 \exp(-\alpha h) \sin^2(\Phi_F h) \sin^2(\pi n w_1/P). \tag{13.8}$$

The last formula defines the radiation intensity in the diffraction orders beginning with $n = 1, 2, 3, \ldots$; for the diffraction orders $n = -1, -2, -3, \ldots$ the relation $I_{x.n} = I_{x.-n}$ is valid.

In thin plates or films of uniaxial magnets, a maze-like rather than a stripe domain structure usually exists. This circumstance does not affect the relationships presented above; the diffracted light simply forms concentric circles instead of a linear diffraction pattern, the radiation intensity in the first order being equal to $I_1 + I_{-1}$, and that in the second order being equal to $I_2 + I_{-2}$, etc.

If we consider the propagation of light that is polarized along the x-axis, i.e., perpendicular to the system of the stripe domains (figure 13.11), we will obtain the same pattern: in the zero diffraction order the light is polarized in the same direction as the incident radiation, and in other diffraction orders it is polarized perpendicularly. As orthogonally polarized beams do not interfere, the phase diffraction grating, which is formed by the system of stripe or maze-like domains, diffracts non-polarized radiation too.

The wall image formation in bismuth-substituted iron garnets associated with the diffraction of light was analysed by Ohkoshi *et al* (1995) using a magnetic phase grating model. Small wall widths and large Faraday rotations yield wall images with high contrast in both bright-field and dark-field configurations.

The diffraction efficiency of magnetooptical deflectors (figure 13.12(a)), which are based on the diffraction of light by the domain structures, is determined from the expression (when $w_1 = P/2$) (Scott and Lacklison 1976)

$$\eta_d = I_{+1}/I_0 = 4\pi^{-2} \sin^2(\Phi_F h) \exp(-\alpha h). \tag{13.9}$$

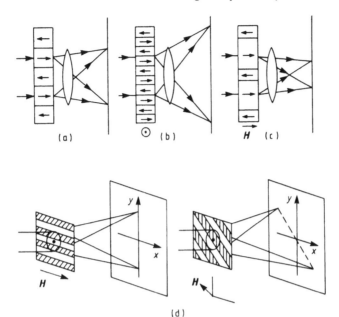

Figure 13.12. Magnetooptical deflectors, where: the magnetic field is zero (a); the field is applied in the plane of the plate (b); and the field is applied perpendicular to the plate (c). Also shown is a two-dimensional magnetooptical deflector (d) (Scott and Lacklison 1976).

Here η_d is the ratio of the intensity in the first diffraction order to the intensity of the incident radiation. In the literature, diffraction efficiency is often defined as a ratio of the total light intensity in the $+1$ and -1 diffraction orders to the intensity of the incident radiation; this value, of course, is twice the one defined by formula (13.8).

As for every magnetooptical device, at given values of the Faraday rotation Φ_F and optical absorption α, which characterize the magnetooptical properties of the material utilized, there is an optimal thickness h_{opt} for the deflector too. Differentiating expression (13.8), which defines the light intensity in the first diffraction order, with respect to h, and assuming that $w_1 = w_2 = P_0/2$, we find

$$h_{opt} = (\Phi_F)^{-1} \tan^{-1} F_1. \tag{13.10}$$

Substitution of (13.10) into (13.8) gives the light intensity in the first diffraction order:

$$I_+ = I_- = I_0 \pi^{-2} 4 \sin^2\{\tan^{-1} F_1 \exp(-F_1^{-1} 2 \tan^{-1} F_1)\}. \tag{13.11}$$

The dependence of the efficiency of a magnetooptical deflector for the sum of the first two diffraction orders on the magnetooptical figure of merit of a magnetooptical material is shown in figure 13.2 (curve 3).

In order to control the angle of diffraction of the light, it is necessary to change the period of the domain structure. This can be achieved in either of two ways. When an external magnetic field is applied along the easy magnetization axis in the films with uniaxial magnetic anisotropy (the field is applied perpendicular to the plane of the film), the domains with magnetic moments directed along the field expand, and those with magnetic moments directed oppositely contract. In the first stage, the period of the domain structure does not change (see figure 7.1(b)); therefore the diffraction angles are also almost unchanged—only redistribution of the light intensities among the diffraction orders takes place; in particular, even diffraction orders appear, and the intensity in the first diffraction order decreases, proportionally to $\sin^2(\pi w_1 / P)$.

Further increase of the magnetic field results in rapid increase of the period of the domain structure; the width of the domains with the magnetization oriented along the field also grows rapidly, while the width of the domains with the opposite orientation of the magnetization weakly decreases (see figure 7.1(c)). In this stage of the restructuring of the magnetic structure, the angles of diffraction of the light decrease. This process of structural change of the domain structure can be treated as a variation of symmetry of the diffraction grating.

When the magnetic field is applied in the plane of the film, the period of the domain structure decreases (figure 13.13, curve 1), but the diffraction grating, in the first approximation, remains symmetric, because domains with opposite orientations of magnetization remain equal in width. The existence of a cubic component of the magnetic anisotropy in the films with substantial uniaxial magnetic anisotropy results in asymmetry of the domain structure, which depends on the orientation of the magnetic field in the plane of the samples with respect to the crystalline axes. This decrease of the period of the domain structure increases the angles of diffraction of the light, because the grating period decreases when the field is applied in the plane of the film.

In this case, the change of the diffraction angles is not accompanied by the appearance of even diffraction orders. At the same time, the increased field in the plane of the film results in a decrease of the diffraction efficiency. The reason for this is that the magnetization vectors in the neighbouring domains deviate from the vertical in the direction of the magnetic field in the plane, and therefore the resultant Faraday rotation—which is proportional to the cosine of the angle between the direction of propagation of the light and the orientation direction of the magnetic moment—decreases.

Controlling the stripe structure using the field in the plane of the film can provide two-dimensional deflectors of optical radiation. Diffracted beams always remain in the plane that is perpendicular to the surface of the magnetic field and to the direction of the stripe magnetic structure. When the direction of the magnetic field in the plane of the film changes, the stripe domain system will rotate, following the direction of the applied field. This results in two-dimensional deflection of the radiation (figure 13.12(d)).

We should note that neither acoustooptical, nor electrooptical light deflectors possess this unique feature. In epitaxial iron garnet films, in which the growth component of the magnetic anisotropy substantially exceeds the cubic component, when a field comparable to the field of the uniaxial magnetic anisotropy is applied in the plane of the film, the domain structure period decreases twofold. In epitaxial iron garnet films with predominantly cubic magnetic anisotropy, e.g. in yttrium iron garnets, grown on the substrates oriented in the (111) plane, far more substantial variations of the domain structure period are attainable. The ratio of the periods can vary over the range 5–7, the domain width being changed from 3 to 22 μm (Johansen *et al* 1971).

Figure 13.13 shows the dependence of the diffraction angle and of the period of the domain structure on the magnetic field, which is applied in the plane of the film of $(BiNdYb)_3Fe_5O_{12}$ composition, in which the magnetization is declined away from the normal into the plane of the film. After application of a magnetic field exceeding the field of the uniaxial magnetic anisotropy, a system of parallel stripe domains with width, in the absence of a bias field, of $w = 0.65$ μm were formed in the film. At the wavelength $\lambda = 0.63$ μm, the first-order diffraction angle was 30°. When a field of 400 Oe was applied in the plane of the film, the change of the diffraction angle reached 20° (Numata *et al* 1980).

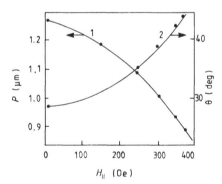

Figure 13.13. The dependence of the period of the stripe domain structure (curve 1) and of angle of diffraction of the light (curve 2), for $\lambda = 0.63$ μm, on the magnetic field in the plane of the film of $(BiNdYb)_3Fe_5O_{12}$ composition, with $P_0 = 1.28$ μm (Numata *et al* 1980).

Numata *et al* (1980) used a double-layer epitaxial film that was grown on a substrate of GGG oriented in the (111) plane as an example of a two-dimensional deflector of optical radiation. The first layer, of composition $(BiSmLuCa)_3(FeGe)_5O_{12}$, had significant growth-induced uniaxial magnetic anisotropy ($4\pi M_s = 285$ G, $h = 3.4$ μm). The film was grown by liquid-phase epitaxy. A layer of yttrium iron garnet 0.88 μm thick was grown on the film by vapour-phase epitaxy.

When a field exceeding the saturation field of the domain structure was applied perpendicular to the surface of this double-layer structure, if there was a component of the magnetic field in the surface of the film, then, in the lower epitaxial layer, a regular system of stripe domains formed, which was oriented along the field. For the sample described, the domain width in the absence of the field was $w = 3.8$ μm. The structure of the stripe domain system did not depend on the azimuthal angle in the plane of the film, and, by rotating the field in the plane of the film, it was possible to obtain a symmetric grating of stripe domains for any azimuthal angle. In the films with uniaxial magnetic anisotropy, the presence of a cubic component hampers the homogeneous rotation of the stripe domain system by the field that is applied in the plane of the film.

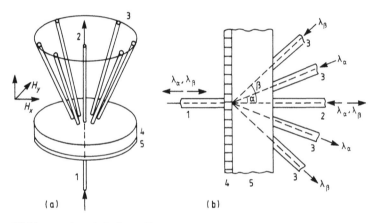

Figure 13.14. (a) A switch for a fibre communication line based on a magnetooptical deflector. The components are: the input (1); the central output (2); the switched output (3); fibres, and the substrate (4); and the epitaxial iron garnet film with a stripe domain structure (5). (b) A multiplexer for optical signals based on a magnetooptical deflector. The components are: the input (1); the central (2), and side (3) output fibres; the epitaxial film with a stripe domain structure (4); and the substrate (5) (Sauter *et al* 1981).

Figure 13.14(a) shows a device that is designed for the commutation of optical signals propagating in fibre communication lines (Sauter *et al* 1981). The light that exits from the lower fibre is diffracted by the stripe domain structure, with the period 3.4 μm. In order to increase the diffraction efficiency of the device, Sauter *et al* used an epitaxial film of bismuth-substituted iron garnet with a large Faraday rotation. In the device, the zero diffraction order was picked up by the central output fibre, which was located on the same axis as the input fibre. The beam of the first diffraction order could be sent by the deflector to any of six side fibres inclined to the vertical at the angle that corresponded to the first-order diffraction.

The diffracted beam was switched between the fibres by rotation of the stripe domain system, the transition from one position to another being caused by application of a magnetic field pulse of less than 1 μs duration to the stripe domain system. The output signal had a signal-to-noise ratio of more than eight. This commutator of fibre-optic communication lines has significant advantages over existing ones: because there are no moving components, the system functions in the $\lambda > 1$ μm wavelength region, which is the most suitable for communication lines, and energy is consumed only when the orientation of the stripe domain system is switched, and is not expended at any other time. The system provides transmission of information both ways: from the lower fibre to the side ones, and back.

Figure 13.14(b) shows a multiplexing/demultiplexing system for optical signals of different wavelengths (Sauter *et al* 1981). For a given period of the domain structure, the diffraction angle depends of the wavelength of the light, in accordance with the relation (13.6), so each of the beams excites the desired fibre, which is positioned at the angle that corresponds to the first-order diffraction of the appropriate wavelength.

The diffraction efficiency of magnetooptical deflectors based on epitaxial iron garnet films can be substantially increased if the film is placed in a laser resonator. However, in this case the Q-factor of the resonator falls abruptly; that is why practical demonstrations of this system are still questioned. Creation of an optical system that consists of two coupled resonators solves the problem for the effective magnetooptical deflector for the infra-red spectral band (Krawczak and Torok 1980). Figure 13.15 is a diagram of such a system.

By choosing the appropriate transmission coefficient for the middle semitransparent mirror, one can achieve a situation in which the laser resonator supports a stable generation of radiation, and in which a system of standing waves appear in the deflector resonator, which significantly increase the diffraction efficiency of the system. If the reflection coefficient of the middle mirror is low, and the attenuation caused by the introduction of the magnetic film into the resonator exceeds the gain in one pass of the radiation, there is no generation. If the reflection coefficient of the intermediate mirror is optimal, the intensity of the optical radiation in the deflector resonator will be substantially lower than that of the optical radiation in the laser resonator. In an experiment using a double-cavity laser, a diffraction efficiency of 10% was achieved at the wavelength $\lambda = 1.06$ μm when a 20 μm epitaxial iron garnet film—which in the ordinary regime provided a diffraction efficiency of less than 0.5%—was used (Krawczak and Torok 1980).

13.6 Spatial filtering of optical signals

Diffraction of light by domain structures is often used in the study of thin magnetic films—in particular, for measurements of the domain structure parameters of films with very low contrast of the domain structure, which corresponds to diffraction efficiencies near 10^{-3}–10^{-6}. Let us assume that light incident on the magnetic film is linearly polarized along the y-axis.

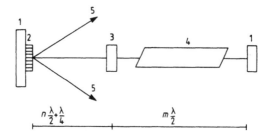

Figure 13.15. The intra-resonator magnetooptical deflector. The components are: 1: reflectors; 2: a magnetic film with a stripe domain structure; 3: a semitransparent mirror; 4: the active element of the laser; and 5: the diffracted light (Krawczak and Torok 1980).

If the diffraction efficiency of the object is in the range 10^{-3}–10^{-6}, almost all of the light that passes through the magnetic film corresponds to the zero diffraction order, which has a polarization coinciding with the original one. Crossed dichroic polarizers provide attenuation of the light at a level of about 10^{-2}–10^{-4}. At the same time, all of the information on the domain structure is contained in the diffraction orders that have polarization directed along the x-axis, i.e., perpendicular to the polarization of the incident radiation. At these diffraction efficiencies, the intensity of the zero diffraction order, even after attenuation by a polarizer in the crossed position, is several orders of magnitude higher than the intensity of the radiation in the first diffraction orders.

When the domain structure is observed using the polarizing microscope, the low contrast of the pattern is caused by the masking background that is associated with radiation polarized along the y-axis, which bears no information about the object. The method of spatial filtering allows one, in principle, to totally eliminate the y-component of the radiation by blocking the zero diffraction order; thus utilization of the spatial filter in a polarizing microscope substantially increases the contrast when the domain structures are observed. In this case, the spatial filter is simply a non-transparent circle of appropriate diameter, which is located in the centre of the rear focal plane of the objective (Henry 1976).

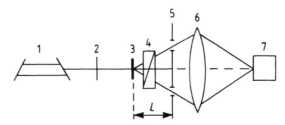

Figure 13.16. A schematic diagram of the spatial filtering of the diffraction orders.

Spatial filtering is suitable for measurements of periods of domain structures of micrometre and submicrometre dimensions, without employing optical polarization microscopes. Figure 13.16 shows a schematic diagram of the set-up for measurement of the domain width in thin magnetic films with uniaxial magnetic anisotropy by the diffraction method. The laser beam (1) passes through the modulator (2), and the magnetic film (3), which contains a maze-like domain structure, and is diffracted by the latter. The zero diffraction order is first attenuated by the analyser (4), which has its transmission axis directed perpendicularly to the plane of polarization of the incident beam. The analyser is necessary for the reduction of the intensity of the scattered light. Radiation of the zero order, which still passes through the crossed polarizer, is totally blocked by the spatial filter (5), which is a ring slot in a non-transparent screen. The diffracted light is gathered by the lens (6) into the photodetector (7).

Moving the spatial filter along the axis, which coincides with the laser beam path, one can find a position in which the diffracted cone of the first order will pass through the ring slot. Knowing the radius of the spatial filter R and the distance L from the filter to the film, one can find the diffraction angle θ, which is defined by the relation

$$\tan \theta = R/L.$$

The period of the domain structure P, and hence the width of the stripe domain $w = P/2$, are determined from the diffraction angle found using relation (13.6) for the first diffraction order:

$$P \sin \theta = \lambda.$$

The other magnetic parameters of the film are determined as follows (Johnson *et al* 1980). From the measured thickness, h, of the film (h is measured independently by standard methods), and P_0 (P_0 is the period of the equilibrium domain structure in the absence of a bias field), using the relation (7.4), the characteristic length of the material, l, is calculated. Although the formula (7.4) was derived for stripe domain structure, experimental data show that it can be applied for maze-like domain structure too (Henry 1976, Johnson *et al* 1980).

In the following stage, the period of the domain structure, P, is measured at some bias field, H. From the known values of P and H, and from the calculated value l, using the relations (7.2) and (7.3), and the expression

$$h^{-1}l = \pi^{-3}h^{-2}P^2 \sum_{n=1}^{\infty} n^{-3}[1 - (1 + P^{-1}2\pi nh)$$
$$\times \exp(-P^{-1}2\pi nh)\sin^2(\pi n(1 + M_s^{-1}M)/2)]$$

one can determine the ratio M/M_s of the magnetization of the sample to the saturation magnetization by the iteration method. Then, from the known values of M/M_s, l, h, P, and H, and from the relations (7.2) and (7.3), the value of $4\pi M_s$ is determined.

Using the calculated value $4\pi M_s$ and the independently measured exchange constant A (the value of A is usually found from the results of the measurement of the Curie temperature of the sample), and the known values of l and h, the Q-factor of the material and the collapse field of the bubble magnetic domain H_0, the uniaxial magnetic anisotropy constant K_u, and the field of the uniaxial magnetic anisotropy H_a are determined from the formulae (Henry 1977)

$$Q = (32\pi A)^{-1}(4\pi M_s l)^2$$
$$H_0 = 4\pi M_s[1 + (4h)^{-1}3l - (3l/h)^{1/2}]$$
$$K_u = A^{-1}(\pi l M_s^2)^2 \qquad H_a = M_s^{-1} 2K_u.$$

The spatial filtering method can be successfully used to enhance the signal-to-noise ratio when measuring the coercive force or recording the map of magnetic defects of a magnetic film (Henry 1976, 1977). On the basis of the method of spatial filtering, a set-up for measurement of the periods of the submicrometre domain structures was created (Kotov *et al* 1982). The main features of the set-up were the use of the spatial filter, and a photodetector in the form of a ring. As a result, the set-up was significantly simplified, and the measurement limits were expanded to 0.9–15 μm. The measurement error at the domain width $w \approx 0.5$ μm was $\pm 2\%$.

Spatial filtering of the diffracted beams makes it possible to measure the Faraday rotation of thin magnetic films of uniaxial materials with stripe or maze-like structure, which are in the demagnetized state. The light intensities in the zero-order beams and in the diffracted beams are determined from the expressions (13.7) and (13.8). Measuring the intensity ratio of the two orthogonal polarizations:

$$\sum_{n=1}^{\infty}(I_{x,n} + I_{x,-n})/I_{y0} = 8\pi^{-2}\tan^2(\Phi_F h)\sum_{n=1}^{\infty} n^{-2}$$

one can determine the Faraday rotation Φ_F. In the experimental set-up, a calibrated 10^1–10^3 times attenuation of the zero-order intensity I_{y0} with following measurement of the signal intensity ratio is envisaged (Kotov *et al* 1981a, b, c).

13.7   Magnetooptical transparencies and displays

A two-dimensional magnetooptical cell matrix based on magnetooptical materials can perform as a spatial or time light modulator (magnetooptical transparency), or a display, depending on the particular application conditions. When the matrix is used as an intermediate information carrier in an optical information-processing system, in spatial filtering, in an optical processor, or in the formation of pages of optical memory, it is called a magnetooptical transparency, or a spatial–time optical modulator. If the matrix of magnetooptical cells is used as an output device for the representation of information, it is usually called a display.

Magnetooptical displays are the devices that are designed for representation of the text or graphic information when the operator observes the panel directly, or the information is transferred to a screen with magnification.

Currently the most promising materials as regards magnetooptical displays and transparencies are epitaxial films of bismuth-substituted iron garnets. The pattern is observed due to the Faraday rotation when the structure works in transmittance mode, and due to the doubling of the Faraday rotation when the light reflects from the coating deposited on the opposite side of the structure. Information in magnetooptical displays of this type is presented in the form of magnetic domains, which are contained in the matrix with oppositely directed magnetizations, or using local magnetooptical cells that have two fixed magnetization states (figure 13.17).

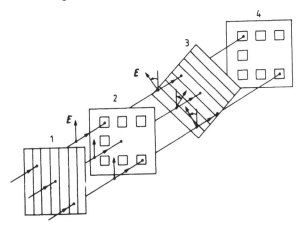

Figure 13.17. Magnetooptical displays based on domain structures or island cells. The components are: 1: the polarizer; 2: the magnetooptical medium; 3: the analyser; and 4: the screen.

As the magnetic domains, or the cells with differently oriented magnetizations with respect to the direction of propagation of the light, rotate the plane of polarization of the linearly polarized light due to the Faraday effect, clockwise or anticlockwise, then placing an analyser, which transmits light with one direction of polarization and blocks light with the other direction of polarization, beyond the magnetooptical transparency, on the screen or directly beyond it, one can obtain the picture of the symbol presented. Because the analyser is positioned at the crossing point with respect to the plane of polarization in the 'closed' information cell, the intensity of the light that passes through the 'open' information cell can be described by the expressions that were obtained for the calculation of the efficiency of the optical system working as an optical switch.

One of the main parameters of a magnetooptical display is the contrast of the picture, i.e., the ratio of the intensity that passes through the 'open' cell to the intensity of light passing through the 'closed' cell. When dichroic polarizers are used for magnetooptical displays based on epitaxial films of bismuth-substituted iron garnets, for monochrome radiation with the wavelength $\lambda = 0.57$ μm, a contrast of 180:1 can easily be reached, and, if certain additional measures are taken, the contrast can exceed 2000:1 (Lacklison *et al* 1975, Paroli 1984).

We should consider in more detail the factors that restrict the maximum contrast of a magnetooptical display. If white light is used, the main factor will be the dispersion of the Faraday rotation. Because of this, the analyser can be positioned for maximum blocking of just one wavelength. The spectral characteristics of the transmission coefficient of a magnetooptical display are shown in figure 13.18(a) (Lacklison *et al* 1975). The optimal result is achieved when the analyser is positioned such that light with the wavelength $\lambda = 0.57$ μm is blocked. If the wavelength of the light is changed in either the short-wavelength or long-wavelength direction, the light intensity that passes through the 'closed' cell increases.

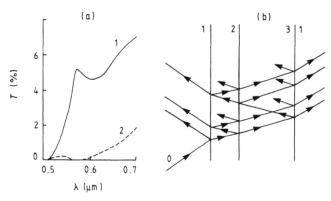

Figure 13.18. (a) The transmission spectrum of a magnetooptical cell based on a film of composition $Bi_{1.1}Sm_{1.9}Fe_{3.9}Ga_{1.1}O_{12}$: open (curve 1) and closed (curve 2) states (Lacklison *et al* 1975). (b) The reflected beams defining the contrast of the magnetooptical transparency. Key: 1: air; 2: the epitaxial film; 3: the substrate; and 4: the incident beam.

When the object is illuminated with white light, a colour contrast is observed between the cells with opposite orientations of magnetization. An 'open' cell has yellow coloration, and a 'closed' one has a red or green hue, depending on the position of the analyser. A four-layer structure, in which the rotation angle of the plane of polarization can take any of five discrete values ($-4\Phi, -2\Phi, 0, +2\Phi,$ and $+4\Phi$), by combining different states of the cells in the layers, one can obtain green, red, and yellow coloration (Balbashov and Chervonenkis 1979).

When monochrome radiation is used, the maximum achievable contrast is restricted by the following factors. Consider a single-layer epitaxial structure, which consists of a non-magnetic GGG substrate and an epitaxial film of a bismuth-substituted iron garnet (figure 13.18(b)). For simplicity, let us consider only beams I, II, III, and IV in the output of the structure. The main beam has the intensity

$$I_0(1 - R_{12})(1 - R_{23})(1 - R_{13})\exp(-\alpha h)$$

where R_{12}, R_{23}, and R_{13} are the reflection coefficients of the air–epitaxial film, substrate–epitaxial film, and substrate–air interfaces, respectively, and its plane of polarization is rotated with respect to the incident beam by the angle $\Phi = \Phi_F h$. We should note here that, for a beam incident at an angle to the interface between the two media in the general case, the transmitted beam has elliptical polarization. For instance, deviation of the direction of incidence of the beam by several degrees from the normal to the interface is equivalent to rotation of the plane of the incident beam by several minutes of arc.

Beam II has the intensity

$$I_0(1 - R_{12})R_{23}R_{12}(1 - R_{23})(1 - R_{13})\exp(-3\alpha h)$$

and the angle of rotation of the plane of polarization $\Phi = 3\Phi_F h$. For beam III we can write

$$I_0(1 - R_{12})(1 - R_{23})R_{13}R_{23}(1 - R_{13})\exp(-\alpha h)$$

and the angle of rotation of the plane of polarization is $\Phi_F h$. Finally, the intensity of beam IV is

$$I_0(1 - R_{12})(1 - R_{23})^3 R_{13}R_{12}(1 - R_{13})\exp(-3\alpha h)$$

and the total angle of rotation of the plane of polarization $\Phi = 3\Phi_F h$. These formulae do not take into account the ellipticity of the light, which arises because of magnetic circular dichroism, and oblique-incidence beams due to finite divergence of the light beam.

It is important that for most of the radiation sources, the spatial coherence of the beams I, II, and III, IV is distorted. Adding beams I and II results in elliptical polarization of the light at the output. The same is true for beams III and IV. The radiation is additionally depolarized because of the magnetic circular dichroism, final beam divergence, and induced birefringence in the film and the substrate. The ellipticity, though not very large, can reduce the contrast from the theoretically achievable value, which is approximately equal (when film polarizers are used) to 10^3–10^4:1, to be compared with the experimentally observed ratio of 180:1 (Paroli 1984). Ellipticity can, in principle, be avoided by means of an optical compensator, the simplest of which is the quarter-wave plate, so the contrast can be increased, and can reach 2000:1 (Paroli 1984).

However, beams III and IV substantially complicate the situation. When beams III and IV pass through the substrate twice, they lose their coherence with beams I and II. This results in the appearance of a radiation component with polarization that is orthogonal to the initial polarization of the incident beam; this component leads to a background that cannot be avoided by using an optical compensator. In order to increase the optical contrast, the air–epitaxial layer and air–substrate interfaces should be antireflection coated, which, in the ideal case, completely suppresses beams II, III, and IV. When an epitaxial structure with films on both sides of the substrate is used, even antireflection coating does not prevent additional beams that reflect from the epitaxial layer–substrate interface. The problems as regards the antireflection coating of these interfaces (film–substrate), and the experimental results, are considered in section 13.9.

Existing magnetooptical displays can be divided into two classes of organization.

To form the picture, displays of the first type employ bubble domains, which are moved along the permalloy pattern by a magnetic field that rotates in the plane of the plate, i.e., the principle is similar to that on which bubble-memory devices are based. The translation elements form a sequential register, which substantially increases the display fill time.

An epitaxial film of composition $Bi_1Tm_2Fe_4Ga_1O_{12}$, having the nominal bubble dimension 7 μm, was used in one such device (Lacklison *et al* 1977). Characters were displayed using the shift registers, in which the character was formed from 10×7 or 7×5 domains. The data were recorded into a register with a 36×36 format (capacity 1.3 kbit), with the clock rate of 5 kHz. In order to avoid shadowing the domains by the permalloy elements, after the register was filled the magnetic field in the plane of the sample was fixed in a position that located the bubble in the gaps between the permalloy elements.

In another device, an epitaxial film of composition $(GdTm)_3(FeGa)_5O_{12}$, with the nominal bubble diameter 8 μm, was used. The total capacity of the display was 4096 bits, and the bubble movement structure comprised 16 parallel shift registers, each being 'folded' four times. With a clock rate of 100 kHz, the display fill time was 2.7 ms. Additional structures for restriction of the bubble position were formed by local variations of the film magnetization through the propagation structure. The additional structures increased the bubble diameter fourfold in the data-display position; the domain preserved its circular form and position, which substantially increased the brightness of the picture (Paroli 1984).

Bubble-based magnetooptical displays did not find broad practical application, because they required permanent magnets and coils of specific design in order to form the planar rotating field. In addition, the desired contrast for the displayed information required a high illumination level of the display, which resulted in overheating of the device. Increase of the display capacity leads to unacceptable growth of the data-recording time, because the structures were organized as sequential shift registers.

An entirely different approach to magnetooptical display design is that of the formation of the matrix of magnetooptical cells with arbitrary access, by means of a system of vertical and horizontal current buses. The cells can be from several micrometres to hundreds of micrometres in size, and can be formed as mesa structures, as isolated islet structures, or by local diffusion or local laser annealing.

A magnetooptical matrix based on islet cell structure has been used in magnetooptical displays and printers. An epitaxial film of the composition $(BiGdTm)_3(FeGa)_5O_{12}$, with the easy magnetization axis oriented perpendicular to the plane of the film, had a specific Faraday rotation of $\Phi_F = 4$ deg μm^{-1} at the wavelength $\lambda = 0.56$ μm (the bismuth content was at about 1.1 atoms per formula unit). The gallium and gadolinium content in the film was chosen such that the compensation point was near room temperature. A non-magnetic garnet of complex composition, with an increased unit-cell parameter, was used as the substrate. The epitaxial film, which was 5 μm thick, was etched down to the substrate to form square cells, 65×65 μm, with the pitch between the cells equal to 125 μm. The current buses run between the cells. The structure was placed between polarizers, which blocked the light when the film was magnetized in one of the directions. When the display was illuminated with white light, the contrast was 20:1 (Hill and Schmidt 1982).

The data were recorded and stored in the following way. At room temperature the magnetization of the material was near zero, so the anisotropy field was very high. The coercive force at the compensation point also grew abruptly; therefore, at room temperature the cell conserved the state with the given magnetization direction for an indefinitely long time. When the cell was heated—say, to 100–200 °C—the magnetization grew, while the coercive force dropped abruptly, and the anisotropy field also dropped. In this state the cell was remagnetized by a magnetic field pulse of 15 μs duration and nearly 100 Oe strength. The data cell was remagnetized in two steps. First, the seed of reversed magnetization appeared, then the floating domain turned into a normal domain, and finally the domain expanded to the size of the data cell.

The cell can, in principle, be heated by optical radiation, but the required intensity is unacceptable from the practical point of view. For local heating of the film, a system with resistive straps turned out to be very effective. A resistor strap, which covers the corner of the magnetooptical cell, is formed between the current buses. When the current pulse runs through the strap, the magnetooptical material is heated, so local application of a magnetic field pulse forms a domain of reversed magnetization. The power release on the strap is 30 mW. The whole switching procedure takes about 20 μs.

In the set-up with the resistor straps, there is a problem of cross-coupling, because, when the two current buses are connected to the pulse generator, the pulse leaks via the neighbouring strips. In order to avoid the leakage, which substantially increases the power consumption and the probability of erroneous cell switching, a compensating potential should be applied to the other current buses. Stabilization of the matrix performance with different numbers of cells, depending on the number of simultaneously switched elements, is discussed by Hill (1980).

This approach was employed in industrial fabrication of high-speed printers.

The single module consists of two lines, each containing 256 cells, which are mutually shifted by half of the period; the printer consists of five modules. In the unit, the light from a halogen lamp is delivered to the cells by polymer fibres, and, after spatial filtering by a system of optical isolators, is focused on the photosensitive material, like in standard copying machines. The lamp, of 100 W power, provides 0.2–0.4 μW power at each point of the pattern; the radiation is concentrated in the 0.53–0.65 μm spectral band. The resolution is 12 points mm^{-1} (up to 4000 points per line), and the data transmission rate is 1 Mbit s^{-1} at the cell switching frequency of 2 kHz. The entire assembly, without the fibres, is $45 \times 80 \times 220$ mm in size (Hill and Schmidt 1982).

The main advantages of the device are: information storage without power consumption; lack of moving parts; high integration (fivefold with respect to light-emitting diodes); high spatial resolution; high printing rate (2000 characters per second); and comparatively low power consumption.

One promising type of display is the magnetooptical indicator. Such devices employ bismuth-substituted iron garnet films with the composition that corresponds to the compensation point, which have large irregular ('rag') domain structures (Paroli 1984). At the operating temperatures, the demagnetizing field turns out to be lower than the coercive force in the films; therefore the domains formed, of arbitrary configuration (usual size: up to 1 cm), can exist in the metastable state for indefinitely long times. At the same time, the configuration of the domains can be controlled using the current buses, by applying magnetic fields of the order of 10 Oe (Paroli 1984). Studying such 'rag' domains existing in iron garnet films, for which the condition $l > h$ is fulfilled, where l is the so-called characteristic length of the bubble film material, and h is the thickness of the sample, have shown that the strength of the applied magnetic field which is necessary to move the domain walls in such films is between 4 and 6 Oe for films with l between 0.1 and 10 μm (Kotov *et al* 1986a, b).

Like in liquid-crystal indicators, all ten digits are formed using a combination of seven segments. Thermal treatment provides the films with a positive constant of uniaxial magnetic anisotropy. The film structure ensures stable domain seeding, and formation of the desired symbols, on running the 20–100 mA current pulses of 250–30 μs duration through the electrodes. The information is erased by an external magnetic field source. The advantages of the indicators are: lack of constant bias fields; rather high contrast; considerable variety of the displayed symbol sizes; and relatively simple formation schemes for the data arrays (Paroli 1984).

When magnetized to saturation, a sample of homogeneous magnet with $H_a - 4\pi M_s > H_s$ can preserve the single-domain state for an indefinitely long time, provided that there are no microdefects, which are the seeding sources of domains with reversed magnetization. In the islet structures of high-quality iron garnet films, the single-domain state can linger, with an element size up to 3.5 cm, and $4\pi M_s \approx 300$ G. Switching of an individual cell from one state to another requires application of a magnetic field that exceeds $H_a - 4\pi M_s$ normal to the sample surface (Paroli 1984).

This principle for information storage also applies to magnetooptical transparencies, which comprise 48×48 or 64×64 cell arrays; 11×11 mm samples having data capacities of 16 kbits have been demonstrated. Random-access information storage in the devices is provided by two systems of perpendicular current buses, which are located between the magnetooptical cells. The cells are switched by simultaneous application of current pulses to both horizontal and vertical buses. Because the magnetooptical cell switching has a threshold, an appropriate choice of the current amplitude switches only one cell: the one that lies at the intersection point of the buses. For reliable cell switching, the deviation of the switching thresholds throughout the entire transparency should not exceed 25% (Paroli 1984).

There are several types of organization of current buses.

First, the current pulses were applied to pairs of adjacent horizontal and vertical buses, the current flowing in neighbouring buses in opposite directions. Only the cell that lay where the fields from the horizontal and vertical buses added was switched. In order to lower the switching threshold, and to reduce the deviation of the switching fields from one cell to another, local implantation or local annealing in the edge areas of the magnetooptical cells was proposed (MacNeal *et al* 1983). Local annealing provides a controllable reduction of the cell switching fields, from several thousands to 100–200 Oe. The cells are reliably switched when current pulses of 100 mA amplitude are applied. To switch the selected cell, one should apply the current pulse to one horizontal and one vertical bus.

When a cell is locally ion implanted, it switches in the following way. First, a floating domain forms in the implanted zone; then, the domain grows though the entire film thickness; and, finally, the entire cell remagnetizes via domain wall displacement. If the material utilized has a domain wall velocity ~ 10 m s^{-1}, the switching of a cell of size 50 μm takes approximately 5 μs; if the ultimate velocity reaches 10^3 m s^{-1}, the switching time must reduce to 50 ns. The domain in the ion-implanted region is formed by a current pulse of 200 ns duration. The transparency is filled with a clock rate of 1 MHz; the domains which appear in the ion-implanted area are expanded, and the cells are remagnetized by the biasing field, which is created by a common coil. The deviation of the switching field was $\pm 25\%$ from the nominal value.

In comparison with magnetooptical transparencies with resistor heating, the above-described device has a shorter cell switching time, and a wider temperature interval (operation at temperatures up to 100 °C was achieved); the contrast reaches 20:1 with white light and 100:1 with monochromatic light. The mirror coating allows one to use the transparency with this cell organization in reflected light (Paroli 1984).

The question of cell switching of the magnetooptical transparency demands special discussion. The theory predicts that, in order to switch a uniformly magnetized cell of the transparency, one must apply a magnetic field $H_a^* =$

$H_a - 4\pi M_s$. For usual magnetooptical materials, H_a^* reaches several thousand oersteds; if the switching is provided by the x- and y-buses, then a 100 μm cell requires a current of several amperes, which prevents practical implementation of the magnetooptical transparencies.

Real switching fields can be substantially lower, because the switching process begins at local inhomogeneities, which exist at the edge of the cell in particular. The deviation of the switching thresholds from one cell to another grows abruptly in this case.

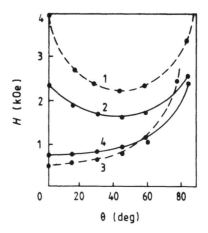

Figure 13.19. The dependence of the switching field H_s on the θ-angle for a square cell of a magnetooptical transparency. The relative lattice deformations in the ion-implanted layer are: 1: 0.00; 2: 0.05; 3: 0.10; and 4: 0.15% (MacNeal *et al* 1983).

For arbitrary field orientation, the switching field H_s of a homogeneously magnetized film is defined by the formula

$$H_s(\theta) = H_a^*(\cos^{2/3}\theta + \sin^{2/3}\theta)^{-3/2} \qquad (13.12)$$

where θ is the angle between the normal to the plane of the film and the direction of the applied field. The switching field attains its maximum at the angles $\theta = 0$ and 90°, and its minimum at $\theta = 45°$. Figure 13.19 shows the dependence of the switching field H_s on the θ-angle for a square cell of a magnetooptical transparency, with side 109 μm (MacNeal *et al* 1983). To form the cells the pits were etched down to the substrate. The $(BiTm)_3(FeGa)_5O_{12}$ epitaxial film had the following parameters: $H_a = 4400$ Oe, $4\pi M_s = 94$ G, $h = 7.6$ μm, and $w = 7.6$ μm. At $\theta = 0°$, the switching field was 3950 Oe; at $\theta = 45°$, it was reduced to 2200 Oe. The dashed line 1 in the figure corresponds to expression (13.12) at $H_a^* = 4350$ Oe. Implantation of B^+ ions, which results in an ~0.05% lattice deformation, lowers H_s, but the character of the $H_s(\theta)$ curve does not change (curve 2). Increase of the implantation dose and increase of the corresponding lattice deformation to 0.1% lead to a substantial change in the dependence of the switching field on the θ-angle (curve 3). The minimum of the switching field, $H_s = 570$ Oe, corresponds to $\theta = 0°$. Curve 3 can be approximated by the expression $566/\cos\theta$. Growth of the lattice deformation level to 0.15% leads to a slight increase of the switching field (curve 4).

In the non-implanted cell, the switching field is determined by coherent rotation of the magnetization, which begins when the bias field is higher than $H_s(\theta)$, which is defined by formula (13.12). For low implantation doses in the ion-implanted layer, the uniaxial anisotropy field decreases, which decreases the switching threshold but preserves the dependence on the θ-angle, which is defined by expression (13.12). In this case the cell switches in the following way. In the ion-implanted region, coherent rotation of the magnetization creates a 180° domain wall, which moves toward the interface between the implanted and non-implanted areas under the influence of the magnetic field.

The gradient of the uniaxial magnetic anisotropy (this is the decisive factor) at the interface creates an energy barrier, which prevents movement of the domain wall. At low ΔK_u (low implantation doses), the 180° domain wall overcomes this barrier, and pushes into the non-implanted region. The switching is complete when the domain wall reaches the substrate. In this case the switching field is evidently determined by coherent rotation of the magnetization.

At higher ion-implantation doses, H_a^* decreases, reaching the state in which the 180° domain wall, which formed in the ion-implanted region, cannot pass the interface between the implanted and non-implanted regions. The switching field in this case is defined by the pressure that must be created by the external magnetic field in order to push the domain wall into the non-implanted layer. The field which is required for the cell switching is defined by the energy gradient of the domain wall $\nabla\sigma_w$. Assuming that the H_a changes only slightly in the domain wall width, and disregarding the demagnetizing fields, we can write the approximate expression

$$\nabla\sigma_w = M_s \nabla H_a \Delta$$

where $\Delta = (A/K_u)^{1/2}$ is the domain wall width parameter. Equalizing $|\nabla\sigma_w|$ with the external magnetic field pressure $2M_s H \cos\theta$, we find

$$H_s = (2\cos\theta)^{-1} \nabla H_a \Delta. \tag{13.13}$$

Formula (13.13) satisfactorily describes the angular dependence $H_s(\theta)$ when the lattice deformation is 0.1% or more.

This consideration shows that the switching field of the transparency cell must drop if one manages to reduce $|\nabla H_a|$. The reduction can be achieved by having a more gradual change of the lattice deformation profile in the transition from the implanted to the non-implanted regions. An experiment has shown that multiple implantation, with different ion energies, results in a substantial decrease of the threshold of the switching field; for example, a value of 150 Oe, instead of 1500 Oe, for the non-implanted film was reached, and the deviation of the threshold field lies within 10%. Annealing after the ion implantation leads to similar results (MacNeal *et al* 1983).

13.8   Magnetooptical read heads and creation of a magnetic image

The process of magnetooptical information read-out is in fact a transfer
of information from a magnetic carrier to a recording medium, which has
higher values of the Faraday or Kerr effects, with subsequent magnetooptical
information read-out, using one of the two above-mentioned effects, from
an intermediate medium.   The method of magnetooptical read-out using
intermediate media is useful if the optical characteristics of an information carrier
are not such as to allow read-out directly from the carrier (for example, when the
information pick-up using the Kerr effect does not provide the desired signal-
to-noise ratio).

Nowadays the most promising media for magnetooptical pick-up heads
are epitaxial films of bismuth-substituted iron garnets, having high values of
Faraday rotation. In the double-pass Faraday effect mode (light passes through
the magnetooptical medium twice, reflecting from a dielectric mirror located on
the bottom surface of the film) the films provide values of Faraday rotation up
to several dozen degrees for film thicknesses of the order of 10 μm at the light
wavelength 0.63 μm.

Figure 13.20 shows the process of read-out from the magnetic carrier using
a thin iron garnet film.   Magnetic leakage fields, which are created by the
domain structure of the information carrier, modify the domain structure of
the iron garnet reading film. Under the influence of the magnetic leakage fields,
the domain magnetization in the uniaxial film is oriented upwards, where the
magnetic charges in the information medium below are positive and downwards
in the case of negative magnetic charges (see figure 13.20(a)). Thus, when the
information transfers from the storage carrier to the magnetooptical read-out
medium, the domain walls in the carrier are situated below the domain centres
in the magnetooptical read-out medium, and vice versa.

When the information carrier moves, the related domain structure in the
upper read-out layer is displaced. Variation in the Faraday rotation for the light
that passes through the film and reflects from its bottom surface is converted
into amplitude changes of the read-out signals by an analyser.

One of the most important parameters of the magnetooptical pick-up head is
its minimal period for information read-out. It is evident that the minimal read-
out period is defined by the domain structure parameters of the read-out medium,
and the gap width, because, for effective information transfer, the gap between
the magnetic read-out medium and the information carrier must be smaller than
the domain structure period. Another crucial parameter is the ultimate read-out
frequency, which is determined chiefly by the dynamical properties of domain
walls of the magnetooptical read-out medium.

The main results of the experimental studies of the process of information
transfer from the carrier to the information read-out medium are the following
(Nomura 1985).

The limit for the minimal information-recording wavelength l_r^{min}, which

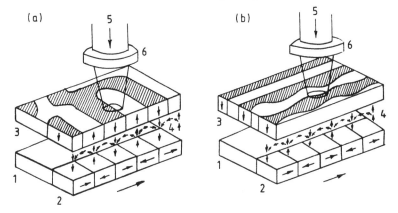

Figure 13.20. Magnetooptical information read-out from magnetic media: using epitaxial iron garnet films with maze-like (a), and stripe (b) domain structure. Key: 1: the storage medium; 2: the track with recorded information; 3: the magnetooptical film; 4: magnetic leakage; 5: the laser beam; and 6: the objective (Nomura 1985).

is defined for undistorted information transfer from the carrier to the magnetooptical read-out medium, is approximately equal to the period P of the maze-like domain structure. Achievement of even lower values of l_r^{min} is possible if the magnitudes of leakage fields are substantial.

The decrease in the vertical component of the magnetic leakage field H_z leads to sinusoidal distortions of the domain structure during the achieving of the maximum wavelengths, which determines the upper limit of the undistorted information transfer. Reduction of l_r^{min} leads to lower values for l_r^{max}; for example, if $l_r^{min} = 4$ μm, the maximum value is limited by $l_r^{max} = 20$ μm.

The ultimate velocity of the domain wall displacement in the case of strong coupling between the information carrier and the magnetooptical read-out medium (high values of the leakage field H_z) is determined by dynamic conversion in the domain wall in a region close to l_r^{min}. However, in the case of weak coupling (low values of the leakage field H_z) in a low gradient of the displacement field, the velocity of the domain wall displacement is insufficient to have an effect.

The method of information transfer is characterized by certain specific processes that are manifested in the joining of information domains to domains in the film periphery, or to domains in the other recording tracks. The presence of the coupling also results in cross-coupling distortions during read-out of a signal that is recorded on a narrow track.

Gusev and Kotov (1991) have analysed the attainable minimum period for information read-out using iron garnet films with stripe or maze-like domain structures. It was shown that, by optimizing the thickness of the film, with $4\pi M_s = 1600$ G and $l = 100$ nm, it is possible to obtain a spatial resolution of about 200 nm when an optimal thickness in the range $h = 100$–200 nm is used.

The saturation field of such a film equals 500 Oe. By optimizing the reading wavelength—for example, using $\lambda = 440$ nm—it is possible to obtain rather a high magnetooptical reading signal (Burkov *et al* 1986).

Nomura (1985) have proposed a method for the information read-out from a magnetic carrier which uses a system of the stripe domains oriented along the recording tracks (see figure 13.20(b)). In this method of information pick-up, the magnetic leakage fields, which are related to the magnetic charges on the carrier surface, modulate the stripe domain width, i.e. magnetic leakage fields that are oriented upwards increase the widths of the domains whose magnetization is oriented along the field and decrease the widths of the domains that are oriented otherwise.

When the carrier is displaced with respect to the magnetooptical head, the area that contains the width-modulated domains follows the displacement of the information carrier. Like in the previous case, the resulting magnetization modification in the laser-illuminated area, which results from the change of the domain area for upward- and downward-oriented domains, leads to modification of the signal amplitude after the light has propagated through an analyser.

In the presence of magnetic leakage from the domain structure of the information carrier, the domain width varies almost sinusoidally with the period, which is determined by the wavelength of the recorded information. If the magnetization depends sinusoidally on the information that is recorded on a magnetic medium (for example, if the information carrier is a magnetic disk), the average magnetic leakage field H_z in the z-direction (normal to the film surface) is described by the expression (Nomura 1985, Tokumaru and Nomura 1986)

$$\bar{H}_z = (2\pi h)^{-1} l_r H_p [1 - \exp(-l_r^{-1} 2\pi h)] \exp(-l_r^{-1} 2\pi S)$$

where S is the gap between the magnetic disk and the iron garnet film surface, and H_p is the magnetic leakage field at the surface of the magnetic disk.

One of the variants of disk memory with a magnetooptical pick-up device for reading information recorded using a magnetic head was described by Nomura (1985). The memory medium is a disk of 35 cm diameter with a CrO_2 magnetic layer. For a rotation speed of the dc motor higher than 1000 revolutions per minute, the disk rotates stably, hovering on the air gap, which is 300 μm thick. The signals were recorded onto the magnetic disk using a magnetic head having a recording track width of 120 μm and a gap width of 1.8 μm. The recording area occupied a surface of 13–15 cm radius. A He–Ne laser was used as the light source; its beam was focused onto the iron garnet film surface by an objective with a focal length of 7.8 cm. Glan–Thompson prisms were used as the polarizer and analyser, and an avalanche photodiode was used as a photodetector. The frequency response of the system, which comprised the photodetector and the amplifier, was linear in the range 0.1–10 MHz.

The magnetooptical pick-up assembly was installed in a holder in front of the lens objective. At a rotation speed exceeding 1000 revolutions per minute, the air gap remained constant due to the air-cushion effect, provided that the dimensions of the magnetooptical element were less than or equal to 1.5×1.5 mm. Technical data related to the magnetic disk, and the parameters of the epitaxial iron garnet film that was used in the magnetooptical pick-up head, are listed below, in table 13.3.

Table 13.3. Parameters of the film used in the magnetooptical read head (Nomura 1985).

Magnetic medium	CrO_2
Residual magnetic induction, B_r (G)	1500
Coercive force, H_c (Oe)	550
Magnetic film thickness (μm)	5
Film composition	$(YSmLuBiGa)_3(FeGe)_5O_{12}$
Saturation magnetization (G)	439
Film thickness, h (μm)	2.7
Domain structure period, P_0 (μm)	4.5

Use of the above-described system demonstrated the possibility of a distortion-free playback for signals having the recording wavelengths $l_r = 6.3$ μm and 50 μm, and for a track width of 15 μm, with the neighbouring tracks being separated by non-modulated stripe domains. Using the results, one can evaluate the laser beam diameter: $d \approx 5$ μm. This value agrees well with the theoretical one:

$$d = (\pi D)^{-1} 4\lambda F \approx 5 \ \mu m$$

where D is the diameter of the incident laser beam (in our case it corresponds to the objective aperture), which was 1.3 mm, $F = 7.8$ mm, and $\lambda = 0.63$ μm.

In principle, it is possible to read out up to the track width that is close to the stripe domain width. The advantage of the method described, in comparison with that based on a maze-like domain structure, is the absence of sinusoidal distortions of the domain structure when the leakage field H_z decreases. The read-out is stable over a wide recording wavelength range from 5 to 200 μm, and the shape of the oscillation is not rectangular even when the recording wavelength is larger than the diameter of the pick-up light beam. The read-out characteristics remain almost unchanged even when the disk rotation speed varies from 16 to 50 m s^{-1}, while in the case of magnetooptical heads with the maze-like domain structure, the amplitude falls abruptly, and the noise grows at $v = 30$ m s^{-1} (Nomura 1985).

In order to increase the spatial resolution, Tokumaru and Nomura (1986) proposed a magnetooptical head which is shown in figure 13.21. Magnetic film with high magnetic permeability provides the penetration of the magnetic flow from the information carrier to the magnetooptical read-out, and the domain width in the magnetooptical film is modulated to the extent that is quite sufficient to detect the magnetization changes using a laser beam. The new design of the pick-up head is expected to provide the information read-out from the carrier with enhanced track densities and higher recording density in the track. In the experiments, a domain width variation of more than 1 μm was detected, while the nominal domain width was 2.5 μm. Calculations show that the system can provide a signal-to-noise ratio of about 75 dB with a laser beam diameter of 2 μm.

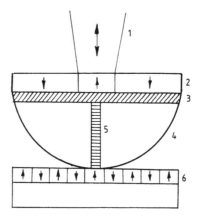

Figure 13.21. A magnetooptical head with magnetic flow closure. The components are: 1: the laser beam; 2: the magnetooptical read-out medium; 3: the mirror layer; 4: glass; 5: a film with high magnetic permeability; and 6: the magnetic information carrier (Tokumaru and Nomura 1986).

In order to overcome the drawbacks which are intrinsic to magnetooptical pick-up devices that use films with maze-like or stripe structure as a recording medium, it is possible to utilize epitaxial bismuth-substituted iron garnet films with so-called in-plane or quasi-in-plane magnetic anisotropy. In these films, the following relation is valid: $H_a \sim 4\pi M_s$ (the uniaxial magnetic anisotropy field is almost equal to the film demagnetizing field). When substrates with (111) orientation are used, the magnetization usually tilts away from the film surface at an angle of 15–25° if the condition $H_k \sim 4\pi M_s$ is satisfied (the angle is measured from the film surface), or, to be more precise, the magnetization orients as close as possible to one of the ⟨111⟩ axes, which are directed at the angle of 19.5° with respect to the film surface.

In the general case, in these films a low-contrast wedge-like domain structure is formed, which may be eliminated by applying a uniform magnetic field with a strength of 10–30 Oe in the film plane. If the condition $H_k \approx 4\pi M_s \approx 500$ Oe is satisfied, the magnetization of this film follows the vertical component of the leakage field for the magnetic tape, which is brought into contact with the film surface. Figure 13.22 shows a diagram of the magnetic leakage fields of a videotape with an epitaxial 4 μm $(BiLu)_3(FeGa)_5O_{12}$ film with quasi-in-plane magnetic anisotropy ($K_u = 5 \times 10^2$ erg cm^{-3}, $4\pi M_s = 200$ G) (Gusev and Kotov 1991).

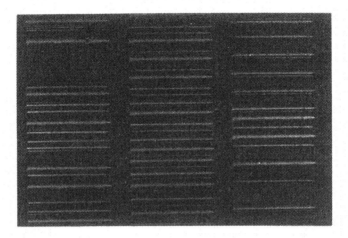

Figure 13.22. A magnetooptical diagram of the information recorded on a videotape using bismuth-substituted iron garnet film with quasi-in-plane magnetic anisotropy (Gusev and Kotov 1991).

Films with quasi-in-plane magnetic anisotropy are free from specific limitations on the period of the read-out information and, as the experimental results demonstrate, provide read-out of information, which is recorded on magnetic tape, with periods from 0.6 μm to several hundred micrometres (Gusev and Kotov 1991).

The domain structure shape-memory effect in YIG:Co, Ge, Ca films studied by Kisielewski *et al* (1995) provides interesting possibilities as regards storing and translating magnetic images from one site to another. This phenomenon is related to magnetization-induced anisotropy, and non-collinear magnetization directions in neighbouring domains.

Polanskii *et al* (1990) and Wijngaarden *et al* (1995) reported observations of magnetic flux structures in high-T_c superconductors. The magnetic flux in a superconductor is mapped by detecting the rotation of the polarization vector of light (Faraday effect) within a magnetooptically active EuSe layer which is evaporated onto the surface of the sample.

13.9 Laser gyroscopes with magnetooptical elements

Thin magnetic films are used in novel laser gyroscope designs, which are used in platform-free navigational systems. The sensitive element of the laser gyroscope is a laser with a ring resonator, where two independent counterpropagating radiation modes are generated. These counterpropagating-mode frequencies are determined by the speed of rotation of the ring resonator in inertial space. The frequency difference of the counterpropagating modes is proportional to the speed of angular rotation (Krebs *et al* 1980).

When the ring laser gyroscope is spinning around the rotation axis, the frequencies of the counterpropagating modes change. For a viewer that rotates with the cavity, it would take longer for the wave that propagates along the direction of rotation to pass the closed loop than for the wave that propagates in the opposite direction. As the velocity of light is constant, this effect could be explained by the viewer as an increase of the total optical path along the direction of rotation, and as a decrease of the optical path in the opposite direction. In the resonant case, the cavity length is a multiple of the light wavelengths, i.e., the change of the optical path leads to frequency shifts of opposite sign for the two counterpropagating modes.

For applications in laser gyroscopes, He–Ne lasers with radiation wavelengths of 0.63 and 1.15 μm are best. The first experiments with laser gyroscopes revealed that their real characteristics differ from what was expected. At low angular velocities, there are: a range of insensitivity and a drift of the zero; instability of the characteristics over time and from one switching to another; and sensitivity to external conditions.

The reason for the region of insensitivity, and for the non-linearity of the characteristics at low speeds of angular rotation, is the coupling of opposite modes, which results from the backscattering by the resonator elements. This coupling leads to mode locking (the capture effect) over a certain range of rotation speed: $-\Omega_L < \Omega < \Omega_L$, where Ω_L is the capture threshold.

In order to minimize the counterpropagating-mode coupling due to the backscattering, the laser gyroscopes utilize multilayer dielectric mirrors (up to twenty layers), which provide reflection coefficients higher than 99.7% and backscattering coefficients lower than 0.05%. Nevertheless, the typical value for Ω_L is of the order of hundreds of degrees per hour. To eliminate the effect of capture, the laser gyroscope's operating point is shifted via an artificially created path difference of the counterpropagating modes, which results in frequency splitting ('substitution') even when the laser gyroscope does not really rotate. The required frequency difference is created either by non-reciprocal optical effects (Faraday or Kerr effects), or by rotating the laser gyroscope with constant speed or by inducing mechanical torsional oscillations.

One of the laser gyroscope designs employs a magnetooptical mirror. In this case the frequency substitution results from the transverse magnetooptical Kerr effect. In a previous design, a Faraday element was used, but a Faraday cell in the resonator requires temperature stabilization, gives rise to light backscattering, and substantially increases the cost of the whole system. One of the drawbacks of this solution is higher loss in the resonator due to radiation absorption by the magnetooptical mirror. A laser gyroscope with a magnetooptical mirror has a temperature sensitivity lower than 0.02 deg h^{-1} K^{-1} and a magnetic sensitivity of 0.03 deg h^{-1} Oe^{-1}, and tolerates accelerations up to 20g.

The Faraday cell was used in the differential laser gyroscope to escape the capture range. The device employs the four-frequency operating mode with circularly polarized radiation, and with two pairs of the counterpropagating modes excited simultaneously. As the polarizations of the gyroscope modes are different, the Faraday cell creates a bias for each of the modes; these biases have the same value but opposite signs. When the output signals are added, the biases that are created by the Faraday cell compensate each other, while the shifts that are induced by the rotation add to one another. Circularly polarized radiation ensures lower counterpropagating-mode coupling due to the backscattering, and, consequently, a lower effect of capture on the laser gyroscope's performance.

Intra-cavity optical elements complicate the laser gyroscope design, and increase its temperature sensitivity; therefore, at present the problem of the design of magnetooptical mirrors with low losses is of high priority. For this reason, a Faraday cell for utilization in a ring laser resonator should satisfy the following criteria.

(i)   The field for the reverse switching of the magnetization is less than 2 Oe.

(ii) The insertion optical losses are less than 0.5%.

(iii) The magnetooptical figure of merit F_1 more than 60°.

Two variants of Faraday elements, based on epitaxial iron garnet films, were proposed for utilization in the ring laser resonators (Whitcomb and Henry 1978). The first structure consists of a one-sided epitaxial iron garnet film of thickness $\lambda/(2n)$ on a GGG substrate, and of two antireflection MgF_2 coatings of thickness $\lambda/(4n)$, which are deposited on the epitaxial layer and on the substrate.

For the Faraday element of the first type, Whitcomb and Henry used epitaxial layers of the composition I: $(YBi)_3(FeGa)_5O_{12}$. Special measures were taken to decrease the optical absorption at the operating wavelength, $\lambda = 1.152$ μm. In the series of procedures in which the epitaxial layers with different contents of Ca^{2+} ions were grown, the absorption coefficient α varied from 410 to less than 10 cm^{-1}. The epitaxial structure parameters are listed below, in table 13.4 (Whitcomb and Henry 1978).

The main drawback of the Faraday element, which is shown in figure 13.23(a), is the reflection losses at the epitaxial film–substrate interface. To avoid the losses, Whitcomb and Henry (1978) fabricated the Faraday element which is shown in figure 13.23(b). The losses at the film–substrate interface were suppressed owing to the sophisticated fabrication technology. First, the iron garnet layers of thickness $\lambda/(4n)$, which correspond to the antireflection coating, were grown epitaxially on the GGG substrate. For the antireflection coating, they used the composition $(YGd)_3(FeGa)_5O_{12}$, in which one can vary the refractive index from 2.00 to 2.12 by controlling the gallium content. When the optimal value of the refractive index was

Table 13.4. Parameters of the magnetooptical structures (Whitcomb and Henry 1978).

Composition	I	II
Refractive index, n	2.19	2.11
Layer thickness, h (μm)	3	3
Switching field, H_s (Oe)	10	10
Faraday rotation, Φ_F (deg cm^{-1})	200	200
Absorption coefficient, α (cm^{-1})	10	10
Wavelength, λ (μm)	1.152	1.152
Transmission, T		0.995

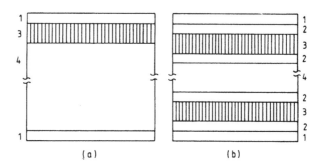

Figure 13.23. Multilayer magnetooptical structures with antireflection coatings for laser gyroscopes. Components: 1: the MgF$_2$ film, $h = \lambda/2$; 2: the paramagnetic iron garnet antireflection coating; 3: the epitaxial iron garnet film; and 4: the substrate (Whitcomb and Henry 1978).

reached, the antireflection coating material was in a paramagnetic state at room temperature.

The magnetooptical layers of the composition II: (YLa)$_3$(FeGa)$_5$O$_{12}$ were grown on the internal antireflection coatings (see figure 13.23(b)). Then the antireflection epitaxial films of paramagnetic iron garnet were grown on the surface of these magnetooptical layers, and finally the system was antireflection coated with MgF$_2$. The structure, which is shown in figure 13.23(b), exhibited extremely low reflection losses (the system transmittance exceeded 0.995).

A magnetooptical element of the first type (see figure 13.23(a)) in a ring laser gyroscope provided 'substitution' that was equivalent to a speed of rotation of 350 deg s^{-1}, which led to the total elimination of the capture effect, the overall losses being lower than 1%, while the losses associated with the magnetooptical element were less than 0.5%. The optical losses which were introduced by the element that is shown in figure 13.23(b) were below 0.1% (Whitcomb and Henry 1978).

13.10 Magnetic field sensors

Ferrimagnetic iron garnet films and bulk single crystals have been investigated as possible current-sensing elements. The Faraday effect within the films permits measurement of the magnetic field or current by a simple polarimetric technique. In some cases the magnetooptical measurement and control systems have various merits compared with the usual magnetic system, including high electrical insulation, strong resistance to electromagnetic induction noise, and non-contact sensing (Ko and Garmire 1995). Magnetooptical field sensors are used to detect the current variations in high-voltage power generating systems. Hence high sensitivity, high accuracy over the operating temperature range, and a high degree of reliability are required for such sensors.

Kamada *et al* (1987) developed magnetic field (current) sensors using the Faraday effect for $Tb_3Fe_5O_{12}$ garnet. The temperature coefficient of the Faraday rotation is a crucial problem when wide temperature operation is required. For example, the temperature coefficient of the Faraday rotation was $+1\%$ for the temperature range from -20 to $+110°C$, and the minimum detectable field was 0.02 Oe for the field range 2–50 Oe.

In Bi-substituted iron garnet films grown by liquid-phase epitaxy, the sensitivity constant S and its temperature coefficient were studied by Kamada *et al* (1987, 1994). The wavelength dependences of the sensitivity constants of $Bi_{1.3}(YLa)_{1.7}Fe_5O_{12}$, in comparison with $(TbY)_3Fe_5O_{12}$, and Flint glass, are given in table 13.5. The temperature sensitivity of Bi-substituted iron garnets can be considerably improved using bilayer films or Bi–Tb iron garnet compositions, as described in section 13.3.

Table 13.5. The sensitivity constants S of $Bi_{1.3}(YLa)_{1.7}Fe_5O_{12}$, $(TbY)_3Fe_5O_{12}$ epitaxial films, and Flint glass (Kamada *et al* 1987).

Wavelength (μm)	S (deg cm^{-1} Oe^{-1})		
	(BiYLa)IG	(TbY)IG	Flint glass
0.8	6.1		8×10^{-4}
1.15	2.0	0.48	
1.3	1.43	0.35	

Kamada *et al* (1994) measured the temperature characteristics of $(BiYGd)_3Fe_5O_{12}$ films, and found that $Bi_{1.3}Gd_{0.43}Y_{1.27}Fe_5O_{12}$ film has the minimum temperature dependence, within 0.5%, over the temperature range from -20 to $+120$ °C

One of the most important applications of magnetooptical field sensors may be that of current measurement for high-voltage conductors. A iron garnet film or crystal is mounted adjacent to a conductor carrying a current. The plane-polarized light from a laser or transmitted through an optical fibre is projected onto the sensing sample, and is reflected by a mirror behind the sample. The angle of rotation of the plane of polarization in the return beam is simply related to the current flowing in the conductor. This method avoids many of the insulation difficulties associated with current measurements for high-voltage conductors, and is suitable for dc or ac operation. At the same time, Bi-substituted iron garnet films with a high degree of temperature dependence of the Faraday rotation are suitable for use as temperature sensors.

The operation of most sensors utilizing thin magnetooptical films is based on domain motion. However, this is not adequate for highly sensitive magnetic field sensors, because of Barkhausen jumps (the amplitude effect) and the domain wall's speed limitation (the frequency effect). Deeter (1995) has analysed the non-linear response phenomenon for magnetic field sensors based on iron garnet thick films and bulk crystals related by their domain structures. It has been shown that differential signal processing resulted in a linear signal for the thick films and a primarily sinusoidal response for the bulk crystals.

Deeter *et al* (1994) have used uniaxial iron garnet films to sense the magnetic fields acting in the plane of the film. A magnetic field applied in the plane of such a film deflects the magnetization of both types of domain toward the applied field, equally. Thus, magnetization rotation, as opposed to domain wall motion, should be the dominant response to an in-plane magnetic field in these films. For magnetic field sensing, magnetization rotation is preferred over domain wall motion, since it does not generally produce hysteresis, and is also typically faster than domain wall motion. Therefore, sensors exploiting these films in a planar geometry (figure 13.4(b)) should exhibit a fundamentally linear response (without Barkhausen jumps), and a better frequency response, up to 1 GHz and higher.

Using planar geometry, Deeter *et al* (1994) attained a magnetooptical sensitivity of 2.5 deg Oe^{-1} for 1.3 μm. They demonstrated the signal linearity, and apparent lack of hysteresis in response to a field amplitude up to 1.5 Oe; at the same time, when the field strength was increased to a field amplitude of 20 Oe, an obvious hysteresis was found. In the case of homogeneous rotation of the magnetization, these restrictions are absent, i.e., the effect of the material's coercivity is minimized. A homogeneous rotation state may be reached by magnetizing the sample up to saturation. Bi-substituted iron garnet films with in-plane and quasi-in-plane magnetic anisotropy are more suitable for these purposes. Because cubic anisotropy prevents homogeneous rotation of the magnetization in the plane of the film, films with minimized cubic anisotropy are the most suitable for use in such magnetooptical field sensors.

Deeter (1995) studied the influence of domain-induced diffraction effects for iron garnet films and crystals on the response functions of magnetic field sensors. It was found that thick films with stripe domain structure produced non-linear response functions, and that the bulk crystals produced qualitatively similar effects. Differential signal processing resulted in a linear signal for thick films and a sinusoidal response for the bulk crystals. These diffraction effects should be considered in the design of iron-garnet-based fibre-optic magnetic field sensors, in which the coupling of light from the film to the fibre could spatially filter out all but the zeroth-order diffracted beam.

A device for the determination of the magnitude and the direction of the in-plane magnetic field strength vector by a magnetooptical method based on Bi-substituted iron garnet film of composition $(BiPrLu)_3Fe_5O_{12}$ was developed and created by Vetoshko *et al* (1991) and Boardman *et al* (1994). Use of the mode of the homogeneous rotation of the magnetization vector of the substituted iron garnet film allowed Vetoshko *et al* to obtain the time response of the system for any magnetic field change over about 2 ns without a reduction of the magnetic material susceptibility. The minimal value of the magnetic field strength measured was 10^{-6} Oe with a spatial resolution of 3 μm. The high sensitivity of the device was achieved by the measurement of the magnetic field in-plane components in orthogonal directions, and by the application of the optical heterodyning technique, in which a single-frequency Ar-ion laser ($\lambda = 514.5$ nm) with a stabilized output was used. To realize the optical heterodyning, as well as to increase the dynamic range of the device, an electrooptical phase modulator was included in the optical set-up.

Kapitulnik *et al* (1994) described polar Kerr effect measurements with an accuracy of 3 μrad, with a spatial resolution of 2 μm, made using a Sagnac interferometer. The interferometric technique provides a number of advantages over conventional polarizer methods, including insensitivity to linear birefringence, the ability to completely determine the magnetization vector in a region, and the ability to sensitively measure magnetooptical effects without applying an external field.

Chapter 14

Magnetooptical memories, disks, and tapes

14.1 Memory addressed by a laser beam

In magnetooptical data storage (DS) with a fixed storage medium the information
is recorded and read by a laser beam. A schematic diagram of such a DS system
is shown in figure 14.1. For information storage, magnetooptical media with
uniaxial magnetic anisotropy ($Q > 1$) and high Faraday or Kerr rotation, in
which the magnetization is oriented perpendicular to the surface of the film, are
employed. Most examples of DS of this type are based on epitaxial iron garnet
films.

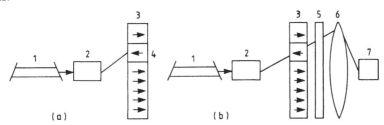

Figure 14.1. A schematic diagram of magnetooptical DS addressed by a laser beam;
recording (a) and reading (b) of the information. Components: 1: the laser; 2: the
deflector; 3: the magnetooptic storage medium; 4: the informational domain (bits of
information); 5: the analyser; 6: the lens; and 7: the photodetector.

Information is recorded using the thermomagnetic method: laser radiation
is focused onto a spot about one micrometre in diameter. When an area
of the storage medium is heated locally, it remagnetizes as a result of the
simultaneous effect of the demagnetizing field and a constant or pulsed bias
field with an intensity of about 100 Oe. The pulsed magnetic field is applied
almost simultaneously with the impact of the laser beam. The information is
read by a laser beam of lower intensity.

Depending on the organization of the information cells, the magnetooptical

311

DS addressed by the laser beam has an information storage density of about 10^6–10^7 bits cm^{-2}; the random-access time is about 0.1–5 μs (Chang *et al* 1965, Paroli 1984). Magnetooptical DS of this kind has substantial advantages over other types: the absence of moving parts; and a high data rate. On the other hand, it is difficult to design DS of this type with high capacity (more than 10^6 bits), because existing deflectors of laser beams give a restricted number of resolved spots.

As in DS on magnetooptical disks, the information can be recorded by heating a local spot of the field to a temperature above the Curie temperature or the compensation temperature. The possibility of thermomagnetic recording on iron garnet film with $4\pi M_s = 36$ G and $T_c = 125$ °C was for the first time demonstrated by Krumme *et al* (1972) and Hill *et al* (1975). In film that was 4 μm thick, mesa structures of square cells with the side 10 μm were formed, the groove depth between the cells being 2 μm, i.e., the film was etched to half of its thickness. The groove width was also 2 μm. The cells were switched by the impact of a laser beam of 6 μs duration focused to a spot 4 μm in diameter. The laser power was 20 mW and a constant bias field with an intensity of 10 Oe was applied during recording.

The possibility of thermomagnetic recording on the epitaxial films with bubble domains that were used for DS was demonstrated by Kaneko *et al* (1983). During the recording, storage, and reading of the information, the film was maintained in a constant bias field H_s, which satisfied the condition $H_2 < H_s < H_0$, where H_2 and H_0 are the strengths of the magnetic field for elliptic instability and the collapse of the cylindrical magnetic domain, respectively.

When a local area of the film is irradiated, in the course of heating this area is remagnetized as a result of the simultaneous effect of the demagnetizing field and the bias field. The size of the bit of information is determined by the diameter of the cylindric magnetic domain which exists in the film at the given bias field, i.e., by the magnetic characteristics of the material. In epitaxial iron garnet films, magnetic domains of diameter 0.2 μm can be achieved (Starostin and Kotov 1982). To stabilize the bubble domain in a particular position, a moderate coercive force is required, which is easily attainable during the epitaxial growth of the film; the distance between the domains can be three times their diameter.

Whereas for the Curie point recording, the film must be heated by 100–300 °C (figure 14.2(a)), for the compensation point recording, the information cell can be switched when it has been heated by just several dozens of degrees. If the compensation point is near room temperature, the magnetization is near zero, and the coercive force is rather high (figure 14.2(b)). As the Faraday effect for iron garnet is not proportional to the resultant magnetization of the material, but is instead a difference between the contributions of octahedral and tetrahedral magnetic sublattices which have opposite signs near the compensation point, the magnetic domains with different magnetization directions of the sublattices will rotate the planes of magnetization of the light in opposite directions even if $M_s \approx 0$. The contrast between the two cells will be rather high. In order to record information at the compensation point, it is enough to heat the film area by 20 °C; a resultant magnetization whose direction can be switched by application of a pulsed bias field will appear.

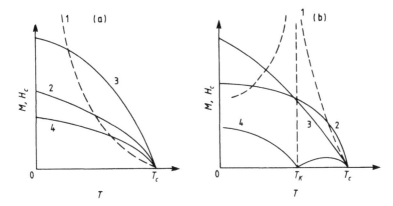

Figure 14.2. The temperature dependencies of the coercive force H_c (curve 1) and the resulting magnetization M (curve 2) as functions of temperature for the cases of Curie point (a) and compensation point (b) thermomagnetic recording. Curves 3 and 4 represent the temperature dependencies of the magnetization for the sublattices (rare earths and transition metals, respectively, for amorphous RE–TM metal films).

In order to form information cells with the area 12×12 μm^2 having local variations of the compensation temperature, the epitaxial film, which had a sputtered silicon thin film between the information cells, was annealed. This treatment created a matrix of magnetooptical cells, where the compensation temperature in the cells was 20 °C higher than in the rest of the array. For an appropriate choice of the composition of the initial film and the annealing conditions, the compensation temperature of the regions between the information cells is near 300 K. In this structure the cells were switched by a laser pulse of 5 μs duration; the laser power was 25 mW, and during the recording a constant bias field of 40 Oe was applied (Krumme and Hansen 1973).

In order to lower the laser power which is used for recording the information, it is desirable to increase the absorption coefficient of the iron garnet in the visible region, which—at the same absorption level—decreases the film thickness. By introducing Co^{2+} ions into the structure, the laser power for recording was reduced from 24 to 4 mW (Inoue *et al* 1983).

Different variants of double-layer structures have been proposed, in which a layer with a higher absorption coefficient (for example, an intermetallic film of Dy–Fe or some other composition) was deposited on the epitaxial iron garnet film; the laser power in this case was reduced to 2 mW with the reading characteristics remaining the same, because the information is read from the side of the iron garnet film with doubled Faraday rotation (the light passes through the iron garnet film, reflects from the metal film or from a special reflecting coating, and again passes through the iron garnet film) (Yokoyama *et al* 1983).

A double-layer structure, which consists of an epitaxial iron garnet film 2 μm thick, which is coated successively with a transparent In_2O_3/Sn_2O_3 conductor layer, with a layer of the photoconductor CdS:Cu that is 4 μm thick, and with a layer of the conductor In_2O_3/Sn_2O_3, lowered the laser power required for the recording by several orders of magnitude. In one case, isolated cells, of area 8×8 μm^2, were formed in the epitaxial film; the distance between the cells was 20 μm (Krumme and Hansen 1973, Krumme and Schmitt 1975, Krumme *et al* 1977).

When the cell in which the bit of information was recorded, was illuminated, the resistance of the photoconductor decreased by about three orders of magnitude. As the cell is simultaneously affected by the pulse of light and by the pulse of current of 30 μs duration, which is applied to the photocell (the voltage on the electrodes is 5 V), it is heated by approximately 20 °C, so when the external magnetic field with intensity 100 Oe is applied to the cell, it switches. The full switching time is defined by the heat dissipation, and is approximately equal to 100 μs. Because the information cell is heated by the ohmic heat of the photoconductor rather than by a laser beam, optical power of only 1 μW is required for the recording of the information.

An organization scheme for a random-access magnetooptical memory, based on a double-layer structure composed of a photoconductor and a magnetooptical material, with blocks of 992 bits each, an access time less than 5 μs, and a total memory capacity of 6.5×10^7 bits was proposed (Hill *et al* 1975). Sixteen planes were intended to be used for storage of the information, each plane containing 8×8 blocks, and each block consisting of 256×256 memory cells. The experimental device had a total informational capacity of 1.5×10^5 bits (Krumme and Schmitt 1975, Krumme *et al* 1977).

14.2 Magnetooptical disks

The computer industry has long dreamed of data-storage systems that combine the major advantages of optical disk recording (gigabyte capacity and non-contact technology) and magnetic recording (erasability, reusability, longevity). The advantage of magnetooptical recording in comparison with conventional magnetic recording is the combination of the merits of magnetic and optical techniques, i.e., cyclability of magnetic media that is in principle unlimited, and high storage density, contactless reading and writing, and removability of optical media (Mansuripur 1994). Erasable magnetooptical recording is widely used today in high-performance disk drives, and is viewed as a possible successor to electromagnetic recording that will be used for mainframe computers, hard disks, and magnetic tape drives. Magnetooptical recording using rare-earth–transition metal alloy films exhibiting large perpendicular magnetic anisotropy offers high-density recording (densities near 10^8 bits cm^{-2})—densities ten times those of high-performance disk drives and up to 100 times those of low-end disk drives, lifetimes longer than 10 years, and data transfer rates up to 1 Mbyte s^{-1} with a typical access time of approximately 40 ms.

The disks are removable like flexible disks, but are rigid, and may be rotated at speeds as high as in conventional winchester technology. The write and erasure operations can be performed by laser light modulation or magnetic field modulation at GaAs laser wavelengths. Since the bit density along a track is compatible with today's winchester technology, the data rates are the same also. Access on a disk is achieved with a moving-head actuator like those in winchester rigid-disk files, but containing an optical head. The access time slightly exceeds that for winchester technology because the weight of the optical head is higher, but approximately 100 tracks which are adjacent to the track being read, each containing about 5×10^5 bits, can be accessed in a millisecond. The fundamental broad-band signal-to-noise ratio (S/N) for existing rare-earth–transition metal media is well above 20 dB (a typical value of the carrier-to-noise ratio (C/N) at a bandwidth of 30 kHz is between 45 and 60 dB, depending on the material used, the write and read power, the mark length, and some other parameters), and therefore adequate for maintaining error rates less than 10^{-12}. Substrate imperfections in optical media lead to error rates of 10^{-5}–10^{-6}, but even these relatively high raw-bit error rates are corrected to 10^{-13} by error-correction codes and the hardware being used in write-once optical disk storage systems.

14.2.1 The disk assembly

In a magnetooptical recording system, writing is achieved by thermomagnetic means, whereas reading is performed using either a polar Kerr (in standard systems) or a Faraday (in experimental systems) magnetooptical effect. The thermomagnetic recording is based on locally heating a small spot with a diameter near 1 μm above the Curie temperature.

The process of the recording and reading of the information onto and from a magnetooptical disk is shown in figure 14.3(a). The laser beam (1) passes through a grating (2), a lens (3), a polarizer (4), a beam splitter (5), and through an objective lens (6) to the disk surface (7). In reality, the construction of the magnetooptical head includes a moving reflecting mirror and a moving objective driven by voice coils, which affords one the opportunity to focus the laser beam onto the surface of the disk, to a spot one micrometre across, and to track the groove. The beam reflected from the non-coated amorphous rare-earth–transition metal film has its plane of polarization rotated by about $0.2°$ to $0.5°$ ($\lambda = 780$ nm, $T = 300$ K) due to the polar Kerr effect.

A real magnetooptical disk presents a multilayer structure (a minimum of three layers in the structure: an antireflecting–protective layer on the substrate, a magnetic rare-earth–transition metal layer, and a dielectric protective layer) including an antireflecting layer, and the real angle of rotation can exceed one degree. At the same time, the reflection coefficient of such a structure is lower than that of an unprotected film. In practice, to characterize the magnetooptical quality of the reading media, the product $\Phi_K^2 R$ is used. Finally the reflected beam passes through the objective lens (6), and is reflected by the beam splitter (5), and then comes through the $\lambda/4$ waveplate (9) to the polarizing beam splitter (10). If the polarizer between the lens L_1 and the beam splitter is set at a $45°$ angle with respect to the propagation axis of the polarizing beam splitter, then signals coming through the cylindrical lenses (11) will reach the photodiode assembly (12), and can be subtracted to give the desired signal without the common-mode noise.

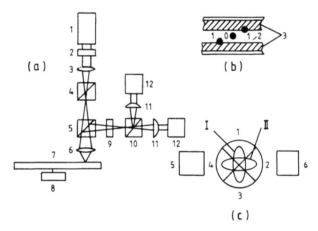

Figure 14.3. (a) A schematic drawing of the optical system for a magnetooptical drive. Components: 1: the laser; 2: the grating; 3: the lens L_1; 4: the polarizer; 5: the beam splitter; 6: the objective lens; 7: the magnetooptical disk; 8: the disk drive; 9: the $\lambda/4$ waveplate; 10: the polarizing beam splitter; 11: cylindrical lenses; and 12: the photodiode assembly. (b) The tracking laser beam spots (first order) and the read–write laser spot (zero order) are shown on the disk groove. (c) The photodiodes P_5 and P_6 for tracking, and the photodiodes P_1, P_2, P_3, and P_4 for focusing and for the signal recording (Kryder 1985, Meiklejohn 1986).

Kryder (1985) and Meiklejohn (1986) discussed magnetooptical disk recording systems which used a GaAs diode laser ($\lambda = 830$ nm) to provide the heat source for writing and the light source for reading information. The bias magnetic field is provided by a coil. The field is generated in the coil by a current pulse in one direction for writing and in the opposite direction for erasing the domain. It should be noted here that, in standard systems, overwriting information needs one rotation of the magnetooptical disk to erase recorded information and another rotation to record new information, which is a disadvantage of such a system.

Before considering the processes of tracking and focusing the laser beam on the surface of the magnetooptical medium, let us consider the construction of the disk. One method of making a magnetooptical disk substrate is to inject a soft plastic material into a die that contains a grooved mould. The spiral grooves are about 70 nm deep, 700 nm wide, and 1.6 μm apart. The plastic used to make the substrate must have a very low birefringence compared with that of a standard read-only audio disk, since here the polarized laser beam passes through the substrate material. Birefringence causes depolarization of the light, and decreases the real signal-to-noise ratio.

In order to prevent degradation of the magnetooptical film, a protective coating is deposited on the substrate. The magnetooptical film (in primary systems, amorphous TbFeCo) is deposited on the groove side of the substrate by sputtering or evaporation to a thickness of about 100 nm. A second protective coating is then deposited over the magnetooptical film.

Focusing the laser beam through the substrate gives an essential advantage compared to focusing the beam through the upper protective layer: in this case the diameter of the laser beam on the surface of the substrate is about 1 mm, and small scratches and dust particles on the surface do not seriously influence the recording signal. For this reason, only one side of the magnetooptical disk can be used to record data. In order to record information on both sides, two disks, each having a magnetic film 100 nm thick, are placed back-to-back, so that the grooved sides are adjacent to each other.

Tracking. As was mentioned earlier, the laser beam passes through a grating. The zero-order beam is used for writing and reading information. Typically, laser powers of 8–10 mW and 2–4 mW on the surface of the magnetooptical medium are used for writing and reading, respectively. The two first-order beams from the optical grating are used for tracking, as shown schematically in figure 14.3(b) (Meiklejohn 1986). The cross-hatched region represents the grooves in the disk. The area between the grooves represents the recording track. The three beams are focused on the disk in such a way that the two first-order beams are slightly displaced from the track in opposite directions, and each of the first-order beams partly covers the recording track and partly the groove.

As we mentioned earlier, the grooves are about 70 nm deep and usually have a pitch of 1.6 μm. The reflected first-order beams, after coming through the objective lens, reflecting from the mirror, and coming through the beam splitter, polarizing beam splitter, and cylindrical lenses, fall on separate photodiodes, P_5 and P_6, as shown in figure 14.3(c). The difference signal from these photodiodes is maintained at zero by the radial movement of a mirror or an objective lens driven by a servodrive, keeping the zero-order write–read beam on the track.

Focusing. In order to maintain the focusing on a rapidly rotating disk, a variety of techniques have been developed. An objective lens which focuses the beams on the magnetooptical recording medium is mounted on a coil–permanent-magnet arrangement to allow movement of the lens normal to the surface of the disk. Usually a cylindrical lens and a quadrant photodiode, to produce an error signal when the medium begins to move out of focus, are used.

The ring photodetector used to read information and to focus laser beams on the surface of the magnetooptical medium consists of four pie-shaped quadrants, as shown in figure 14.3(c). The sum of the output of these four quadrants is the read-out signal. At the same time, the cylindrical lens positioned between the polarizing beam splitter and the photodiode assembly (see figure 14.3(a)) causes an elongated beam to appear on the photodiode when the surface of the magnetooptical medium is out of focus, the direction of elongation depending on whether the objective lens is too far away from or too close to the surface.

As shown in figure 14.3(c), the major axis of the ellipse will be vertical if the medium is too far away, and horizontal if the medium is too close. The focal lengths and optical path lengths between the lens L_1, the objective lens, the disk surface, and the cylindrical lens are chosen so that when the laser beam is properly focused on the disk surface, the reflected light beam is also focused on the cylindrical lens. The beam is circular when the magnetooptical medium is in focus. The difference signal representing the subtraction of the sum of the outputs of photodiodes P_1 and P_3 from the sum of the outputs of photodiodes P_2 and P_4 produces the error signal which is fed to the coil of the servodrive to maintain the focus.

14.2.2 Thermomagnetic recording

The process of local magnetization reversal by thermal cycling of a small area in a thin film of MnBi was first demonstrated by Mayer (1958) using electron beam heating. He named the process 'Curie point writing', and explained it on the basis that when the small area was heated above its Curie temperature, it would regain its magnetization on cooling in the direction of the demagnetizing field produced by the adjacent ferromagnetic material. Since that time, the process of Curie point writing has been applied as a method of storing information in magnetooptical media (Fan and Greiner 1968, Chen *et al* 1968, Lewicki 1969). The mechanism of Curie point writing in thin magnetic films has been studied by Bernal (1971).

The reverse process of Curie point erasure has also been demonstrated (Chen *et al* 1968). In this case, a spot of reversed magnetization is heated above its Curie temperature, and forced to regain its magnetization in the same direction as the rest of the film by the application of an external field, smaller than the coercive field of the film, during cooling.

The main advantages of magnetooptical recording are that the optical recording head is at least 1 mm above the disk surface, and this arrangement eliminates the possibility of head crashes, and also allows the disk to be removed. Rigid plastic disks or glass disks covered with a photopolymerization lacquer with grooved recording tracks on centres 1.6 μm across can be rotated at speeds as high as for conventional winchester drives.

Since the bit density along the track is comparable to that of today's winchester technology, a 12-inch-diameter magnetooptical drive has a user capacity of 2.1 Gbytes, the data transfer rate is 360 kbytes s^{-1}, and the seek time for a third of a full stroke is 80 ms. A 3.5-inch-diameter disk drive has been developed that is intended for the personal computer. Such a magnetooptical disk contains 40 Mbytes of data with an overwrite capability of 1 Mbytes s^{-1}, and the one-third-stroke seek time is 20 ms, and the track-to-track access time is 1 ms (Meiklejohn 1986). At the same time, conventional systems usually use 5-inch-diameter disks.

The magnetooptical recording technology offers for the future a greatly increased data rate over that of the winchester drives, since several tracks can be written and read in parallel with a multichannel write/read optical head. This may be achieved by the use of arrays of laser diodes that are focused onto the disks through one optical system.

The formation of recording marks is a very important aspect of recording. This is particularly true for pulse-width-modulation (PWM) recording, which can achieve double the linear density of pit-position-modulation recording. In PWM the mark becomes teardrop shaped, because of thermal diffusion occurring during the thermal writing process. This creates undesirable pulse distortion in the read-out waveform, which hinders efforts to increase the linear density.

Tanabe *et al* (1995) described the theoretical basis of a write-compensation scheme that is determined directly by the simplified thermal transfer function. It was shown that such a function can be approximated to a first-order exponential filter such as that of an integrating circuit. The write-compensation method is then directly given by the inverse function of the filter.

We now discuss the construction and operation of a typical magnetooptical drive, including the writing and reading process, disk assembly, tracking and focusing of the existing magnetooptical recording materials, the attained and expected bit density and signal-to-noise ratio (S/N), as well as the material stability and future developments.

The writing process. The thermomagnetic recording (writing) process is that in which the recording material is heated above some critical temperature (the critical temperature may be the Curie temperature or a temperature above the compensating point temperature where the coercive force is less than the bias field) by a diode laser beam, and the heated spot is allowed to cool in an external magnetic field or demagnetizing field. To make a thermomagnetic recording, a local point on the magnetic film is heated by the laser beam focused to a spot with a diameter near one micrometre.

The magnetic properties of the materials that may be used for Curie point recording and compensation point recording are shown in figure 14.2. In both cases, the coercive force of the film at room temperature is exceeded by some kOe. The sum of the bias and demagnetizing fields usually does not exceed several hundred oersteds. The coercive force at the heated spot decreases along the H_c-curve (figure 14.2(a)), or increases when the compensation point is approached and then decreases along the H_c-curve (figure 14.2(b)) until it is below the sum of the applied bias magnetic field and the demagnetizing field at some temperature. Because the direction of the bias field determines the direction of the magnetization after the laser pulse is turned off, its direction is opposite to the direction of the initial magnetization.

Writing is accomplished by focusing a moderately high-intensity (typical values 5–10 mW) laser beam onto a diffraction-limited spot on the sample. Typically an objective lens with a numerical aperture of 0.5 is used, so the spot has a submicrometre diameter. It is also possible to write into a randomly magnetized medium, by pulsing the polarity of the magnetic field in synchrony with the laser (Kryder 1985). Erasure may be accomplished either with a magnetic field which is large compared to the room temperature coercivity of the medium, or with the simultaneous application of the laser pulse and a small magnetic field.

The reading process.   When a thermomagnetic recording is made, there are regions of inverted magnetization (usually called pits), representing the thermomagnetically recorded magnetic domains, in the recording track magnetization which is in the initial direction. The direction of the local magnetization can be detected now by means of the polar Kerr effect. It should be noted here that the sense of the rotation depends upon the direction of the magnetization in the magnetic film. The plane of polarization rotates clockwise or anticlockwise, depending on whether the magnetization is parallel or antiparallel to the incident light beam. Because the width of a recorded domain is near 800 nm, the sensing optical beam must be focused onto the surface of the magnetooptical media into a spot with a diameter near to the limit of the optical resolution (near one micrometre). When a linearly polarized light beam is reflected from a vertically magnetized magnetooptical material, there is a rotation of the plane of polarization, and ellipticity appears. Because of the ellipticity, it is necessary that after reflection the light is passed through a $\lambda/4$ waveplate to compensate for any ellipticity introduced by the magnetooptical effect, and passed through a polarizing beam splitter which is set at 45° to the polarization direction of the incident light beam.

In order to avoid loss of recorded information during reading, the intensity of the reading beam must be 1 to 2 mW at the surface of the film. In this case the rise in temperature does not exceed 20–50 °C.

With no magnetooptical rotation of the plane of polarization, the intensities of the light beams at each photodiode are the same; thus the difference signal from the photodiode is zero. With a rotation of the plane of polarization, one photodiode receives more light than the other, and a difference signal results. The polarity of the difference signal indicates the direction of the magnetization.

In order to increase the recording density of magnetooptical disks, Takahashi *et al* (1994) have developed a new method that improves the apparent optical resolution, using a medium that consists of a GdFeCo read-out layer and a DyFeCo recording layer. The magnetization of the read-out layer is designed to be in-plane magnetization at room temperature, and to become perpendicularly oriented at high temperatures. Therefore, this medium shows a polar Kerr effect only in areas heated up by a read-out beam, whereas the other areas of the medium are masked by an in-plane magnetization. Using this medium, a single bit at the centre of the read-out beam spot can be detected without cross-talk from the neighbourhood.

14.2.3 Direct overwriting in magnetooptical recording

In the conventional magnetooptical recording process using beam modulation, two passes of the laser beam over the medium are needed to write data. A first pass is needed to erase the medium, and a second pass (with the bias field reversed) is needed to write the data. Direct overwriting is vital if the new generation of magnetooptical disk drives are to achieve shorter access times and higher data rates. For this reason, substantial efforts have been directed towards the development of an efficient and practical method for achieving direct overwriting on magnetooptical media.

Several promising methods for overwriting directly have been suggested, one of which is field modulation with the recording head (Greidanus *et al* 1989, Madison and McDaniel 1990). It was found that two conditions must be met to get well-formed marks (Madison *et al* 1990). First, the switching time needs to be short in comparison to the mark transit time. Second, the field strength needs to be larger than the saturation field.

These two conditions are not independent. If the switching time is very short and the field strength large, then the mark shape is determined by the temperature profile. The signal amplitude resulting from low writing fields is determined by the amount of mixed magnetization in the marks, the size of the marks, and the inter-symbol interference. The signal amplitude from high-field writing is determined by the mark size and inter-symbol interference only.

One possible direct overwriting scheme was demonstrated by Shieh and Kryder (1986, 1987). They showed that domains can be written in certain rare-earth–transition metal alloys, such as GdTbFeCo, with zero magnetic bias field, using relatively high-energy laser write pulses (e.g., 100 ns, 12 mW). With the bias field zero, the domains can then be erased using laser pulses with shorter time duration or lower power.

Rugar *et al* (1988) demonstrated a method for direct overwriting in TbFe using a static magnetic field. The method relies on the domain collapse effect of the thermally induced wall energy gradient. Micrometre-size domains can be overwritten to form submicrometre-size domains using short (950 ns) laser pulses. Complete erasure of micrometre-size domains can be achieved by firing short laser pulses in succession as the medium is moved slowly with respect to the optical beam. The direct overwriting in single-layer structures based on Gd–Tb–Fe–Co and Gd–Tb–Fe films with C/N ratios at the level 32–36 dB, and with a level of 42 dB or better possible, was demonstrated by Schultz *et al* (1993).

Following the limited success of direct overwriting experiments for a single-layer medium, schemes have been investigated that are based on both magnetic field modulation and light intensity modulation using films of two or more layers. Schemes based on light intensity modulation utilize multilayer films, and enable one to eliminate the field switching required in magnetic field modulation. At the same time, a large initializing magnetic field increases the complexity of the drive design, and affects the stability of the written marks.

Saito *et al* (1987a) demonstrated direct overwriting by light power modulation for magnetooptical multilayered media. Using a four-magnetic layer structure (SiN/GdFeCo/TbFeCo/GdFeCo/DyFeCo/SiN), Saito *et al* (1992) demonstrated a C/N ratio of 53 dB at 0.75 μm mark length. A method based on modulation of the light intensity method for direct overwriting using a magnetooptical disk with exchange-coupled rare-earth–transition metal quadrilayered film has been developed by Fukami *et al* (1990, 1991) and Nakaki *et al* (1992). The film consists of a memory, and writing, switching, and initializing layers. The initializing layer is used to initialize the writing layer, and the switching layer is used to switch off the coupling between the writing layer and the initializing layer above the Curie temperature of the switching layer. These two layers have made direct overwriting with only a conventional bias magnet possible. A C/N ratio above 47 dB was obtained using this structure. Kaneko (1992) demonstrated that direct overwriting by light intensity modulation and super-resolution using magnetic quadrilayer structures enable a linear density of 0.3 μm per bit to be achieved. Matsumoto *et al* (1994) proposed a pulse-train recording method for direct overwriting on exchange-coupled five-layer magnetooptical disks.

It is difficult to produce the structure with the quadrilayer magnetic film in large-scale manufacturing, due to extremely severe constraints imposed on the large number of parameters of each layer. Therefore, an ideal manufacturable system would have a simple disk structure, wide power margin, and high writing sensitivity.

Hatwar *et al* (1994) proposed a structure which consists of two magnetic layers (memory and reference: for example, $Tb_{24}Fe_{62}Co_9$ with thickness 130 nm, $T_c = 250$ °C, $T_{comp} = 30$ °C, and $H_c > 13$ kOe at room temperature; and $Tb_{27}Fe_{64}Co_9Zr_{10}$ with thickness 40 nm, $T_c = 120$ °C, and $H_c < 3$ kOe at room temperature, respectively) exchange coupled through an intermediate magnetic layer (for example, $Gd_{35}Fe_{22}Co_{43}$ with thickness 10 nm and $T_c = 400$ °C), and is suitable for direct overwriting by light intensity modulation using just one conventional bias magnet.

The memory and the reference layers have a relatively large perpendicular anisotropy, while the intermediate layer is magnetically soft. The memory layer has a Curie temperature that is greater than or approximately equal to the reference layer Curie temperature. The intermediate layer has a higher Curie temperature than either the memory layer or the reference layer. The room temperature coercivity of the memory layer is higher than the room temperature coercivity of the reference layer. The resulting magnetization of the coupled layers exhibits two compensation points. It is possible to use a two-layer structure consisting of only a memory layer and a soft magnetic underlayer, exchange coupled to each other. The thickness and composition of the reference layer are chosen such that the combined system has a compensation temperature above the ambient temperature but below the write temperature. The structure proposed as a result of the experimental investigation demonstrated a respectable direct overwriting performance of 48 dB using a 3.5 μm mark size, 30 kHz bandwidth, and 7 m s^{-1} disk velocity.

14.2.4 The signal-to-noise ratio

To discuss the signal-to-noise ratio for magnetooptical disk memory, let us first consider an ideal magnetooptical system where the limiting source of noise is shot noise in the differential photodiode detectors. The current produced by the differential photodiode detectors is

$$i_s = \zeta N P_i \Phi_K \sin 2\alpha$$

where ζ is the sensitivity of the photodiodes (typical value $\zeta = 0.3$ A W^{-1}), P_i is the average light intensity in watts incident on the magnetooptical medium during the reading process, R is the reflectivity of the medium, Φ_K is the Kerr rotation angle, and α is the angle between the axes of the polarizer and analyser (here $\alpha = \pi/4$).

The shot-noise-limited S/N ratio is then

$$(S/N)_{shot} = 10 \, \log(2\zeta P_i R \sin^2 \Phi_K / eB) \tag{14.1}$$

where e is the electronic charge in coulombs, and B is the bandwidth of the system in hertz. If typical values are inserted in (14.1), we get $(S/N)_{shot} = 38$ dB (Kryder 1985).

The signal-to-noise ratio can be increased by 4 dB by placing a non-reflecting coating on the opaque magnetooptical medium. Using a quadrilayer magnetooptical structure consisting of the dielectric overlayer, the partly transparent magnetooptical film, the intermediate dielectric layer, and the reflector, it is possible to increase the signal-to-noise ratio by 8.5 dB (Mansuripur 1990, 1994). It should be noted here that the more complicated quadrilayer structure may require better manufacturing control of the technological processes in order to achieve the same yield of usable disk as the simpler non-reflecting coating structure.

The signal-to noise ratio at operating bandwidths is not usually measured directly, but rather is calculated from the measured carrier-to-noise ratio, C/N. The usual bandwidth used in the C/N measurements is 30 kHz. In order to convert the result of the C/N measurement into a value for S/N at a higher frequency, a bandwidth correction must be made. The relation between the signal-to-noise ratio and the carrier-to-noise ratio is given by

$$(S/N)_s = (C/N)_c - 10\log(f_s/f_c)$$

where f_s is the operating frequency desired for $(S/N)_s$ (the signal-to-noise ratio is obtained at a bandwidth f_s) and f_c is the measurement frequency of $(C/N)_c$. If $f_c = 30$ kHz and $f_s = 20$ MHz, the value of $10\log(f_s/f_c) = 27$ dB.

Using a typical value of $(C/N) = 55$ dB measured at $f_c = 30$ kHz, one can obtain the calculated value of S/N at 20 MHz: $55 - 27 = 28$ dB (Meiklejohn 1986). A signal-to-noise ratio of about 25 dB at a bandwidth of 20 MHz is considered more than adequate for a magnetooptical disk system (Freese and Takahashi 1985).

There are other sources of noise besides the shot noise that bring the calculated value of S/N = 38 dB down to the experimental value of S/N = 28 dB. Some of these are laser noise, electronic noise, and media noise. One possible media noise source is the irregularity of the written domain. If the periphery of the domain is ragged instead of circular, there will be some noise introduced into the system, as well as bit shift. In order to obtain circular domains, the magnetooptical materials must have a substantial domain wall stiffness. The desirable material parameters are large domain wall energy density, small magnetization, and a small ratio of domain diameter to film thickness (Shieh and Kryder 1986).

A disk with a high value of C/N shows regular domains, while a disk with a low value of C/N indicates domains with irregular boundaries. The unwritten noise is lower than the written noise, but follows the same trend. This indicates that mark edge roughness is not the only medium-related noise source. Unwritten noise could result from unerased domains present in the disk at the time of manufacture, as well as from variations in reflectivity, surface roughness, etc (Hatwar *et al* 1990).

14.2.5   Magnetooptical recording materials

The materials used for magnetooptical recording must be selected on the basis of their ability to support submicrometre domains, to provide adequate magnetooptical effects for a good signal-to-noise ratio, to be resistant to corrosion, and to have magnetic parameters which remain stable during aging. For magnetooptical recording media, the following properties are required.

(i)   A large uniaxial (perpendicular) magnetic anisotropy K_u to realize the perpendicular magnetization configuration ($K_u < 2\pi M_s^2$).

(ii) A large product $M_s H_c$ to ensure high recording density.

(iii) A Curie point T_c of 150–180 °C to ensure thermal stability of the recorded information, and to provide reasonable writing laser power.

(iv) Very low medium noise.

(v) A large domain wall stiffness in order to minimize recorded media noise caused by the irregularity of the written domain. The desirable material parameters are a large domain wall energy density, a small magnetization, and a small ratio of domain diameter to film thickness.

(vi) Magnetooptical Kerr rotation Φ_K as large as possible, to provide a high signal-to-noise ratio of the read-out signal.

(vii) High corrosion resistance.

Many different materials have been considered because of their promise as potentially useful memory media, including MnBi, and related compounds, rare-earth–transition metal alloys, PtMnSb, and bismuth-substituted iron garnets. However, for MnBi, and sputtered films of bismuth-substituted iron garnets, it is observed that the media noise was too high due to the polycrystalline structure. In amorphous rare-earth–transition metal films the magnetooptical effects are intrinsically small; nevertheless, due to their low media noise, such materials now represent better magnetooptical recording media. Some properties of these materials are discussed later in this section, but their optical and magnetooptical properties were treated in chapter 10.

14.2.5.1 Binary rare-earth–transition metal alloys

GdCo was one of the first materials suggested for thermomagnetic recording (Chaudhari *et al* 1973). Films were examined for compensation point writing and erasing. The spectral dependence of the polar Kerr rotation of this material was reported by Višňovský *et al* (1976).

GdFe has H_c-values lower than desired for a good magnetooptical recording material, resulting in instability of small domains in these films (Meiklejohn 1986).

TbFe has been widely used as a model material in studying thermomagnetic recording in amorphous rare-earth–transition metal films. But this material has some drawbacks, because of the irregularity of its thermomagnetically recorded domains. Rugar *et al* (1988) demonstrated thermomagnetic direct overwriting in TbFe using a thermally induced domain wall energy gradient.

Other binary rare-earth–transition metal alloys. The Kerr rotation angles of sputtered films of Gd–Co, Tb–Co, Dy–Co, and Nd–Co were measured for a wide range of rare-earth content (3–45%) at room temperature, at $\lambda = 780$ nm. The compositional dependence of Φ_K was analysed using mean-field theory, and the relation between the Kerr rotation and the Co-sublattice moment M_{Co} was clarified (Honda and Yoshiyama 1988). Many other binary alloys, such as DyFe, HoCo, HoFe, and GdY, have been investigated, but none of them appears to be useful for thermomagnetic recording (Meiklejohn 1986).

14.2.5.2 Ternary rare-earth–transition metal alloys

GdTbFe films show more promise as a viable magnetooptical recording material than the binary alloys (Meiklejohn 1986).

GdTbCo appears to be one of the better materials for thermomagnetic recording. The stability of domains written in this material is excellent, and the material has a large value of the product $M_s H_c = 8 \times 10^4$ Oe^2 (Meiklejohn 1986).

GdFeCo is suitable for Curie point thermomagnetic recording for Fe-rich compositions, but has too high a Curie point when the cobalt content increases. The H_c-values are too low for high-quality thermomagnetic recording material to be obtained. GdFeCo alloys that contain only a small amount of iron are similar to GdCo alloys, and can be used for compensation point recording. The addition of small amounts of bismuth increases the value of the Kerr rotation up to 0.48° (Meiklejohn 1986).

TbFeCo. Amorphous TbFeCo-based alloy films are of great interest as regards erasable high-density information storage. Films have been obtained by evaporation and sputtering. The TbFeCo films showed a polar Kerr rotation angle Φ_K as large as 0.43° at a wavelength of 830 nm (Ito *et al* 1990). Carrier-to-noise ratios of 50 and 55 dB have been obtained with a bandwidth of 30 kHz (Meiklejohn 1986). Yamamoto *et al* (1987) attained a signal-to noise ratio of 27 dB using TbFeCo. For TbFeCo film compositions near the compensation point, Hatwar *et al* (1990) observed carrier-to-noise ratios (bandwidth: 30 kHz) greater than 60 dB.

Iijima *et al* (1989) have studied the magnetic and magnetooptical characteristics of indium-alloyed TbFe amorphous films. It was shown that alloying with In drastically decreases the compensation temperature, and increases the Kerr rotation by more than 20%. Ternary alloys such as $(GdFe)_{1-y}M_y$ (M = Bi, Sn, Pb) have been prepared to investigate the influence of third components on the Faraday and polar Kerr rotation, and saturation magnetization (Hansen and Urner-Wille 1979, Urner-Wille *et al* 1980, Hartmann *et al* 1984).

Several other ternary alloys have been investigated, such as GdFeBi, GdFeSn, GdFeY, GdFeB, GdCoB, and DyFeCo (Meiklejohn 1986).

DyFeCo. Raasch (1993) compared the recording characteristics of different magnetooptical disks based on $Dy_{29}Fe_{57}Co_{14}$, $Tb_{24}Fe_{70}Co_6$, $Gd_{17}Tb_7Fe_{76}$, and Pt/Co multilayers ($14 \times [Co(0.31$ nm)/Pt(1 nm)]) plus a Pt(1 nm) underlayer. The measurements were performed using a magnetic field modulation recorder with an 820 nm semiconductor laser. It was found that, among the disks tested, that based on $Dy_{29}Fe_{57}Co_{14}$, with $T_c = 490$ K and $T_{comp} = 370$ K, showed the highest write and erasure sensitivity (<8 mW), the highest carrier-to-noise ratio (≥ 55 dB) and the lowest difference between the domain stability and the erase threshold (<2 mW) up to 22 m s^{-1} disk velocity. Pt/Co disks showed less suitable results, as their values of C/N did not exceed 46 dB for erasure fields ≤ 400 Oe and write powers ≤ 10 mW.

14.2.5.3 Quaternary rare-earth–transition metal alloys

The quaternary alloys GdTbFeCo, GdTbFeCo, and GdFeCoBi were investigated (Meiklejohn 1986). Shieh and Kryder (1987) demonstrated a direct overwriting scheme using GdTbFeCo film.

NdDyFeCoTi magnetooptical recording media having a high performance and good reliability have been prepared using an alloy target formed by the casting method. A magnetooptical medium having a composition of $(Nd_{0.2}Dy_{0.8})_{25}(Fe_{0.58}Co_{0.38}Ti_{0.04})_{75}$ has demonstrated a carrier-to-noise ratio over 50 dB (Shimoda *et al* 1987).

NdTbFeCo and PrTbFeCo. Hansen *et al* (1994) investigated the magnetization, Curie temperature, uniaxial anisotropy, coercivity, and Faraday and Kerr rotations as functions of composition and temperature for amorphous rare-earth–transition metal alloys of compositions $RE_{1-x}Fe_x$ and $RE_{1-x}Co_x$, with RE = Pr, Nd and $0 < x < 1$, and the four-element composition of TbNdFeCo or TbPrFeCo, prepared by coevaporation, and intended to provide a new generation of magnetooptical disks for short-wavelength recording.

The spectral variation of the Kerr rotation was measured for some alloys. The magnetization data indicate a strong dispersion of the Pr and Nd subnetworks and of the Fe subnetwork for Fe-rich alloys. The latter gives rise to a maximum of the Curie temperature at around $x = 0.7$, and to very low T_c-values for Fe-based alloys with $x > 0.9$.

The magnetooptical effects result from both the light rare earths and the transition metal alloys, giving rise to significantly larger rotations in the visible region, as compared to the alloys containing heavy rare earths. Recording experiments on standard disks with a TbNdFeCo layer do indeed reveal good write/erase and read performances. The carrier-to-noise ratio ranges between 49 and 56 dB for wavelengths between 458 and 820 nm. Disks based on RE–TM layers containing Pr reveal a lower write/read performance, and appear less suitable for magnetooptical recording. The optical and magnetooptical constants of PrTbFeCo films for wide spread of compositions were investigated by Carey *et al* (1994a, b).

14.2.5.4 NdTbDyFeCo/DyFeCo ultra-thin bilayers

In order to improve thermomagnetic recording behaviour and to obtain a high magnetic field sensitivity, a NdTbDyFeCo/DyFeCo ultra-thin bilayer was produced and investigated by Kawase *et al* (1994). It is known that the magnetic field sensitivity is limited by subdomain formation in domains written at low field. Therefore a high magnetic field is required in order to make the magnetization within the written domain uniform. In the case of RE–TM amorphous films, the subdomain formation is dependent on the concentration of the rare-earth metal, because it determines a demagnetizing field near the Curie temperature. Furthermore, the surface layer, which has a higher magnetization and a lower perpendicular anisotropy than that of the bulk region, is influenced by the concentration of rare-earth metal. Using the bilayers developed provides one with the opportunity to reduce the demagnetizing field, and modify the surface layer. The main layer was a $Nd_6Tb_5Dy_{16}Fe_{58}Co_{15}$ film with a thickness of 20 nm. The ultra-thin layer was of $Dy_{30}Fe_{35}Co_{35}$, with thickness varying from 0.5 to 7 nm. Only the DyFeCo layer has a compensation point. It was found that, in the case of the single main layer having no ultra-thin layer, an applied magnetic field of 250 Oe is required to saturate the carrier-to-nose ratio.

14.2.5.5   Co/Pt multilayers

Lin and Do (1990) achieved a C/N ratio of 59 dB with a reading laser ($\lambda = 820$ nm) power of 1 mW for a 2 MHz carrier at 10 m s^{-1}, and 64 dB with 3 mW read power for a 2.5 MHz carrier at 20 m s^{-1}, using a film structure consisting of a $15 \times [Co(0.32$ nm$)/Pt(1.15$ nm$)]$ multilayer, and an 80 nm silicon nitride enhancement layer on a non-grooved glass disk. The coercivity of the film was about 1.8 kOe, and the Curie temperature was about 330 °C. Hashimoto *et al* (1990) reported a value of the C/N ratio of 50 dB for a $Co(0.45$ nm$)/Pt(1.25$ nm$)$ multilayered disk at a bit length of 5 μm. A 2P (photopolymer) glass substrate with a double-layer disk structure was used.

14.2.5.6   Oxide films for magnetooptical disk memory

Thermomagnetic recording has been performed on some polycrystalline media— spinel $CoFe_2O_4$, hexagonal ferrite $Ba(FeCoTi)_{12}O_{19}$, and bismuth-substituted iron garnet $(BiRE)_3(FeM)_5O_{12}$—which have been prepared by sputtering or pyrolysis on glass substrates (Abe and Gomi 1987). Since they transmit light well, Faraday rotation is used for magnetooptical read-out. The perpendicular magnetic anisotropy is obtained via crystalline anisotropy and stress-induced anisotropy.

Compared to current amorphous rare-earth–transition metal films, oxide films, as magnetooptical recording media, have the following advantages: a strong magnetooptical effect, especially when enhanced by particular ions such as Bi^{3+}, Co^{2+}, and Ce^{3+}; high corrosion resistance; and good reproducibility in film preparation. However, oxide films have counterbalancing disadvantages; media noise due to grain boundaries, and related domain size and wall irregularity; poor squareness of the magnetic hysteresis loop, which reduces the read-out signal-to-noise ratio; and low thermomagnetic recording sensitivity (Abe and Gomi 1987).

With the aim of reducing the critical diameter of the written domain in garnet films prepared by pyrolysis on glass substrates, double-layer bismuth-substituted films were prepared and investigated by Nakagawa *et al* (1994). The critical diameter of the written domains was reduced to 0.2 μm, and a decrease in the noise level of the read-out signal was attained. A value of C/N of 55 dB was achieved for this double-layer film by using an Ar-ion laser of wavelength 515 nm, while the value of C/N was 52 dB for the single-layer garnet films. H_c and K_u for the double-layered film are 1.7 and 1.4 times as large as those for the single-layered film, respectively, and the critical diameter of the written domain is 0.7 times the size of the value for the single-layer film. Double-layer iron garnet films were prepared on glass substrates by Itoh and Nakagawa (1992), and a C/N ratio as high as 53 dB, at a wavelength of 515 nm, was achieved. The double-layer film consisted of an underlayer fabricated by pyrolysis, and a top layer deposited by sputtering. The underlayer was designed to accommodate the small grain size via doping with Rb.

Kawamura *et al* (1995) studied the domain formation and bit shapes for single- and double-layer polycrystalline iron garnet films intended for thermomagnetic recording. It was found that the written bit shapes are strongly related to the domain growth pattern in these films. The two films show the same grain size, of about 50 to 100 nm, but double-layer films give more regular bit shapes than single-layer films.

14.2.5.7 *Materials with potential application for recording in the blue spectral range*

Fu *et al* (1995) investigated magnetooptically the figures of merit (FOM) of: bilayers of $(BiDy)_3(FeGa)_5O_{12}$, 9.5 nm thick, and Co, 0.5 nm thick (the sample consists of ten bilayers with nominal film thicknesses of 100 nm), sputtered onto a $Gd_3Ga_5O_{12}$ substrate, and crystallized under conditions of rapid thermal annealing; MnBi film; multilayered Co/Pt film; amorphous TbFeCoTa film; and fcc Co, in the wavelength range 400–550 nm, and found that $FOM(MnBi) \approx FOM(garnet) > FOM(PtMnSb) > FOM(Co/Pt) \approx FOM(Co) > FOM(TbFeCoTa)$.

Miyazawa *et al* (1993) studied a complex exchange-coupled magnetic multilayer: NdDyTbFeCo/NdTbFeCo/NdDyTbFeCo, with a total magnetic film thickness of 20 nm, in which a layer with a high Kerr rotation angle in the blue spectral range was sandwiched by layers with large coercivities. This magnetic multilayer demonstrated higher recording sensitivity compared with a conventional magnetic single layer and showed a C/N ratio of 47 dB at the wavelength 532 nm. More detailed information concerning multilayer systems and superlattices is presented in chapter 12.

High-density recording with a visible-region laser diode ($\lambda = 680$ nm) used in the direct overwriting of a magnetooptical disk based on quadrilayer magnetic film was investigated by Tsutsumi *et al* (1993). High recording sensitivity was obtained, and the write power was estimated to be about 10 mW at 22 m s^{-1} linear velocity. A C/N ratio of more than 46 dB was obtained at 0.66 μm mark length, and 0.5 μm/bit recording with a bit error rate less than 10^{-5} was achieved for a 1.2 μm track pitch.

Zeper *et al* (1992) compared the recording and reading performance of

(Co/Pt)- and GdTbFe-based disks at wavelengths of 820, 647, and 458 nm. At long wavelengths, the C/N ratio of the GdTbFe disks was higher due to a lower noise level. At the wavelength of 458 nm, the performance of Co/Pt was better, by 3 dB, than that of GdTbFe; this was related to a higher Kerr rotation level for Co/Pt. The C/N ratio for a Co/Pt disk was 51.5 dB for the longest domain period (domain length ≈ 2.5 μm), and still 40 dB for a period of 0.75 μm (≈ 0.38 μm domains). The threshold for writing was the same for the Co/Pt as for the GdTbFe disk, and varied from 2.7 to 3.8 mW as a function of the domain period.

The Pt/Co multilayer disk and the GdFeCo/TbFeCo double-layer disk were investigated as regards use in high-density recording, using a low-noise and high-power green laser, by Kaneko *et al* (1993). A high C/N ratio of 45 dB at a mark length of 0.4 μm was obtained for both disks, with a track pitch of 0.9 μm. The results demonstrate the possibility of areal densities more than three times higher than those for standard magnetooptical disks.

14.3   Magnetooptical tape

Needs in information storage are constantly increasing. Two main parameters are dominant for the tape recorder domain: capacity, and data flow rate. The optical disk allows considerable surface density of information. However, for high-volume needs, disks have a major drawback, and require big and costly systems like the 'juke-box'. Today, magnetic tape is the only medium allowing high-volume information storage. So, the possibility of combining the high density of optical storage with the volume of a tape seems attractive: high density and high recording speed should be achievable, with the advantage of no contact between the recording medium and the writing/reading head.

Dancygier (1987) developed an experimental sample of magnetooptical tape based on the following structure: Kapton tape (the substrate); AlN (a protective layer); Tb_xFe_{1-x} (a Tb-rich film); Tb_xFe_{1-x} (an Fe-rich film); and AlN (another protective layer), and demonstrated the existence of a magnetic anisotropy, perpendicular to the film plane, for TbFe deposited on Kapton. The sensitive layer has a bilayer structure which yields a Kerr rotation as high as $0.6°$. The thickness of the TbFe layer was 50 nm, and the Kapton tape was 25 μm thick. Kapton was used as the substrate material because of its very good thermal properties. The major drawbacks of Kapton are its poor surface state, and its water content. The water problem was solved by the deposition of the AlN protective layer, which is very efficient for this purpose. The problems of surface states were resolved by a very strict selection of the Kapton sample, and an appropriate cleaning process. The magnetic and magnetooptical parameters are presented in table 14.1.

The smallest magnetic domains obtained were 3 μm in diameter. The Kerr rotation angles were as high as $0.6°$ for a sample with the Tb content $x = 0.257$. The increase of the Kerr rotation is related to interference phenomena in the

Table 14.1. Magnetic and magnetooptical parameters of samples deposited on glass and Kapton tape (Dancygier 1987). (In this table H_{cK} denotes H_c measured on the basis of the Kerr effect.)

x	Substrate	M_s (emu cm^{-3})	Φ_K (deg)	H_{cK} (Oe)	K_u (10^5 erg cm^{-3})
25.7	Kapton	89	0.6	1500	7.1
25.7	Glass	125	0.68	2000	7.2
27	Kapton	112	0.48	1300	4.5
27	Glass	188	0.5	1480	5.6
28.4	Kapton	157	0.24	795	2.6
28.4	Glass	225	0,24	1100	2.9
29.7	Kapton	220	0	—	8.3
29.7	Glass	319	0	—	9.2

AlN protective layer. The high coercive field prevents erasure, and gives a good stability to the recorded bits. The Curie temperature for $x = 0.257$ is 150 °C. Static pulse recording experiments have been carried out down to a few tens of nanoseconds. A simulation of the movement of the magnetooptical tape has been made, reaching 1 Mbit s^{-1} per channel. So, the validity of the magnetooptical tape concept has been demonstrated.

In order to achieve a high volume of information recording on the magnetooptical tape, a special multitrack read-out head is necessary. A new magnetooptical multitrack read-out head for magnetic tape was investigated by Maillot *et al* (1994), and high-track-density magnetooptical read-out of magnetic tapes was demonstrated. The magnetic contribution to the track-to-track cross-talk is experimentally measured through the optical scanning of the transducer, and it was found that the anisotropy of the soft poles leads to a strong decrease of the magnetic cross-talk. The optical contribution to the cross-talk was also calculated. These results enable one to foresee that this magnetooptical thin-film technology will be able to accommodate very high track densities, up to the optical resolution limit.

Chapter 15

Integrated magnetooptics

One of the main objectives of integrated magnetooptics—a novel intensively developing branch of applied magnetooptics—is the transmission and processing of optical signals propagating in thin-film magnetooptical waveguides, in which the waveguiding and controlling topological elements and structures are formed using modern methods of thin-film deposition, lithography, and laser annealing.

Epitaxial iron garnet films are widely used in integrated optics elements and devices. They are highly transparent in the wavelength range 1.2–5 μm and exhibit strong magnetooptical properties. Application of magnetooptical films of transparent iron garnet as magnetooptical waveguides was first reported in 1972. Light waveguiding was observed in a film of composition $Y_3Fe_{4.3}Sc_{0.7}O_{12}$, which was grown on a GGG substrate oriented in the (111) plane. A He–Ne laser beam was prism coupled to the film; the insertion losses were 8 dB cm^{-1} at $\lambda = 1.152$ μm and less than 3 dB cm^{-1} at $\lambda = 1.523$ μm (Tien *et al* 1972a).

Epitaxial iron garnet films for integrated magnetooptics should satisfy several specific criteria. On the one hand, the magnetization should be oriented in the plane of the film, with minimal anisotropy and coercive force, in order to ensure rotation of the magnetization vector under magnetic fields of several oersteds. Optical losses result from material absorption of the waveguide, substrate, and upper layer, and from scattering at the interfaces and inside the waveguiding layer. The scattering leads to mode coupling with other guided or radiational modes. The main parameter of the magnetooptical waveguide is the Faraday rotation of the material.

A large number of iron garnet compositions applicable for use in thin-film waveguide design have been proposed and investigated. The most promising structures are epitaxial iron garnet films containing Bi and Pr ions. Incorporation of Pr ions led to in-plane anisotropy, while Bi ions enhance the Faraday rotation of the material. The final goal of integrated magnetooptics is to design an integrated optic scheme that includes light sources and detectors, as well as elements for transmission and processing of optical signals using magnetooptical effects. In this respect, work that is devoted to the epitaxial growth of semiconductor GaInAs–InP heterolasers (Razeghi *et al* 1986) on garnet substrates is of considerable interest.

15.1 Thin-film magnetooptical waveguides

As we have already mentioned, a crucial parameter of the thin-film waveguide is its level of optical losses, which are associated with the material absorption of the waveguide, substrate, and upper layer, as well as with the interface and volume scattering. The scattering leads to the light coupling into other guided and radiational modes. At given operating conditions (the wavelength of light, waveguide mode type, quality of epitaxial layer surface treatment, epitaxial growth peculiarities, and other factors) various mechanisms result in different losses depending on the mode number and the mode type. Figure 15.1 shows the dependence of the total loss and the contributions of different mechanisms (at $\lambda = 1.152\ \mu$m) on the mode number for a magnetooptical waveguide based on epitaxial iron garnet film (Daval *et al* 1975).

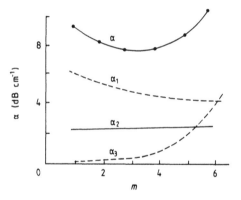

Figure 15.1. Losses for modes of different orders: α: total losses; α_1: the absorption in the film; α_2: the scattering in the film; and α_3: the scattering at the interfaces (Daval *et al* 1975).

Absorption losses slightly decrease with the mode number, because more and more light wave energy is localized outside the absorbing material of the waveguide layer. In the iron garnet epitaxial films, at wavelengths of $\lambda = 1.064$ and $1.152\ \mu$m, the absorption of the waveguide layer is usually high. Scattering losses in the material are low, and independent of the mode index.

Numerous experimental results demonstrate that the scattering of light at the interfaces contributes substantially to the overall losses. In some cases, scattering losses can be reduced by additional surface polishing (Daval *et al* 1975). The losses related to the interface scattering are inversely proportional to the waveguide layer thickness, because for thick waveguides the number of reflections from the interfaces per unit length is lower than that for thin waveguides.

Another principal parameter of the magnetooptical waveguide is its Faraday rotation. A large number of iron garnet compositions for thin-film magnetooptical waveguides have been proposed and investigated (Prokhorov *et al* 1984, Zvezdin and Kotov 1988). The most promising waveguide structures are epitaxial iron garnet films containing Pr and Bi ions. Incorporation of the Pr ions leads to in-plane anisotropy, and Bi ions give rise to the Faraday rotation of the material.

The additional optical absorption caused by impurities incorporated in the iron garnet films during the epitaxial growth process can be reduced considerably by control of the growth parameters, or by annealing treatments. Dotsch *et al* (1992) reported the absorption level 0.3 cm^{-1} for as-grown bismuth- and aluminium-substituted iron garnet films at 1.3 μm and 0.2 cm^{-1}—for films annealed in hydrogen atmospheres.

For $Bi_{0.9}Lu_{2.1}Mg_{0.02}(FePt)_{4.98}O_{12}$ epitaxial layers grown from a Bi_2O_3 melt on GGG substrates of (111) orientation and designed for magnetooptical waveguide fabrication, the absorption coefficient α was decreased to 2 cm^{-1} at $\lambda = 1.152$ μm and to 0.13 cm^{-1} at $\lambda = 1.317$ μm by incorporation of MgO into the melt (Tamada *et al* 1987, 1988).

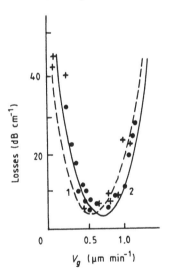

Figure 15.2. The dependence of optical losses on the growth rate, for $\lambda = 1.15$ μm. Curve 1: $Y_{1.55}Gd_{0.45}Fe_{4.1}Ga_{0.9}O_{12}$ (+). Curve 2: $Y_{1.55}Gd_{0.45}Fe_{4.2}Ga_{0.8}O_{12}$ (•) (Meunier *et al* 1987).

It was established that α does not depend on such technological parameters as the growth temperature and the duration of the growth process, and the substrate rotation speed. However, the ferromagnetic resonance width is sensitive to the growth conditions. A minimal value of $\Delta H = 0.7$ Oe was obtained for growth parameters that ensure matching of the substrate and film lattice parameters at the growth temperature $T_p = 830$ °C.

Measurements of optical losses of epitaxial films, which were designed for optical stripe waveguides, revealed the dependence of optical losses on the epitaxial growth temperature (figure 15.2). The difference between the lead and platinum concentrations $|c_{Pb} - 2c_{Pt}|$ turned out to be responsible for the optical losses (table 15.1) (Meunier *et al* 1987).

Table 15.1. The dependence of optical losses and the difference between the concentrations of Pb and Pt ions on the growth rate for epitaxial $Y_{2.55}Gd_{0.45}Fe_{4.2}Ga_{0.8}O_{12}$ film at $\lambda = 1.152$ μm (Meunier *et al* 1987).

Growth rate (μm min^{-1})	$|c_{Pb} - 2c_{Pt}|$	Optical losses for the TE$_0$ mode (dB cm^{-1})
0.40	15×10^{-4}	15
0.52	12×10^{-4}	9.4
0.69	7×10^{-4}	7
0.83	1×10^{-4}	5.4
0.87	10×10^{-4}	7

Ion beam etching of epitaxial Bi-substituted sputtered iron garnet films of composition $Bi_{0.9}Gd_{2.0}Ca_{0.02}Pb_{0.08}Fe_{4.95}Pb_{0.05}O_{12}$, with Faraday rotation $\Phi_F = -2800$ deg cm^{-1} at $\lambda = 1.15$ μm, of 0.9 μm thickness, for 19 minutes in an Ar atmosphere, leads to nearly the same decrease of optical absorption as does the reducing thermal treatment in $Fe(OH)_2$ solution. The films were etched at an Ar pressure of 2×10^{-3} Torr, while the residual pressure was 10^{-4} Torr, the etching rate being 5 nm min^{-1}. More extended treatment results in increase of the losses. For optimal conditions, the losses at $\lambda = 1.52$ μm fell from 70 dB cm^{-1} ($\alpha = 16$ cm^{-1}) to 12 dB cm^{-1} ($\alpha = 3$ cm^{-1}) (Santi *et al* 1987).

Annealing of the Bi-substituted iron garnet epitaxial films in an oxygen atmosphere containing several volume per cent of ozone at 700 °C with subsequent annealing at 500 °C in pure oxygen proved to be rather efficient for diminishing optical losses in the infra-red spectral region, and for narrowing the ferromagnetic resonance (Tamada and Saitoh 1990). The process resulted in a drop in optical losses in the $(BiLuYMg)_3(FePt)_5O_{12}$ layer that was 90 μm thick, with a few magnesium impurities, to less than 1 dB at $\lambda = 1.3$ μm. Simultaneous narrowing of the ferromagnetic resonance (FMR) from 1.7 to 0.8 Oe at 9 GHz was also observed. Tamada and Saitoh explain the fall in optical losses and the FMR narrowing by a decrease in the Fe^{2+}-ion concentration in the iron garnet film during the annealing.

Figure 15.3 shows the dependence of the optical absorption α at $\lambda = 1.15$ μm for an as-grown (curve 1) epitaxial $(BiYLuMg)_3(FePt)_5O_{12}$ film on the parameter

$$R_6 = c_{MgO}/(2c_{Fe_2O_3} + c_{MgO})$$

(the molar ratio of the concentration of magnesium oxide to the total concentration of the Fe_2O_3 and MgO components in the melt), and the corresponding dependence after annealing in an atmosphere that contained several per cent of ozone (curve 2). Layers of the Bi-substituted garnets were grown from the lead-free melt.

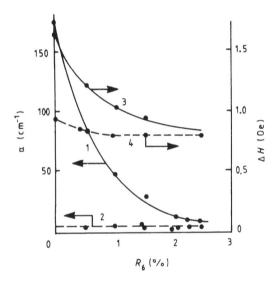

Figure 15.3. The dependence of the optical absorption coefficient α at 1.15 μm for as-grown films (curve 1), and films after annealing (curve 2), and of the resonance width ΔH at 9 GHz for as-grown films (curve 3), and annealed films (curve 4) on the parameter $R_6 = c_{MgO}/(2c_{Fe_2O_3} + c_{MgO})$ (Tamada and Saitoh 1990).

The results demonstrate the high technological adaptability of annealing in ozone and oxygen atmosphere as regards decreasing losses. Magnesium doping of Bi-substituted iron garnet layers without subsequent annealing enables fabrication of films with low optical losses to be achieved only at a fixed concentration of magnesium in the film, which is rather difficult to maintain in mass production.

Annealing in an ozone and oxygen atmosphere leads not only to reduction in the absorption, but also to narrowing of the FMR band. Figure 15.3 presents the dependence of the decrease of the resonance width ΔH on the R_6-parameter for the films before (curve 3) and after (curve 4) additional annealing. The drop in ΔH from 1.7 to 0.8 Oe is promising as regards utilization of Bi-substituted iron garnet films in the design of highly effective devices based on the interaction of optical waveguide modes with magnetostatic waves (Tamada and Saitoh 1990).

15.2   Anisotropic magnetooptical waveguides

Non-degenerate mode propagation in an isotropic magnetooptical waveguide complicates the mode conversion. However, the conversion can be reinforced by optical anisotropy of the substrate. In particular, phase synchrony of the TE and TM modes can be furnished in an anisotropic waveguide grown on a gyrotropic substrate. This structure guides the degenerate modes, because the upper layer and/or the substrate are anisotropic, while the mode coupling is controlled using gyrotropic properties of the substrate. As growth of thin films with the desired parameters and low optical absorption is an unwieldy problem, utilization of gyrotropic substrates is favourable as regards controlling the waveguide's optical parameters.

Let us consider how light propagation in the waveguide can be controlled by the substrate properties (Chung and Kim 1988). We assume that there are no optical losses in all three media, and that the film is isotropic, and the cladding is anisotropic in the general case, but the crystalline axis is orientated so that the non-diagonal elements of the permittivity tensor are zero, and the substrate is anisotropic and gyrotropic.

In this case the mode conversion occurs at the substrate–film interface, while at the film–upper-layer interface there is only a phase shift. Indeed, the eigenmodes for the magnetooptical medium are not in fact purely of TE or TM type, and after a TE (TM) mode reflects from the film–substrate interface, there is always a TM (TE) component in the reflected wave. The two eigenmodes have different decay constants in the substrate. It should be mentioned that the higher the non-diagonal component δ of the permittivity tensor and the deeper the field penetration into the substrate, the stronger the mode conversion. Therefore, when a gyrotropic substrate is used, the mode propagation angle should be chosen close to the critical value $\theta_{cr} = \sin^{-1}(n_2/n_3)$.

For the subsequent analysis, it is useful to introduce the mode conversion matrices. For the path 1–2–3 (figure 4.1) we obtain

$$\begin{pmatrix} E_3^{TE} \\ E_3^{TM} \end{pmatrix} = (R) \begin{pmatrix} E_1^{TE} \\ E_1^{TM} \end{pmatrix}.$$

Here

$$(R) = \begin{pmatrix} \exp i\varphi_{12}^{TE} & 0 \\ 0 & \exp i\varphi_{12}^{TM} \end{pmatrix} \begin{pmatrix} \exp i 2b_e h & 0 \\ 0 & \exp i 2b_m h \end{pmatrix} \begin{pmatrix} r_{ee} & r_{em} \\ r_{me} & r_{mm} \end{pmatrix}.$$

The r-matrix describes the reflection of light from the film–substrate interface, r_{ee} and r_{mm} are the reflection coefficients of TE and TM modes, and r_{em} and r_{me} are the mode conversion coefficients (Smolenskii *et al* 1976a). The $(\exp i\varphi)$-matrix determines the reflection at the film–upper-layer interface, while the $(\exp i 2bh)$-matrix takes into account the phase shift that occurs during propagation along the 1–2–3 path (figure 4.1).

Sequentially multiplying the R-matrices, we obtain the characteristic length l_0 of the total mode conversion under phase synchronization:

$$l_0 = \pi h \tan(\theta_r / r')$$

where $r' = |r_{em}| = |r_{mr}|$.

15.3 Induced optical anisotropy

The mode conversion efficiency of an isotropic magnetooptical waveguide having a Faraday rotation of the order of 100 deg cm^{-1} does not exceed several per cent. Nevertheless, in epitaxial iron garnet waveguides much higher values of the conversion might be reached. For example, Smolenskii *et al* (1976b) reported 40% efficiency of the TE$_6$ → TM$_6$ and TE$_7$ → TM$_7$ conversion in a homogeneous magnetooptical waveguide when they applied a 400 Oe longitudinal field. In the experiment they used an $Y_3Sc_{0.37}Fe_{3.61}Ga_{1.02}O_{12}$ epitaxial layer 9 μm thick, grown on a GGG substrate, which had the saturation magnetization $4\pi M_s = 500$ G.

The principal feature of so strong a mode conversion is its non-reciprocity (Smolenskii *et al* 1976a). For the TE$_6$ → TM$_6$ mode conversion the coefficient changed twofold when the external field reversed. For the TE$_{11}$ → TM$_{11}$ mode conversion, the ratio of conversion coefficients before and after the magnetization switching reached a value of 8, but the overall efficiency was only 2%. The results can be explained by induced optical anisotropy in the waveguide film and in the substrate. Optical anisotropy in the substrate is also responsible for the non-reciprocity of the mode conversion.

Later studies revealed induced optical birefringence in most of the iron garnet films grown by liquid-phase epitaxy. In general, there are two factors that contribute to the film birefringence: the first one is attributed to mechanical tensions, which result from the mismatch of the unit-cell parameters of the film and the substrate; while the second is associated with the film growth process, and is proportional to the magnetic growth anisotropy. A high growth-induced optical anisotropy was observed for the Bi-substituted films, which had a typical value of $\Delta n = 5 \times 10^{-4}$ at $\lambda = 1.15$ μm (Ando *et al* 1985a, b).

If the epitaxial film is annealed for several hours at the temperature 1150 °C, both the growth-induced uniaxial magnetic anisotropy and the growth-induced optical anisotropy almost disappear (figure 15.4).

The mode conversion coefficient can be controlled by regulating the waveguide layer birefringence. Annealing of an epitaxial film of composition $(BiGdLu)_3(FeGa)_5O_{12}$, with the $\Phi_F = 820$ deg cm^{-1}, at a temperature $T = 1150$ °C for seven hours resulted in an increase in the TE$_0$ → TM$_0$ mode conversion coefficient from 50% to almost 100%.

A typical dependence of the entire birefringence Δn, and the contribution that is associated with the mismatch of the unit-cell parameter of the substrate

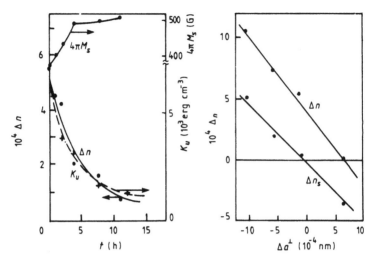

Figure 15.4. The dependence of the optical birefringence Δn, the uniaxial magnetic anisotropy K_u, and the saturation magnetization $4\pi M_s$ on the annealing time t. Also shown is the dependence of the total birefringence Δn, and the stress contribution Δn_s, which is associated with the mismatch of the substrate and film lattice parameters, on the mismatch parameter $\Delta a^{\perp}$ in $(BiGdLu)_3(FeGa)_5O_{12}$ films (Ando *et al* 1985a, b).

a_s and film a_f on the value of $\Delta a = a_s - a_f$ for epitaxial bismuth-substituted iron garnet films is presented in figure 15.4 (Ando *et al* 1985a, b).

Along with the annealing, the phase mismatch of TE and TM modes may be controlled by growing multilayer epitaxial films with a low refractive index difference (Wolfe *et al* 1987). Precise adjustment of the value of $\Delta\beta$ for the films is furnished by etching the top layer or top sputtering a dielectric film, which modifies the tension of the waveguide layer.

15.4 Waveguides with high Faraday rotation

Mode conversion in gyrotropic epitaxial Bi-substituted iron garnets was studied by van Engen (1978) and Kotov *et al* (1979). Figure 15.5 shows the calculated TE–TM conversion coefficients versus the waveguide layer thickness for a number of magnitudes of the Faraday rotation. At $\Phi_F = 1000$ deg cm^{-1} and the waveguide thickness $h > 4$ μm, the TE$_0$–TM$_0$ conversion efficiency exceeds 90%, and is much lower for the higher-order modes.

For epitaxial $(YbPr)_{2.65}Bi_{0.35}Fe_{3.8}Ga_{1.2}O_{12}$ film with $h = 4.13$ μm and $\Phi_F = 510$ deg cm^{-1} at $\lambda = 1.15$ μm, a TE$_0$–TM$_0$ mode conversion efficiency of 43% was observed experimentally (Kotov *et al* 1979). For epitaxial $Bi_{0.68}Pb_{0.02}Tm_{2.3}Fe_{3.8}Ga_{1.2}O_{12}$ film with $h = 4.13$ μm and $\Phi_F = 1300$ deg cm^{-1}, the TE$_6$–TM$_6$, TE$_7$–TM$_7$, and TE$_8$–TM$_8$ mode conversion efficiencies were 80–90%, while the optical losses were 25–30 dB cm^{-1} at

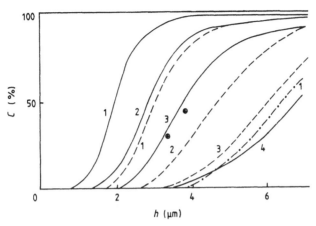

Figure 15.5. Mode conversion coefficients versus film thickness for TE_0-TM_0 (solid curves), TE_1-TM_1 (dashed lines), and TE_2-TM_2 (chain line) conversion, where $\Phi_F = 1760, 920, 460, 115$ deg cm^{-1} (curves 1–4 respectively). $\oplus$: experimental values (Kotov *et al* 1979).

$\lambda = 1.15$ μm. The high mode conversion coefficients could be attributed to the induced birefringence, with $\Delta n = 5.5 \times 10^{-4}$ (van Engen 1978).

15.5 Quadratic magnetooptical effects

Quadratic magnetooptical effects are described by a permittivity tensor containing slightly modified diagonal and non-diagonal elements. Under certain conditions they might lead to a decrease of the phase mismatch of the TE and TM modes, and even to mode degeneracy. Quadratic (-in-magnetization) effects might also result in TE–TM mode coupling in certain configurations. Since for iron garnets, the contribution of these effects may be comparable with that of the linear (-in-magnetization) effects, one can expect a significant influence of the quadratic effects on the propagation of light in gyrotropic waveguides.

Quadratic magnetooptical effects result in the following important features. Quadratic contributions modify the diagonal elements of the permittivity tensor and, therefore, under certain conditions, degeneracy of the TE and TM modes is possible. For yttrium iron garnet, $\Delta n = 6.3 \times 10^{-5}$. Even for these values of Δn, mode degeneracy is possible in sufficiently thick films; for the low-order modes the waveguide thickness should be near 7 μm. Quadratic magnetooptical effects might result in TE–TM mode coupling in the transverse geometry, i.e. when magnetization is in the plane of the film and perpendicular to the direction of propagation of the light. For yttrium iron garnet, the quadratic element of the permittivity tensor $\varepsilon_{xy}^{(2)} \simeq 10^{-5}$, which is only one order of magnitude lower than the linear component δ ($\delta = 3.4 \times 10^{-4}$) (Pisarev *et al* 1969, Hepner *et al* 1975).

These speculations were experimentally verified for an epitaxial gadolinium–gallium iron garnet magnetooptical waveguide grown on a (111)-oriented GGG substrate (Hepner *et al* 1975). He–Ne laser radiation of wavelength $\lambda = 1.152$ μm propagated along the $[\bar{1}10]$ axis (the z-axis). The radiation was coupled in and out using a rutile prism. The film refractive index and thickness, which were determined by measuring the mode excitation angles, were 2.15 and 4.5 μm, respectively.

During the measurements a low-magnitude field component H_x was applied to the sample in order to prevent the appearance of a z-projection of the magnetization. The losses for the TE$_0$ and TM$_0$ modes at the operating wavelength were 5 dB cm^{-1}. For the experiment the TE$_0$–TM$_0$ conversion efficiency for the Faraday effect configuration reached 75% at 6 mm conversion length. In the Gd$_{0.5}$Y$_{2.5}$Fe$_{4.1}$Ga$_{0.9}$O$_{12}$ film that was 5.6 μm thick, 90% mode conversion was achieved in the same geometry. The optimal field magnitude H_x was 1100 Oe. Mode conversion was absent at $H_x = 0$ and for $H_x > 1500$ Oe.

As the expected magnetic birefringence was unlikely to cause such high mode conversion coefficients, the effect was measured in a longitudinal field. To explain the results, it is necessary to assume the presence of growth-induced or stress-induced birefringence with $\Delta n = 1.1 \times 10^{-4}$. Using the expression for the mode conversion coefficient:

$$C(z) = \frac{K^2}{\Delta\beta^2/4 + K^2} \sin^2\left[\left(K^2 + \frac{\Delta\beta^2}{4}\right)^{1/2} z\right] \tag{15.1}$$

where $C(z)$ is the mode conversion coefficient, and K is the coupling coefficient, and matching the experimental curves with the calculated ones, we get $\Delta\beta/k = 4.7 \times 10^{-5}$, and $\Phi_F = 130$ deg cm^{-1} in the Faraday effect geometry and $\Phi_L = 75$ deg cm^{-1} in the Cotton–Mouton configuration.

$\Delta\beta/k$ is lower than the calculated value for an isotropic waveguide: it takes the value $\Delta\beta/k = 1.6 \times 10^{-4}$. On the basis of these results, we can conclude that growth-induced optical anisotropy with $\Delta n = 1.1 \times 10^{-4}$ leads to suppression of the phase mismatch of coupled modes.

The magnetic linear birefringence phenomenon (the Cotton–Mouton effect) is used for efficient waveguide mode conversion in iron garnet films that are grown with misorientation of the normal to the film plane with respect to the (111) axis (Ando *et al* 1985a, b). In such films, the magnetization has an in-plane component, which results in magnetic linear birefringence. Experimental results for the TE$_0$–TM$_0$ mode conversion, which are compared with the calculated data for the Faraday and the Cotton–Mouton configurations, are illustrated in figure 15.6.

Comparison of the magnetic linear birefringence $\Phi_L = 9.0 \pm 2.0$ rad cm^{-1} for the films studied with $\Phi_L = 2.8$ rad cm^{-1} for Gd$_3$Fe$_5$O$_{12}$ single crystal demonstrates the considerable influence of the bismuth ions on the magnetic linear birefringence of iron garnets in the $\lambda = 1.15$ μm wavelength region.

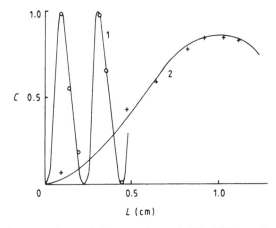

Figure 15.6. A comparison of the experimental TE_0–TM_0 mode conversion with calculated values for the Faraday (curve 1) and Cotton–Mouton (curve 2) geometry (Ando *et al* 1985a, b).

15.6 Sputtered epitaxial Bi-substituted iron garnet films

One of the most impressive advances of modern technology is the development of the method of epitaxial growth of Bi-substituted yttrium iron garnet films by ion sputtering of ceramic targets onto a heated garnet substrate. Reactive ion beam sputtering produced the epitaxial Bi-substituted iron garnets with the best magnetooptical properties in the visible spectral range (Okuda *et al* 1987, 1990).

Incorporation of Bi ions modifies the Faraday rotation as a linear function of the Bi content; for the $Bi_x RE_{3-x} Fe_5 O_{12}$ composition at $x = 1$, the proportionality coefficient is 2.1 deg μm^{-1} at $\lambda = 0.63$ μm (Hansen and Krumme 1984). In the films that were grown by liquid-phase epitaxy, a raising of the bismuth content to $x = 2.3$ ($a_f = 1.255$ nm) was achieved, the Faraday rotation being $\Phi_F = -4.8$ deg μm^{-1} for $\lambda = 0.63$ μm (Hansen *et al* 1984).

Fabrication of films with the ultimate Bi concentration ($Bi_3 Fe_5 O_{12}$) by rf diode sputtering (Okuda *et al* 1990) revealed non-linear growth of the Faraday rotation when the Bi content exceeded $x = 2$. The $Bi_3 Fe_5 O_{12}$ film shows giant Faraday rotation in the ultra-violet and visible regions, which amounts to 2.6×10^5 deg cm^{-1} at 540 nm and 8×10^4 deg cm^{-1} at 633 nm.

Experiments revealed that the main decisive factors for the epitaxial process are the substrate temperature and the oxygen partial pressure. Within the region of epitaxial growth, decrease in the oxygen flow results in higher optical absorption and conduction; at the same time, the Faraday rotation grows with the oxygen flow. Table 15.2 (Okuda *et al* 1987) summarizes the main structural, optical, and magnetooptical parameters of the sputtered epitaxial films. The highest Faraday rotation of $\Phi_F = -5.5$ deg μm^{-1} for $\lambda = 0.63$ μm was achieved for this set of samples; the sample with the slightly lower $\Phi_F = -5.25$ deg μm^{-1} exhibited the highest magnetooptical figure of merit $F = \Phi_F/\alpha = 17.1$ deg, which makes these films attractive as regards various applications in magnetooptical devices.

Here a_s and a_f are the substrate and film unit-cell parameters, and $\Delta a^\perp = a_s - a_f^\perp$ is the substrate–film lattice mismatch parameter; also, V_d is the deposition rate, O_2 is the oxygen flow, and T_s is the substrate temperature. The Faraday rotation and optical absorption were measured at the wavelength $\lambda = 0.63$ μm, the substrate orientation was (111), GCGMZ is $(GdCa)_3(GaMgZr)_5O_{12}$, and NGG is $Nd_3Ga_5O_{12}$. The films of compositions $Bi_2Re_1Fe_{3.8}Al_{1.2}O_{12}$ (Re = Y, Dy, Tb) and $Bi_2Y_1Fe_{5-2x}Co_xGe_xO_{12}$ with $x = 0$, 0.5, 0.8 were sputtered.

Table 15.2. Structural, optical, and magnetooptical parameters of sputtered epitaxial films (Okuda *et al* 1987).

Φ_F (deg μm^{-1})	4.5	4.6	5.25	5.5
α (cm^{-1})	3820	3430	3060	6320
Φ_F/α (deg)	11.8	13.4	17.1	8.7
Substrate	GCGMZ	GCGMZ	GCGMZ	NGG
a_s (nm)	1.2495	1.2448	1.2495	1.2509
a_f (nm)	1.2543	1.2513	1.2554	1.2546
$\Delta a^\perp$ (nm)	0.0088	0.0119	0.0107	0.0068
h (μm)	1.05	1.30	0.85	1.05
V_d (μm h^{-1})	0.08	0.12	0.05	0.08
O_2 (cm^3 s^{-1})	0.2			
T_s ($^\circ$C)		555	600	555

The lowest substrate temperature at which the epitaxial sputtering took place was approximately 380 °C. The x-ray data have shown that the films that were grown at lower substrate temperatures were amorphous. The films that were grown on the NGG substrates with relative lattice mismatch between the substrate and film lower than 0.5% had a mirror-smooth surface; the domain structure was the same as in the high-quality films grown by liquid-phase epitaxy; and the coercive force for $\Delta a/a < 0.1\%$ was less than 5 Oe. The films had uniaxial anisotropy:

$$K_u^s = -\frac{3}{2}\lambda\sigma = -\frac{3}{2}\lambda\frac{E(a_s - a_f)}{a_f(1 - v)}$$

which was induced by mechanical tensions. Because of the difference in sign of the magnetostriction constant λ, for $(BiDy)_3(FeAl)_5O_{12}$, $K_u^s > 0$, while for $(BiTb)_3(FeAl)_5O_{12}$ films, $K_u^s < 0$. Cobalt-substituted films (with Co contents up to $x = 0.8$) on (111) NGG and (111) GGG substrates exhibited strong uniaxial anisotropy (up to 10^5 erg cm^{-3}).

Chapter 16

Integrated magnetooptical devices

In this chapter particular schemes for and examples of functional elements and devices, as well as design prospects for the future, are described. Special attention is paid to a new branch of integrated magnetooptics at the junction of microwave techniques and integrated optics, which is based on waveguide mode interaction with magnetostatic waves. Also we consider the achievements and contemporary state of material design for integrated magnetooptical devices based on interactions with magnetostatic waves.

In order to achieve high performance of integrated optic devices designed for optical modulation in thin-film magnetic layers, non-reciprocal mode conversion, and interaction with magnetostatic waves, the iron garnet films should exhibit low optical losses, strong magnetooptical characteristics, and, in some cases, should have low ferromagnetic resonance linewidths.

16.1 Waveguide magnetooptical modulators

The first integrated magnetooptical device that was based on waveguide mode conversion was the switch modulator of optical radiation with a meander-like driving electrode (Tien *et al* 1972a, b, 1974). The device employed an epitaxial $Y_3Fe_{3.5}Sc_{0.4}Ga_{1.1}O_{12}$ film 3.5 μm thick that was grown on a (111)-oriented GGG substrate. The magnetization vector lay in the plane of the film as a result of an appropriate choice of the lattice mismatch parameter. The field giving rise to the magnetization rotation in the plane of the film did not exceed 1 Oe; the magnetization was $4\pi M_s = 600$ G, and the anisotropy field for tuning the magnetization vector out of the plane of the film was 620 G.

The design of the device is depicted in figure 16.1. The light was coupled in and out by the prism couplers; the driving electrode, which had a period of $p = 2\pi/\Delta\beta = 0.25$ cm, where $\Delta\beta = 24.7$ rad cm^{-1}, provided the change in the coupling between the TE and TM modes (when $M \parallel x$ the mode conversion was maximal while when $M \parallel y$ there was no mode conversion). The magnetization vector was controlled using two meander-like electrodes, which were positioned

345

perpendicularly to each other. The theoretical evaluation of the conversion efficiency was 62%; a value of 52% was obtained experimentally.

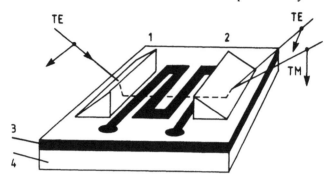

Figure 16.1. Thin-film magnetooptical modulators of laser radiation. The components are: 1: input and output prisms; 2: the meander circuit; 3: the epitaxial iron garnet film; and 4: the substrate (Tien *et al* 1972a, b).

The light was modulated in the following way: a constant bias field was applied at an angle of $45°$ with respect to the x- and y-axes. When the magnitude and the direction of the current in the driving electrodes varied, the component of the magnetization vector that was parallel to the propagation direction of the light altered—that is, the coupling between the TM and TE modes was modified. The ultimate modulation frequency (300 MHz) was limited by the detector response time.

In another case, Tseng *et al* (1974), in order to create a periodical change of the magnetization direction, used a structure that was formally equivalent to a set of small magnets creating an amplitude- and sign-modulated external magnetic field, which also provided considerable mode coupling.

For this modulator a $Gd_{0.5}Y_{2.5}Fe_4Ga_1O_{12}$ film was grown on a (111)-oriented GGG substrate. The film thickness was $h = 6.4$ μm, the saturation magnetization was $4\pi M_s = 200$ G, and the index of refraction was $n = 2.14$ for He–Ne laser radiation of wavelength $\lambda = 1.15$ μm, which was coupled in and out of the film by diffraction gratings. For the modes with $n = 3$, the phase velocity difference was $\Delta\beta = 68 \pm 11$ rad s^{-1}; therefore, the permalloy structure period was 450 μm.

A permalloy film of thickness $h = 0.35$ μm was sputtered onto the sapphire substrate, and then the permalloy stripes, 450×900 μm^2, were fabricated using the routine photolithographic process. The structure was pressed against the magnetic dielectric film. The interaction length reached 1 cm; the optimal mode conversion, which reached 80%, took place when a 1 Oe longitudinal field or a 2.4 Oe transverse field was applied. In the same work, 70% modulation was achieved at 60 Hz frequency, and, after upgrading the bias field system, the frequency response limit was raised to 300 MHz.

Periodic alteration of the magnetization direction can also be used for TE–

TM mode phase matching rather than for phase velocity matching, which we considered in previous cases (Henry 1975). The proposed structure is illustrated in figure 16.2. In the first variant (see figure 16.2(a)), the magnetization changes from longitudinal (first region) to transverse (second region) along the *y*-direction, for which there is no coupling between the TE and TM modes. In the regions of longitudinal magnetization the modes converted, while transverse magnetization leads to mode isolation and accumulation of the phase difference, so there is none of the reverse energy transfer that takes place when the magnetization is uniform.

In the second variant (figure 16.2(b)), the magnetization periodically alters from longitudinal to vertically transverse. As in such configurations the coupling of TE and TM modes takes place for both magnetization directions, in order to isolate them a stripe domain system is created in the regions of the second type. Because of the alternating magnetization orientation in this region, there is no directional energy transformation between the TE and TE modes. To create the periodic alteration of the magnetization orientation, ion implantation or local annealing is proposed.

Metal overlayers, which are deposited onto the waveguide film, may serve as filters which transmit only TE modes. The TM-mode attenuation for aluminium-coated films exceeds 60 dB mm^{-1}, while for the TE modes, the losses are less than 1 dB mm^{-1} (Prokhorov *et al* 1984).

Controlling magnetization orientation in the plane of the film, one can use a metal-coated magnetooptical waveguide to fabricate an amplitude magnetooptical modulator for TE modes. When the magnetization orientation is in the plane of the film and perpendicular to the direction of propagation of the light, there is no TE–TM mode coupling, and the TE radiation propagates in the metal-coated magnetooptical waveguide film almost without losses. When the magnetization in the film was oriented along the propagation direction, the TE mode converted into a rapidly attenuated TM mode, the energy coupling of the TE and TM modes being independent of the sign of the mode coupling coefficient. An appropriate choice of the metal stripe length provides the desired TM-mode attenuation.

The element can also operate as an optical gate, which transmits or blocks TE-mode radiation, depending on the magnetization orientation in the plane of the film with respect to the direction of propagation of the light. If the magnetization is oriented at some angle to the direction of propagation of the light, the TE-mode modulation is provided by the driving signal that changes the angle between the magnetization direction and the direction of propagation of the light. The disadvantage of this modulator is the substantial signal attenuation in the linear region of modulation characteristics. In order to create periodic alteration of the magnetization direction, it was proposed that one could utilize ion implantation or local film annealing (Tolksdorf *et al* 1987).

A magnetically alterable phase grating integrated coupler–modulator was proposed and formed, utilizing the magnetic stripe domains in Bi-substituted

I   II   I   II   I

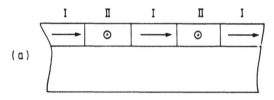

(a)

I   II   I   II   I

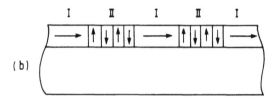

(b)

Figure 16.2.   Magnetooptical waveguides with periodically altered magnetization directions. The arrows show the magnetization direction (Henry 1975).

iron garnet film, by Sauter *et al* (1977). The epitaxial iron garnet structure is bonded to an rf-sputtered waveguide, with an epitaxial layer adjacent to the waveguide. Laser light entering the substrate and film was diffracted by the phase grating formed by stripe domains of the iron garnet film, and is coupled to the waveguide if the coupling condition is satisfied. In experiments, the light was coupled to a glass waveguide, and 72% intensity modulation was achieved.

16.2   Waveguide non-reciprocal elements

One of the important objectives of integrated optics is the creation of non-reciprocal devices for optical isolators and gyroscopes. The principles of non-reciprocal element design for integrated optics were considered by Wang *et al* (1972), and Yamamoto and Makimoto (1974). In a non-reciprocal waveguide, the mode coupling coefficient depends on the propagation direction.

    An anisotropic magnetooptical waveguide that consists of a magnetic film and an anisotropic upper layer can, under certain conditions, provide the desired non-reciprocity (Yamamoto and Makimoto 1974, Zvezdin and Kotov 1988, Prokhorov *et al* 1984).

    The non-reciprocity of mode conversion up to 7 dB was experimentally demonstrated by Smolenskii *et al* (1977). A $(BiTm)_3(FeGa)_5O_{12}$ epitaxial iron garnet film was brought into optical contact with a titanium-diffused lithium niobate waveguide. Thus a double-layer waveguide, or a system of two coupled waveguides, depending on the gap between the waveguide and the film, was formed. The mode conversion took place in the gyrotropic film, which was magnetized along the direction of propagation of the light. The mode coupling coefficient depended on the direction of propagation of the light and the gap between the waveguide and the film, which was determined by the pressure on the film, and on the coupling length of the waveguide and the film. A conversion non-reciprocity of 7 dB at the wavelength of 1.15 μm was obtained for a 3 mm contact spot and under appropriate pressure.

Apart from single-section devices, non-reciprocal mode conversion may be achieved in so-called cascade converters, which consist of two sections that are formed in the same epitaxial film (Wang *et al* 1972, Yamamoto and Makimoto 1974). In the devices, in one section the external magnetic field and the film magnetic moment are both oriented in the direction in which the light propagates (the Faraday geometry), while in the second one the magnetization is perpendicular to the direction of propagation of the light in the plane of the film (the Cotton–Mouton configuration). The waveguide modes are phase matched due to the deformations of the film, which result from the substrate and film lattice mismatch.

Using the Jones matrix formalism for light propagation in the double-section configuration, for the forward direction we have

$$\mathbf{T}^+ = \mathbf{T}_1^+ \mathbf{T}_2^+ = i \begin{pmatrix} -1 & 0 \\ 0 & 1 \end{pmatrix}$$

and for the reverse direction of light propagation we have

$$\mathbf{T}^- = \mathbf{T}_1^- \mathbf{T}_2^- = i \begin{pmatrix} 0 & 1 \\ 1 & 0 \end{pmatrix}.$$

Here $\mathbf{T}^\pm$, $\mathbf{T}_1^\pm$, and $\mathbf{T}_2^\pm$ are the conversion matrices for the entire device, and for the first and second sections, respectively, '+' indicates the forward direction of light propagation, and '−' the opposite direction. Thus, utilization of two sections in a non-reciprocal device can, in principle, result in complete mode conversion for the forward-propagating mode and lack of conversion for the reverse one.

The experimental device employed a 5 μm epitaxial $(YGd)_3(FeGa)_5O_{12}$ film grown on a (111)-oriented GGG substrate. The light propagated in the $\langle 110 \rangle$ direction. The Faraday section length was 3.5 mm, while the section with transverse magnetization was 12 mm long. Non-reciprocal mode conversion in the iron garnet epitaxial films was also observed when the magnetic field was switched into the direction that was opposite to that of the collinear orientation of the magnetization and propagation of light in the $Y_3Fe_{3.6}Sc_{0.4}Ga_1O_{12}$ film. This effect was associated both with reciprocal conversion in the anisotropic substrate, which was induced by mechanical tensions, and with non-reciprocal conversion in the gyrotropic film (Smolenskii *et al* 1976b).

Even higher non-reciprocity of the conversion coefficients ($C^+ = 0.93$, $C^- = 0.09$) was observed for the magnetooptical waveguide that was based on epitaxial iron garnet in a non-uniform magnetic field. A theoretical study of the waveguide mode propagation in a five-layer system that consisted of two isotropic waveguides coupled via a gyrotropic or/and anisotropic medium revealed that the structure can be used as an optical circulator, which would not require mode division at the input and output, and as a gate that would not require mode filtering (Prokhorov *et al* 1984).

Non-reciprocal mode propagation can be implemented also in leaky waveguides (Yamamoto and Makimoto 1977). In lithium-niobate-clad yttrium iron garnet film the leakage loss was 10 dB cm^{-1} at $\lambda = 1.15$ μm. In such a device, the waveguide mode is converted into a radiation mode, and its energy scatters into the adjacent media. In this case there are no strict requirements for the mode synchronization, conversion length, film thickness, and optical contact quality. Leakage-type gates operate at any propagation length, i.e. double length leads to double the non-reciprocity effect. However, absolute non-reciprocity cannot be achieved in such elements, because of the exponential character of the mode conversion dependence. High optical losses for the forward-propagating mode (approximately 10 dB at 10 dB non-reciprocity) can also be attributed to the drawbacks of this device.

The waveguide structure that is shown in figure 16.2(b) can also be utilized in the design of non-reciprocal waveguide mode conversion devices (Henry 1975). Non-reciprocal elements, together with metal-clad mode filters, are useful for mode selection devices. The devices transmit only one type of waveguide mode; for example, they guide TE modes, and do not guide TM modes. A metal cladding, which is deposited onto the waveguide layer, or a metal substrate, can be used as filters transmitting only TE modes.

Wolfe *et al* (1985) proposed a scheme for an effective non-reciprocal element with periodical alteration of the sublattice magnetization in an epitaxial iron garnet film containing 1.3 gallium atoms per formula unit (a composition that corresponds to compensation of the resulting magnetic moment). The sublattice magnetization was periodically altered using local laser annealing (figure 16.3). The device provided non-reciprocal rotation of the polarization plane by $\pm 45°$ at $\lambda = 1.45$ μm without any ellipticity of the outlet light (i.e. in absence of modes of another type), the forward-to-backward mode attenuation ratio being equal to 500.

An analysis of the optical radiation propagation in a thin-film magnetooptical waveguide, which employed the Poincaré sphere for representation of the polarization states, revealed that maximum non-reciprocity of the mode conversion can be provided by inducing periodical regions of a constant length, which is equal to half of the waveguide birefringence period, in which the Faraday rotation alters its sign, and a final element of a length that is a quarter of the birefringence period in the waveguide. The number of elements of length $p/2$ is determined by the waveguide Faraday rotation, and the phase difference $\Delta\beta$ of the waveguide modes. If the ratio $\Phi_F/\Delta\beta$ satisfies the following condition:

$$\tan^{-1}\left(\frac{2\Phi_F}{\Delta\beta}\right) = \frac{90°}{2n+1} \tag{16.1}$$

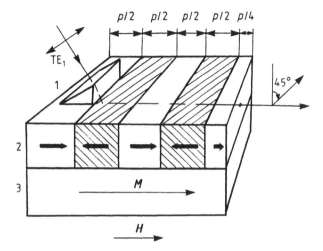

Figure 16.3. A non-reciprocal magnetooptical element. Components: 1: input prisms; 2: an epitaxial iron garnet film with a periodic change of direction of its magnetization; and 3: the substrate (Wolfe *et al* 1985).

where n is an integer, then n elements of length $p/2$ and one element of length $p/4$ ensure rotation by 45°. In the convertor with $n = 3$, which is shown in figure 16.3, an input TE mode (with the plane of polarization parallel to the plane of the film) is transformed into a hybrid mode that corresponds to a +45° angle between the plane of polarization and the normal to the plane of the film. Reversing the magnetic field leads to rotation of the plane of polarization by −45°.

The periodical modulation of the magnetization direction was created as follows. An epitaxial bismuth–gallium iron garnet film with the gallium content corresponding to the sublattice magnetic moment compensation was slowly cooled after the growth, which was equivalent to low-temperature annealing. This procedure shifted the equilibrium between gallium ions in tetrahedral and octahedral sublattices towards higher gallium content in the tetrahedral sites; therefore, the resulting magnetization was determined by the octahedral sublattice.

During the high-temperature annealing, the equilibrium shifts towards higher gallium content in the octahedral sublattice; the magnetization is defined by the tetrahedral sublattice. Local high-temperature laser annealing forms the periodical structure with reversed sublattice magnetization, which naturally results in alteration of the sign of the Faraday rotation.

The experimental $(YBi)_3(FeGa)_5O_{12}$ sample, 2.8 μm thick, was grown on a (111)-oriented GGG substrate. The film had substantial magnetic anisotropy; its magnetization was oriented normally to the film surface. After growth, the film was exposed to high-temperature annealing in a nitrogen atmosphere, followed by slow cooling. The high-temperature annealing eliminated uniaxial growth-induced anisotropy, and compressing tensions in the plane of the film resulted in a negative magnetoelastic component of the magnetic anisotropy; therefore, the resulting magnetization was in the plane of the film. The Faraday rotation was equal to 140 deg cm^{-1} at the wavelength 1.5 μm. The birefringence period for the system of waveguide modes was 1.84 mm, which, according to equation (16.1), leads to $n = 5$ (Wolfe *et al* 1985).

Tunable KCl:Tl^0 laser radiation was coupled into the film using a rutile prism; the output radiation was detected at the polished edge of the film. The magnetization orientation was switched using a small permanent magnet. The film was annealed in air, using continuous argon laser radiation. The laser radiation was focused by a lens with a focal length of 20 cm onto a spot of diameter 40 μm. The film was moved at the speed of 2 cm s^{-1} in 120 μm steps between the runs. The laser power was 0.85 W, which is 10% lower than the film damage threshold. This procedure formed annealed stripes 0.92 μm wide, separated by virgin areas of the same width. The final 0.46 μm element was not annealed.

Figure 16.4 shows the dependence of the angle of polarization of the output radiation on the distance between the prism edge and the edge of the waveguide. In the uniform film (without annealing), small oscillations of the plane of polarization of the radiation with respect to the initial state, which had the same period as the waveguide mode birefringence, were observed. At the wavelength 1.45 μm, the output radiation was linearly polarized at 45°. When the magnetization was altered, the polarizer decreased the output light intensity by a factor of 500, which corresponded to 27 dB optical losses. The non-reciprocity of the device was confirmed when the light beam which was polarized at 45° was coupled in through the polished edge of the waveguide and propagated in the opposite direction. The radiation, which was coupled out of the waveguide by a rutile prism, corresponded to the initial TE$_0$ mode only when the film's magnetic moment was altered (Wolfe *et al* 1985).

Among integrated optical devices, the single-mode non-reciprocal waveguides attract special interest. A single-layer iron garnet epitaxial film grown on a GGG substrate supports a single-mode propagation (there are only two zero-order TE$_0$ and TM$_0$ modes in the waveguide) at layer thicknesses below one micrometre, which is unacceptable for a number of practical applications. Various multilayered structures are currently under study.

Hernandez *et al* (1987) showed theoretically that in a three-layer single-mode waveguide structure (air–waveguide layer–substrate) it is impossible to obtain high non-reciprocity by optimizing the propagation distance of the light and the magnetization orientation. In a four-layer structure (air–waveguide layer–sublayer–substrate), appropriate adjustment of the difference in refractive index Δn between the waveguide layer and the sublayer, of the waveguide thickness h_2, of the coupling length L, and of the orientation of the magnetization of the waveguide layer can provide 100% mode conversion for the light

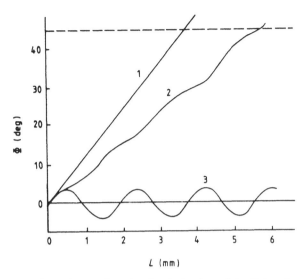

Figure 16.4. The dependence of angle of polarization of the light, Φ, at the output of the waveguide with respect to the polarization state at the input. Key: 1: bulk garnet; 2: the laser-annealed iron garnet film; and 3: the as-grown film (Wolfe *et al* 1985).

that propagates in the forward direction, and no conversion for the light that propagates in the opposite direction.

The problem of mode propagation in a one-section non-reciprocal device based on an epitaxial bismuth-substituted iron garnet film in a hybrid elliptically polarized mode approximation was considered theoretically by Ueki and Miyazaki (1987).

Taki and Miyazaki (1987) experimentally fabricated a single-element non-reciprocal element based on $Bi_1Y_2Fe_4Al_1O_{12}$ film sputtered on a GGG substrate. The optical parameters of these films are listed in table 16.1. In the experiment, the light was coupled in and out of the waveguide using rutile prisms. Two permanent magnets oriented the film magnetization in the optimal direction, which provided maximum mode conversion non-reciprocity. It is known that the effect is achieved when, for a given propagation direction, the contributions of the Cotton–Mouton and the Faraday effects to the mode conversion coefficient are equal. The maximum ratio of the intensities for the forward and backward directions, I_F/I_B, was 13 dB at the conversion length $L = 2.4$ mm, the overall propagation losses being 5.5 dB (2.3 dB mm^{-1}).

The processes of reciprocal and non-reciprocal mode conversion in waveguides were analysed theoretically by Imamura *et al* (1987). The authors determined the relationship between geometric parameters of the mode converter section and the magnetization orientation, which provided 100% conversion non-reciprocity of the forward and backward mode propagation directions for the bismuth-substituted iron garnet epitaxial films.

Table 16.1. Optical ($\lambda = 1.15$ μm) and magnetic properties of the film (Taki and Miyazaki 1987). See the text for details.

Refractive index, n	Birefringence, Δn	Specific Faraday rotation, Φ_F (deg cm^{-1})	Absorption coefficient, α (cm^{-1})	Coercive force, H_c (Oe)
2.18	1.28×10^{-3}	540	2	3

Friez *et al* (1987) proposed a non-reciprocal stripe waveguide structure. The structure is a two-layer epitaxial film (the top medium was $Gd_{0.45}Y_{2.55}Fe_{4.2}Ga_{0.8}O_{12}$, a film $h = 3.7$ μm thick; the bottom layer was $Gd_{0.45}Y_{2.55}Fe_{4.1}Ga_{0.9}O_{12}$, 2.9 μm thick; the refractive indexes were $n_1 = 2.1565$ and $n_2 = 2.1511$, respectively). For the degenerate TE- and TM-mode propagation in the structure, Friez *et al* proposed compensating for the effective refractive index difference of the TE and TM modes with the birefringence that was induced by the substrate and film lattice parameter mismatch. They calculated optimal waveguide layer parameters, as well as the geometric dimensions of the waveguide that would provide the single-mode degenerate TE and TM zero-order mode propagation. Experimentally fabricated stripe waveguides exhibited losses lower than 5 dB cm^{-1}.

Tolksdorf *et al* (1987) developed multilayered and 'hidden' stripe magneto-optical waveguides for non-reciprocal waveguide devices. A four-layer waveguide structure with a symmetric single-mode waveguide was prepared using two epitaxial growth processes. During the first one a layer L_4 (the absorbing layer) of cobalt-substituted iron garnet was grown on the GGG substrate. Then the film was epitaxially grown from a cobalt-free melt. In this process the layer refractive indexes were modified by controlling the substrate rotation rate. With $\Delta T = 53$ K supercooling, and the substrate rotation rate $\omega = 30$ rpm, the L_3 layer of yttrium iron garnet, of composition $Y_{2.93}Pb_{0.07}Fe_5O_{12}$ (the cladding layer), was grown; then the rotation speed was increased to 160 rpm, resulting in modification of the L_2 layer composition to $Y_{2.91}Pb_{0.09}Fe_5O_{12}$ (the core layer); and then, to grow the L_1 layer (another cladding layer), the rotation speed was decreased back to its initial value $\omega = 30$ rpm.

The upper transition layer, which was formed during the final stage of the epitaxial process, was removed by etching in phosphorous acid heated up to 130 °C. An additional layer of cobalt-substituted iron garnet was necessary for the suppression of higher-order modes, which can be excited in a three-layer structure. In this waveguide, single-mode propagation in the L_2 layer was confirmed experimentally. Guided modes of different polarizations were phase matched by a combination of the birefringence that was induced by the

Table 16.2. Optical properties of epitaxial films (Tolksdorf *et al* 1987).

Layer No	Refractive index	Thickness, h (μm)	Lead content	Relative lattice mismatch, $\Delta a^{\perp}/a$
1	2.2200	3.3	0.066	+0.11
2	2.2236	3.7	0.090	−0.28
3	2.2200	3.3	0.066	+0.11
4	2.2190	4.4	0.040	+0.47

substrate and the film lattice mismatch $\Delta a^{\perp} = (a_s - a_f^{\perp})/a = -2 \times 10^{-4}$, and also the additional birefringence resulting from external mechanical tensions. By changing the lead-ion content in the layers, Tolksdorf *et al* could obtain the desired refractive indices. Table 16.2 summarizes the optical properties of the epitaxial films.

The three-layer waveguide structure was grown from a bismuth-containing melt. In this case, the index change $\Delta n > 10^{-2}$ could be achieved by controlling the substrate rotation speed. For the films that were grown using this method, the Faraday rotation reached $\Phi_F = -1300$ deg cm^{-1} at $\lambda = 1.3$ μm, with relatively low optical absorption ($\alpha = 0.2$ cm^{-1}). Fabrication of the three-layer waveguide structure in a single epitaxial process has substantial advantages over a multiple-stage epitaxial process, because it avoids irregular sublayers and contamination at the interfaces. Control of the substrate rotation rate ensures high reproducibility and accuracy of the refractive index profiles.

Efficient connection with optical fibres requires stripe magnetooptic waveguides. Nowadays, the technology of stripe waveguide fabrication is well developed (Tolksdorf *et al* 1987). The L$_3$ and L$_4$ layers are grown from bismuth-containing melts at $\Delta T = 37$ K and $\omega_3 = 60$ rpm; then the rotation rate is increased to $\omega_2 = 90$ rpm for growing the subsequent layers of thicknesses $h_3 = 5$ μm (L$_3$) and $h_2 = 4$ μm (L$_2$). The difference between the refractive indexes of the layers was 3×10^{-3}, i.e., the same as for single-mode optical fibres. The stripe waveguide is formed in the L$_2$ layer using conventional lithography and etching of SiO$_2$ in a HF solution. Then the L$_1$ layer is grown on this structure, its properties being identical to those of the L$_3$ layer. The final step was the growth of the cobalt-substituted L$_0$ layer to absorb the higher-order waveguide modes.

Dotsch *et al* (1992) have proposed a new type of optical waveguide isolator based on the TM$_0$ mode propagating in a stripe waveguide which is divided by compensation walls. In such a waveguide, the intensity distributions are different for forward and backward propagation. The backward mode can thus be eliminated by localized optical absorption. It is demonstrated that

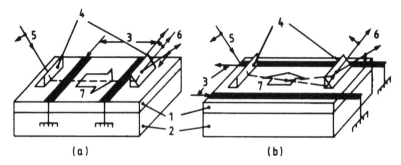

Figure 16.5. Typical experimental configurations for studying collinear (a) and non-collinear (b) interaction of waveguided optical modes with the MSW. Key: 1: the film; 2: the substrate; 3: microstripes; 4: rutile prisms; 5: the input light beam; 6: output light beams.

optical modulators can be realized using dynamical mode conversion induced by domain and domain lattice resonances. Dynamical conversion efficiencies of 10% at frequencies up to 3 GHz are measured. Further enhancement of the non-reciprocal effect for TM-mode propagation was attained using an iron garnet double-layer film exhibiting the opposite direction of Faraday rotation (Wallenhorst *et al* 1995).

Mizumoto *et al* (1993) described an interferometric waveguide optical isolator, employing a non-reciprocal phase shift. Here, the non-reciprocal phase shift means that the phase change that the light wave experiences differs according to the propagation direction. An optical interferometer is composed of branching devices, non-reciprocal phase shifters in two arms, and a reciprocal phase shifter in one of the arms. The isolator has the advantage of not needing phase matching and complicated magnetization control.

16.3 Waveguide magnetic field sensors

Deeter *et al* (1994) described a magnetooptical magnetic field sensor based on uniaxial bismuth-substituted iron garnet films in an optical waveguide geometry. In the optical waveguide geometry, these films exhibit large values of saturation Faraday rotation, which in turn lead to high sensitivity. For example, at the wavelength 1.3 μm, for an optical path 1 cm long, it is possible to obtain a resulting Faraday rotation of more than 1000°. Of course, in order to realize such a rotation, it is necessary to achieve a very accurate matching of the phase velocities of the TE and TM modes, or single-mode propagation with a high level of absorption or radiation losses for any other mode, or to use very thick films—with thicknesses of some hundreds of micrometres.

In films with uniaxial magnetic anisotropy, the stripe or maze domain structure favours magnetization rotation as the primary magnetization process if the value of the external magnetic field is small compared with the saturation field. The magnetization rotation process is significantly faster than the domain wall motion, which is the primary magnetization process for bulk iron garnet crystals. On the basis of the proposed structure of a magnetic field sensor, Deeter *et al* demonstrated a virtually flat frequency response from dc to at least 1 GHz. For iron garnet films with in-plane magnetic anisotropy—for example, Pr-containing films—a saturated in-plane magnetic field of 10 Oe leads to another type of efficient waveguide field sensor. In such a configuration, the sensitivity is determined by the relationship of the measured field to the saturation field, and the value of the saturation Faraday rotation.

16.4 Interactions of optical waveguide modes and magnetostatic waves

There is an extremely promising trend in integrated magnetooptics which is based on waveguide mode conversion as a result of interactions with surface and volume magnetostatic waves (MSW). Typical MSW frequencies are 1–20 GHz, and typical wavelengths range from 10 to 10^3 μm. The MSW can be easily excited in magnetic films, and exhibit relatively low attenuation in materials with narrow FMR lines (yttrium iron garnet, lithium spinel, and others). Using this phenomenon, it is possible to create optical modulators operating at frequencies of 1–20 GHz, narrow-band tunable optical filters, non-reciprocal optical devices, and circulators for fibre communication systems (Tsai and Young 1990, Matyushev *et al* 1995).

Figure 16.5(a) shows typical experimental configurations for the study of the interaction of the waveguide mode with the MSW. This interaction is in fact the stimulated Mandelstamm–Brillouin scattering of light, which propagates in the waveguide, by the MSW. The main manifestations of the scattering are mode conversion, light modulation, and beam diffraction by the MSW.

One should differentiate between collinear and non-collinear light scattering (figures 16.5(a), 16.5(b)). The theory for collinear light scattering in a planar magnetooptical waveguide was developed by Guliaev *et al* (1985). Collinear scattering by the MSW was studied experimentally by Fisher *et al* (1982); they observed and studied optical mode conversion ($\lambda = 1.15$ μm) that was induced by surface MSW, which were excited in an epitaxial iron garnet film of thickness 11.7 μm, in the frequency range from 3.5 to 6.5 GHz, when the light waves propagated in the same direction as the MSW (or in the opposite direction); the interaction length was 7 mm.

The mode conversion due to the scattering by the MSW reaches its maximum when the phase synchronism condition is satisfied:

$$\omega' = \omega \pm \Omega \qquad k' = k \pm q$$

where ω and Ω are the frequencies of the light and the MSW, respectively, and k and q are their wave vectors; the primed symbols refer to the scattered wave. At the UHF signal power of 5 mW, the waveguide mode conversion efficiency was 0.02%, and it grew to 4% when the UHF signal power increased to 1.7 W.

Rutkin *et al* (1985) observed two regions of rather effective mode conversion when they studied the interaction of waveguide modes with surface MSW, in collinear propagation, in an yttrium iron garnet film 5.4 μm thick, having the FMR linewidth $2\,\Delta H = 0.9$ Oe, which was grown on a (111)-oriented GGG substrate. The first region, in which the mode conversion was the strongest, corresponded to the phase synchronism condition for collinear interaction of the waveguide mode; the linearly excited maximum mode conversion was 0.04%. By varying the tangential component of the magnetizing field, Rutkin *et al* were able to tune the conversion maximum in the 2–4 GHz frequency range. In the second area the most probable cause of the mode transformation was the scattering of light by parametric excited spin waves.

In an experiment set up to study collinear low-loss propagation of optical and magnetostatic waves in $Bi_{0.9}Lu_{2.1}Mg_{0.02}(FePt)_{4.98}O_{12}$ films, the mode conversion efficiency was $R = 2.6\%$ at the conversion length $L = 1$ mm, and the UHF power was $P = 280$ mW, while for $Y_3Fe_5O_{12}$ films, $R = 0.55\%$ at $L = 8$ mm and $P = 175$ mW (Tamada *et al* 1987). Using the following expression for the mode conversion coefficient:

$$C(Z) = \frac{K^2}{K^2 + (\Delta\beta/2)^2} \sin^2 \left[\left(K^2 + \left(\frac{\Delta\beta}{2} \right)^2 \right)^{1/2} Z \right]$$

where $\Delta\beta = \beta_{TE} - \beta_{TM} - q_M$, $K \sim (\Phi_F/2)\sin\theta_m$, β_{TE} and β_{TM} are the TE- and TM-mode propagation constants, respectively, q_M is the wavevector of the magnetostatic wave, θ_m is the precession angle of the magnetostatic wave, and K is the coupling constant, we can see satisfactory agreement of the theoretical estimates with the experimental data.

Nakano and Miyazaki (1987) investigated experimentally the optical mode conversion occurring as a result of interaction with magnetostatic waves for stripe domain $Y_3Fe_5O_{12}$ waveguides, which were rf sputtered onto a GGG substrate. Amorphous yttrium iron garnet films were sputtered at the substrate temperature 600 °C, with subsequent annealing for 4 h in an oxygen atmosphere at 1100 °C, the heating and cooling rates being equal to 250 °C h^{-1} and 80 °C h^{-1}, respectively. The propagation losses of magnetostatic waves at 2.3 GHz were 1.63 dB cm^{-1}. The experimentally established conversion efficiency for the stripe domain waveguide was $R = 1\%$ at 2.35 GHz.

Collinear light scattering by the MSW can be employed in the design of optical gates and modulators in the IR spectral band, the frequency response being much faster than that of their acoustooptical counterparts. Double scattering by the two counterpropagating MSW of the $TE_0 - TM_1 = TE_2$ type can be used for convolution of two microwave signals.

Non-collinear light scattering by the MSW offers even richer possibilities (figure 16.5(b)). The phase synchronism condition in this case takes the form

$$\beta_1 - \beta_2 = k_{MSW}$$

where the $\beta_{1,2}$ are the incident and the scattered wave propagation constants, and k_{MSW} is the MSW wavevector. This effect, which is called Bragg diffraction, can be utilized in stripe domain integrated optic filters, deflectors, channel switches, convolvers, microwave-range correlators, and instant spectrum analysers.

Tsai *et al* (1985) achieved experimentally 2.5% TE → TM mode conversion for the transverse propagation geometry of the light with respect to the MSW for 1 W input power at the transducer. The FWHM of the Bragg diffraction was 450 MHz at the frequency 4 GHz ($H = 803$ Oe). The thicknesses of the samples examined ranged from 3 to 14 μm for a 1 mm MSW aperture. By varying the magnetic field in the range from 200 to 1700 Oe, Tsai *et al* could tune the frequency of scattering from 3 to 7 GHz.

The highest TM → TE mode conversion efficiency achieved so far, 47%, was reached in an interaction with backward-propagating volume MSW (BV MSW), for a UHF power of 0.8 W at 4 GHz frequency, and with a 7 mm interaction length (Tamada *et al* 1988); the mode conversion was still fairly effective in the range 3.5–7.5 GHz. A BV MSW was excited in a $Bi_{0.9}Lu_{2.1}Fe_5O_{12}$ film, which was grown from a Bi_2O_3-based melt. Optical losses in the film at $\lambda = 1.3$ μm did not exceed 0.6 dB cm^{-1}—this was achieved by inclusion of MgO in the melt; the Faraday rotation was 0.14 deg μm^{-1}, and the FMR linewidth was 0.7 Oe.

Light scattering by the MSW provides the possibility of designing a spatial magnetooptical Fourier processor, which carries out convolution of two signals. The convolution is furnished in the following way. In the Fourier plane of a point source image, which is located behind the first geodesic lens, there is an integrated lens, which creates the spatial light modulation $g(y)$. In the same plane, a MSW creates the modulation $f_\omega(y - V_g t)$ via Bragg diffraction. The resulting signal in the image focal plane, which is formed behind the second geodesic lens, is proportional to the convolution:

$$\int g(y) f_\omega(y - V_g t) \, \mathrm{d}y. \tag{16.2}$$

MSW-based devices operate approximately one order of magnitude faster than acoustooptical elements, because the group velocity of MSW greatly exceeds the sound velocity ($V_g^{MSW} \sim 10^7$–10^8 cm s^{-1}). As was already mentioned, a single microstripe ensures wide-band excitation of the MSW. Using phased and crossed MSW beams, like in acoustooptics, one can attain wide-band Bragg diffraction (up to 1 GHz and higher). Calculations predict the possibility of creating Bragg diffraction deflectors having several dozens of resolvable spots. An additional advantage of the MSW is that, in order to create wide-band Bragg diffraction, one can apply a non-uniform magnetic field to the microstripe, which tunes the central diffraction band. Therefore, these devices are very promising as regards integrated optics microwave-signal-processing systems in the 1–20 GHz frequency range.

To realize these opportunities, it is necessary to increase the magnetooptical figure of merit of the material—for example, by bismuth substitution in yttrium iron garnet. Theoretical considerations demonstrate that for $Y_{3-x}Bi_xFe_5O_{12}$ films, it is possible to achieve 30% mode conversion at $P = 100$ mW for $x > 1$. Experiments on the MSW propagation in bismuth-substituted iron garnet films confirm that for x ranging from 0.2 to 0.6 the FMR line is not widened.

The propagation of surface and backward volume MSW was studied for $Bi_x(YLu)_{1-x}Fe_5O_{12}$ films, which were grown using liquid-phase epitaxy on (111)-oriented GGG substrates. In the films, the FMR width, which was determined from the attenuation of the surface MSW in the frequency range 3.6–4.0 GHz, was $\Delta H = 1.5$ Oe. The Faraday rotation at $\lambda = 0.63$ μm was 0.9 deg μm^{-1}, which corresponded to a bismuth content of about half of a formula unit. As the observed attenuation does not correlate with the actual bismuth content, bismuth-substituted iron garnet films with even lower MSW losses ($2\,\Delta H = 0.5$ Oe) and higher Faraday rotations can be expected (Bogun *et al* 1985).

Winkler *et al* (1994) studied the dynamic conversion of optical modes in iron garnet films of composition $(YBi)_3(FeAl)_5O_{12}$ grown by liquid-phase epitaxy on (111)-oriented GGG substrates. The films support periodic lattices of parallel stripe domains. A simple strip antenna is used to excite the domain wall resonance and the two branches of the domain resonance in the frequency range up to 5 GHz. The resonance frequencies and the dynamic components of the magnetization are calculated using a hybridization model. Good agreement of the calculated and measured resonance frequencies is obtained if the quality factor of the film is larger than 0.6. The excited-domain resonances cause dynamic conversion of transverse electric and transverse magnetic modes via the Faraday and the Cotton–Mouton effects. The mode coupling and conversion are calculated using perturbation theory. The dynamic conversion efficiencies are measured at the fundamental and the first-harmonic frequencies, and for the zero diffraction order, as functions of the static induction applied in the plane of the film, parallel to the stripes. Conversion efficiencies up to 18% are achieved at a frequency of 2.8 GHz.

To conclude this section, let us summarize the new possible applications areas for the waveguide mode scattering by MSW: multichannel beam switching in integrated and fibre-optic systems, high-speed optical modulators, Fourier analysis of UHF signals, optical wide-band UHF signal processing, and narrow-band optical filtering.

16.5 Materials for waveguide/MSW devices

The main problem in the design of materials for optical signal-processing devices based on the interaction with MSW is that of enhancing the Faraday rotation of the material, while keeping the MSW propagation losses at a minimum. Wolfe *et al* (1987) and Butler *et al* (1990) demonstrated that the MSW propagation losses, which are measured in dB μs^{-1}, lie in the range

$$76.4 \, \Delta H < L \text{ (dB } \mu s^{-1}) < 76.4 \left[1 + \frac{\omega_M^2}{4\omega^2} \right]^{1/2} \Delta H$$

where ΔH is the FWHM of the FMR, ω is the operating frequency, and ω_M is the product of the gyromagnetic ratio and the saturation magnetization. This relationship demonstrates that MSW propagation losses grow with the magnitude of the MSW wave vector k; therefore, using the expression for the MSW propagation losses in the form $76.4 \, \Delta H$ dB μs^{-1} for small k, i.e., for forward-propagating volume MSW, is a valid approximation.

The TE–TM mode conversion efficiency of the MSW (for diffraction by the MSW) in thin-film magnetooptical waveguides is a quadratic function of the specific Faraday rotation at the operating wavelength λ; therefore, despite higher MSW propagation losses, devices that are based on materials with large Faraday rotations, such as bismuth-substituted iron garnets, can exhibit even better performances than those based on yttrium iron garnets.

Wolfe *et al* (1987) reported the results of a study of magnetooptical and UHF properties of $Y_{1.5}Bi_{1.5}Fe_5O_{12}$ films grown by liquid-phase epitaxy on (111)-oriented substrates with a large unit-cell parameter ($a_s = 1.2496$ nm). The films were grown from a conventional lead-containing $PbO–Bi_2O_3–B_2O_3$-based melt at 800 °C. At a frequency of 9.5 GHz, the FMR linewidth ΔH of the films was 1.38 Oe, which corresponded to 105.4 dB μs^{-1} with the approximation of small wave vectors. The measured propagation losses of the MSW, which travelled in delay lines that were based on epitaxial iron garnet films with high bismuth contents, were 8.97 dB cm^{-1} at 9.95 GHz and 19.3 dB cm^{-1} at 19.14 GHz.

Comparison of different results does not lead to any definite conclusion concerning the effect of bismuth content on the FMR width. Santi *et al* (1987) demonstrated an increase of the FMR width from $\Delta H = 0.35$ Oe at $x = 0.41$ to $\Delta H = 0.7$ Oe at $x = 1.03$ for films grown from a lead-containing melt. Also, Bogun *et al* (1985) showed that for $(BiYLu)_3Fe_5O_{12}$ films, with x ranging from 0.24 to 0.52, ΔH remained practically unchanged at 1.5 Oe, while modification of the film growth conditions led to a growth of ΔH up to 2 Oe for $x = 0.38$. Razeghi *et al* (1986) fabricated films with $x = 1.4$ and $H = 0.73$ Oe using a lead-free melt. Ando *et al* (1985a, b), using $Bi_{0.9}(LuYMg)_{2.1}(FePt)_5O_{12}$ films which were grown from a lead-free melt, achieved $\Delta H = 0.7$ Oe.

The available experimental data lead to the following conclusions: good results from the point of view of the FMR linewidth may be achieved using either lead-containing or lead-free melts. So far it is not certain which of the proposed melts is the best for technological processes. In most cases, ΔH grows with bismuth content, but the available data lead to the suggestion that increase in the FMR linewidth may be associated with changes in growth conditions—in particular, with a higher degree of supercooling—rather than with the Bi ions.

References

Abe M and Gomi M 1987 *J. Magn. Soc. Japan. Suppl.* 1 **11** 299–304

Abramishvili V G, Komarov A V, Ryabchenko S M, Derkach B E and Savchuk A I 1987 *Sov. Phys.–Solid State* **29** 644

Afonso C N, Giron S, Lagunas A R and Vicent J L 1981 *IEEE Trans. Magn.* **17** 2849

Afonso C N, Lagunas A R, Briones F and Giron S 1980 *J. Magn. Magn. Mater.* **15–18** 833–4

Agal'tsov A M, Gorelik V S, Zvezdin A K, Murashov V A and Rakov D N 1989 *Lebedev Inst. Lett.* **5** 37

Agranovich V M and Ginzburg V L 1984 *Spatial Dispersion in Crystal Optics and Theory of Excitons* (Berlin: Springer)

Ahrenkiel R K and Coburn T J 1973 *Appl. Phys. Lett.* **22** 340–1

——1975 *IEEE Trans. Magn.* **11** 1103–8

Ahrenkiel R K, Moser F, Carnall E, Martin T, Pearlman D, Lyu S L, Coburn T and Lee T H 1971 *Appl. Phys. Lett.* **18** 171–3

Akhmediev N N, Borisov S B, Zvezdin A K, Lubchanskii I L and Melichov Yu V 1985 *Fiz. Tverd. Tela* **27** 1075–8

Akhmediev N N and Zvezdin A K 1983 *Pis. Zh. Eksp. Teor. Fiz.* **38** 167–9

Aktsipetrov O A, Braginskii O V and Esikov D A 1990 *Sov. J. Quantum Electron.* **20** 259

Alex M, Shono K, Kuroda S, Koshino N and Ogawa S 1990 *J. Appl. Phys.* **67** 4432–4

Altarelli M 1993 *Phys. Rev.* B **47** 579

Ando K, Takeda N, Koshizuka N and Okuda T 1985a *J. Appl. Phys.* **57** 718

——1985b *J. Appl. Phys.* **57** 1277

Angelakeris M, Poulopoulos P, Flevaris N K, Knapek R, Nyvlt M, Prosser V and Višňovský Š 1995 *J. Magn. Magn. Mater.* **140** 579–80

Ankudinov A and Rehr J J 1995 *Phys. Rev.* B **51** 1282

Antonini B, Blank S L, Lagomarsino S, Paoletti A, Paroli P, Scarinci F and Tucciarone A 1980 *J. Magn. Magn. Mater.* **20** 216–9

Antropov V P, Liechtenstein A I and Harmon B N 1995 *J. Magn. Magn. Mater.* **140** 1161–2

Argyres P N 1955 *Phys. Rev.* **97** 334–45

Atkinson R 1977 *Thin Solid Films* **47** 177–86

Atkinson R, Dodd P M, Kubrakov N F, Zvezdin A K and Zvezdin K A 1996 *J. Magn. Magn. Mater.* **156** 169

Atkinson R, Grundy P J, Hanratty C M, Pollard R J and Salter I W 1994a *J. Appl. Phys.* **75** 6861–3

364   References

Atkinson R and Kubrakov N F 1995 *Proc. R. Soc.* A **449** 205–22

Atkinson R, Kubrakov N F, Utochkin S N and Zvezdin A K 1994b *J. Appl. Phys.* **75** 6786–8

Atkinson R and Lissberger P H 1992 *Appl. Opt.* **31** 6076–81

——1993 *J. Magn. Magn. Mater.* **118** 271–7

Azzam R M A 1978 *J. Opt. Soc. Am.* **68** 1756

Bagdasarov H S, Volodina I S and Kolomiitsev A I 1982 *Quantum Electron.* **9** 1158

Balbashov A M and Chervonenkis A Ya 1979 *Magnetic Materials for Microelectronics* (Moscow: Energia) p 216

Bartholomew D U, Furdyna J K and Ramdas A K 1986 *Phys. Rev.* B **34** 6943

Baudelet F, Dartyge E, Fontaine A, Brouder C, Krill G, Kappler J P and Piecuch M 1991 *Phys. Rev.* B **43** 5857

Baumgarten L, Schneider C M, Petersen H, Schafers F and Kirschner J 1990 *Phys. Rev. Lett.* **65** 492

Belov K P, Zvezdin A K, Kadomtseva A M and Levitin R Z 1979 *Usp. Fiz. Nauk* **119** 447

Bernal E G 1971 *J. Appl. Phys.* **42** 3877–87

Betzig E and Trautmann J K 1992 *Science* **257** 189

Bland J R C, Hughes H P, Blundell S J and Jonson N F 1993 *J. Magn. Magn. Mater.* **125** 69–77

Blazey K W 1974 *J. Appl. Phys.* **45** 2273–80

Blume M and Gibbs D 1988 *Phys. Rev.* B **37** 1779

Boardman A D, Voronko A I, Vetoshko P M, Volkovoy V B and Toporov A Yu 1994 *J. Appl. Phys.* **75** 6804

Bogun P V, Gusev M Yu, Kandyba P E, Kotov V A, Popkov A F and Sorokin V G 1985 *Fiz. Tverd. Tela* **27** 2776

Bolotin G A 1975 *Phys. Met. Metalloved.* **39** 731–42

Bolotin G A and Sokolov A V 1961 *Phys. Met. Metalloved.* **12** 6

Bonarski J and Karp J 1989 *J. Phys.: Condens. Matter* **1** 9261

Borovik-Romanov A S, Jotikov V G, Kreines N M and Pankov A A 1977 *Physica* B **86–88** 1275–6

Brandle H, Schoenes J, Hulliger F and Reim W 1990 *IEEE Trans. Magn.* **26** 2795–7

Brandle H, Weller D, Scott J C, Parkin S S P and Lin C-J 1992 *IEEE Trans. Magn.* **28** 2967–9

Breed D J, de Leeuw F H, Stacy W T and Voermans A B 1978/79 *Philips Tech. Rev.* **38** 211–24

Brouder C and Nikam M 1991 *Phys. Rev.* B **43** 3809

Bruno P 1989 *Phys. Rev.* B **39** 865

Buckingham A D and Stephens P J 1966 *Annu. Rev. Phys. Chem.* **17** 399–427

Burkov V I and Kotov V A 1975 *Fiz. Tverd. Tela* **17** 3108–10

——1983 *Proc. Conf. on Physics of Magnetic Phenomena (Tula)* (Tula: Tula State Pedagogical Institute Press) pp 87–8

Burkov V I, Kotov V A, Balabanov D E and Semin G S 1986 *Zh. Tekh. Fiz.* **56** 2073–5

Buschow K H J 1988 *Ferromagnetic Materials* vol 4, ed E P Wohlfarth and K H J Buschow (Amsterdam: Elsevier) pp 493–595

Buschow K H J and van Engen P G 1981 *Solid State Commun.* **39** 1

——1983 *Phys. Status Solidi* a **76** 615–9

——1984 *Philips J. Res.* **39** 82–93

Buschow K H J, van Engen P G and Jongebreur R 1983 *J. Magn. Magn. Mater.* **38** 1–22

Buschow K H J, van Engen P G and Mooij D B 1984 *J. Magn. Magn. Mater.* **40** 339–47

Buss C, Frey R, Flytzanis C and Cibert J 1995 *Solid State Commun.* **94** 543–8

Butler J C, Kramer J J, Esman R D, Craig A E, Lee J N and Ryuo T 1990 *J. Appl. Phys.* **69** 4938–40

Carey R, Gago-Sandoval P A, Newman D M and Thomas B W J 1994a *J. Appl. Phys.* **75** 6789–90

Carey R, Newman D M, Snelling J P and Thomas B W J 1994b *J. Appl. Phys.* **75** 7087–9

Carey R, Newman D M and Thomas B W J 1983 *Thin Solid Films* **102** 245–9

Carey R, Thomas B W J and Bains G S 1990 *IEEE Trans. Magn.* **26** 1924–6

Carnall E Jr, Pearlman D, Coburn T J, Moser F and Martin T W 1972 *Mater. Res. Bull.* **7** 1361–8

Carra P and Altarelli M 1990 *Phys. Rev. Lett.* **64** 1286

Carra P, Harmon B N, Thole B T, Altarelli M and Sawatzky G A 1991 *Phys. Rev. Lett.* **66** 2495

Carra P, Thole B T, Altarelli M and Wang X 1993 *Phys. Rev. Lett.* **70** 694

Chaboy J, Marcelli A, Garcia L M, Bartolomé J, Kuz'min M D, Maruyama H, Kobayashi K, Kawata H and Iwazumi T 1994 *Europhys. Lett.* **28** 135

Chang J T, Dillon J F and Gianola U F 1965 *J. Appl. Phys.* **36** 1110

Chaudhari P, Cuomo J J and Gambino R J 1973 *Appl. Phys. Lett.* **22** 337–9

Chen D, Otto G N and Schmit F M 1973 *IEEE Trans. Magn.* **9** 66–83

Chen D, Ready J F and Bernal G E 1968 *J. Appl. Phys.* **39** 3916–27

Chen L Y, McGahon W A, Shan Z S, Sellmyer D J and Woollam J A 1990 *J. Appl. Phys.* **67** 5337–9

Chen Q B, Wall A and Onellion M 1995 *J. Magn. Magn. Mater.* **139** 171–4

Chetkin M W and Didosjan Yu S 1968 *Laser Unconv. Opt. J.* **44** 12

Chetkin M W, Didosjan Yu S, Ahutkina A I and Chervonenkis A Ya 1971 *Pis. Zh. Eksp. Teor. Fiz.* **13** 3414

Cho J, Gomi M and Abe M 1990 *Japan. J. Appl. Phys.* **29** 1686–9

Choe Y J, Tsunashima S, Katayama T and Uchiyama S 1987 *J. Magn. Soc. Japan Suppl.* 1 **11** 273–6

Chung S K and Kim S S 1988 *J. Appl. Phys.* **63** 5654–9

Clogston A M 1959 *J. Physique Radium* **20** 374–7

——1960 *J. Appl. Phys.* **31** 198S–205S

Coey J M D 1978 *J. Appl. Phys.* **49** 1646

Collins S P, Cooper M J, Brahmia A, Laundy D and Pitkanen T 1989 *J. Phys.: Condens. Matter* **1** 323

Condon E U and Shortley G H 1935 *Theory of Atomic Spectra* (New York: Cambridge University Press)

Daalderop G H O, Kelly P J and den Broeder F G A 1992 *Phys. Rev. Lett.* **68** 682–5

Daboo C, Bland J R C, Hicken R J, Ives A J R, Baird M J and Walker M J 1993 *Phys. Rev. B* **47** 11 852

Dancygier M 1987 *IEEE Trans. Magn.* **23** 2608–10

Daval J, Ferrand B, Geynet J, Challeton D, Peuzin J C, Leclert A and Monerie M 1975 *IEEE Trans. Magn.* **11** 1115–7

Daval J, Ferrand B, Milani E and Paroli P 1987 *IEEE Trans. Magn.* **23** 3488–90

de Groot R A, Mueller F M, van Engen P G and Buschow K H J 1983 *Phys. Rev. Lett.* **50** 2024 de Groot R A, Mueller F M, van Engen P G and Buschow K H J 1984 *J. Appl. Phys.* **55** 2151–4

Deeter M N 1995 *Appl. Opt.* **34** 655–8

Deeter M N, Bon S M, Day G W, Diercks G and Samuelson 1994 *IEEE Trans. Magn.* **30** 4464–6

Dieke G H 1968 *Spectra and Energy Levels of Rare-Earth Ions in Crystals* (New York: Interscience) p 401

Dieke G H and Crosswhite H M 1963 *Appl. Opt.* **2** 675–8

Dillon J F Jr 1958 *J. Appl. Phys.* **29** 539

——1968 *J. Appl. Phys.* **39** 922–9

Dillon J F Jr, Furdyna J K, Debska U and Mycielski A 1990 *J. Appl. Phys.* **67** 4917–9

Dillon J F Jr, Kamimura H and Remeika J P 1966 *J. Phys. Chem. Solids* **27** 1531–49

Dillon J F Jr, van Dover R B, Hong M, Gyorgy E M and Albiston S D 1987 *J. Appl. Phys.* **61** 1103–7

Dionne G F and Allen G A 1994 *J. Appl. Phys.* **75** 6372–4

Donovan B and Medcalf T 1965 *Proc. Phys. Soc.* **86** 1179–91

Doormann J L, Fiorani D, Giammaria F and Lucari F 1991 *J. Appl. Phys.* **69** 5130–2

Dotsch H, Hertel P, Luhrmann B, Sure S, Winkler H P and Ye M 1992 *IEEE Trans. Magn.* **28** 2979–84

Drogemuller K 1989 *J. Lightwave Technol.* **7** 340–6

Druzhinin A V, Lobov I D and Mayevskiy V M 1981 *Pis. Zh. Tekh. Fiz.* **7** 1100–2

——1985 *Pis. Zh. Tekh. Fiz.* **11** 879–82

Druzhinin A V, Lobov I D, Mayevskiy V M and Bolotin G 1983 *Phys. Met. Metalloved.* **56** 58–65

Dudziak E, Bozym J and Pruchnik D 1995 *Acta Phys. Pol.* **87** 563–6

Ebert H 1990 *Habilitation Thesis* University of Munich

Ebert H, Drittler B, Zeller R and Schutz G 1989 *Solid State Commun.* **69** 485

Ebert H, Strange P and Gyorffy B L 1988a *J. Appl. Phys.* **63** 3035

——1988b *Z. Phys.* B **73** 77

Ebert H and Zeller R 1989 *Physica* B **161** 191

——1990 *Phys. Rev.* B **42** 2744

Edelman I S and Baurin V D 1978 *Phys. Status Solidi* a **46** K83–6

Egashira K, Katsui A and Shibukawa A 1977 *Rev. Electr. Commun. Lab.* **25** 163–71

Egashira K and Manabe T 1972 *IEEE Trans. Magn.* **8** 646

Erskine J L 1975 *AIP Conf. Proc.* **24** 190–9

Erskine J L and Stern E A 1973 *Phys. Rev.* B **8** 1239–55

——1975 *Phys. Rev.* B **12** 5016–24

Falicov L H and Ruvalds J 1968 *Phys. Rev.* **172** 508–13

Fan G Y and Greiner 1968 *J. Appl. Phys.* **39** 1216

Feil H and Haas C 1987 *Phys. Rev. Lett.* **58** 65

Fergusson P E, Stafsudd O M and Wallis R F 1977 *Physica* B **89** 91

Ferrand B, Armand M F, Gay J C, Olivier M, Daval J and Milani E 1987 *J. Magn. Soc. Japan. Suppl.* 1 **11** 195–8

Fiebig M, Frolich D, Krichevtsov V V and Pisarev R V 1994 *Phys. Rev. Lett.* **73** 2127–30

Fisher A D, Lee J N and Gaynor E S 1982 *Appl. Phys. Lett.* **41** 779

Freeman A G and Fu C L 1987 *J. Appl. Phys.* **61** 3356

Freese R P and Takahashi T 1985 *Japan. J. Magn.* **85–88** 81–8

Friez P, Machui J and Meunier P L 1987 *J. Magn. Soc. Japan. Suppl.* 1 **11** 385–8

Fu H, Yan Z, Lee S K and Mansuripur M 1995 *J. Appl. Phys.* **78** 4076–90
Fujii Y, Tokunaga T, Hashima K, Tsutsumi K and Sugahara H 1987 *J. Magn. Soc. Japan.*
    Suppl. 1 **11** 329–32
Fukami T, Kawano Y, Tokunaga T, Nakaki Y and Tsutsumi K 1991 *J. Magn. Soc. Japan.*
    Suppl. 1 **15** 293–8
Fukami T, Nakaki Y, Tokunaga T, Taguchi M and Tsutsumi K 1990 *J. Appl. Phys.* **67**
    4415–6
Fumagalli P, McGuire T R, Plaskett T S and Gambino R J 1992 *IEEE Trans. Magn.* **28**
    2970–2
Fumagalli P, Plaskett T S, Weller D, McGuire T R and Gambino R J 1993 *Phys. Rev.*
    Lett. **70** 230–3
Gaj J A, Gałązka R R and Nawrocki M 1978 *Solid State Commun.* **25** 193
——1993 *Solid State Commun.* **88** 923–5
Gambino R J and Fumagalli P 1994 *IEEE Trans. Magn.* **30** 4461–3
Gambino R J, Fumagalli P, Ruf R R, McGuire T R and Bojarczuk N 1992 *IEEE Trans.*
    Magn. **28** 2973–5
Gambino R J and Ruf R R 1990 *J. Appl. Phys.* **67** 4784–6
Gambino R J, Ruf R R and Bojarczuk N 1994 *J. Appl. Phys.* **75** 6871
Gan'shina E A 1994 *Thesis* Moscow State University
Gerrard A and Burch J M 1975 *Introduction to Matrix Methods in Optics* (London: Wiley)
Gibbs D, Harshman D R, Isaacs E D, McWhan D B, Mills D and Vettier C 1988 *Phys.*
    Rev. Lett. **61** 1241
Gibbs D, Moncton D E and D'Amico K L 1985 *J. Appl. Phys.* **57** 3619
Girgel S S and Demidova T V 1987 *Opt. Spectrosk.* **62** 63
Goedkoop J B, Fuggle J C, Thole B T, van der Laan G and Sawatzky G A 1988a *Nucl.*
    Instrum. Methods Phys. Res. A **273** 429
——1988b *J. Appl. Phys.* **64** 5595
Goedkoop J B, Thole B T, van der Laan G, Sawatzky G A, de Groot F M F and Fuggle
    J C 1988c *Phys. Rev.* B **37** 2086
Gomi M and Abe M 1994 *J. Appl. Phys.* **75** 6804
Gomi M, Furuyama H and Abe M 1990 *Japan. J. Appl. Phys.* **29** L99–100
——1991 *J. Appl. Phys.* **70** 7065
Gomi M, Sato K and Abe M 1988 *Japan. J. Appl. Phys.* **27** L1536–8
Gorban' I S, Gumenyuk A F and Degoda V Ya 1985 *Opt. Spectrosk.* **58** 464–6
Graham E B and Raab R E 1992 *Phil. Mag.* B **66** 269
Green A E and Georgiou G 1986 *Electron. Lett.* **22** 1045–6
Greidanus F J A M, Jacobs B A J, Spruit J H M and Klahn S 1989 *IEEE. Trans. Magn.*
    25 3524–9
Gubarev S I 1981 *Sov. Phys.–JETP* **53** 601
Guliaev Yu V, Ignat'ev I A, Plehanov V G and Popkov A F 1985 *Radiotekh. Elektron.*
    8 1522–30
Gusev M Yu and Kotov V A 1991 *Proc. Conf. on Microelectronics (Simpheropol)*
    (Simpheropol: Simpheropolunivpress) p 9
Hannon J P, Trammell G T, Blume M and Gibbs D 1988 *Phys. Rev. Lett.* **61** 1245
Hansen P, Hartmann M and Witter K 1987 *J. Magn. Soc. Japan. Suppl.* 1 **11** 257–60
Hansen P and Krumme J P 1984 *Thin Solid Films* **114** 69
Hansen P, Raasch D and Mergel D 1994 *J. Appl. Phys.* **75** 5267–77
Hansen P, Robertson J M, Tolksdorf W and Witter K 1984 *IEEE Trans. Magn.* **20** 1099

Hansen P and Urner-Wille M 1979 *J. Appl. Phys.* **50** 7471

Hansen P, Witter K and Tolksdorf W 1983 *Phys. Rev.* B **27** 6608

Hartmann M, Hansen P and Willich P 1984 *J. Appl. Phys.* **56** 2870–3

Hashimoto S, Matsuda H and Ochiai Y 1990 *Appl. Phys. Lett.* **56** 1069–71

Hashimoto S, Ochiai Y and Aso K 1989 *Japan. J. Appl. Phys.* **28** L1824–6

Haskal H M 1970 *IEEE Trans. Magn.* **6** 542

Hatwar T K, Genova D and Stinson D G 1990 *J. Appl. Phys.* **67** 5304–6

Hatwar T K, Genova D J and Victora R H 1994 *J. Appl. Phys.* **75** 6858–60

Heim K R and Scheinfein M R 1996 *J. Magn. Magn. Mater.* **154** 141–52

Hennel A M, Twardowski A and Godlewski 1965 *Acta Phys. Pol.* A **67** 313

Henry C H, Schnatterly S E and Slichter C P 1965 *Phys. Rev.* **137** 583

Henry R D 1975 *Appl. Phys. Lett.* **26** 408–11

——1976 *Mater. Res. Bull.* **11** 1295

——1977 *IEEE Trans. Magn.* **13** 1527

Hepner G, Desormiere B and Castera J P 1975 *Appl. Opt.* **14** 1479

Hernandez J, Canal F, Dalmau L and Torner L 1987 *J. Magn. Soc. Japan. Suppl.* 1 **11** 381–4

Hibiya T, Morishige Y and Nakashima J 1985 *Japan. J. Appl. Phys.* **24** 1316–9

Hilfiker J N, Zhang Y B and Woollam J A 1994 *IEEE Trans. Magn.* **30** 4437–9

Hill B 1980 *IEEE Trans. Electron. Devices* **27** 1825

Hill B, Krumme J-P, Much G, Pepperl R, Schmidt J, Witter K and Heitmann H 1975 *Appl. Opt.* **14** 2607–13

Hill B and Schmidt K P 1982 *Philips Electron. Co. Mater.* **4** 169

Hirata T and Naoe M 1992 *IEEE Trans. Magn.* **28** 2964–6

Hiratsuka N and Sugimoto 1986 *Trans. IEE Japan* A **106** 125

Hodges L H, Stone D H and Gold A V 1967 *Phys. Rev. Lett.* **19** 655–9

Honda S, Nawate M, Yoshiyama M and Kosuda T 1987 *IEEE Trans. Magn.* **23** 2952–4

Honda S and Yoshiyama M 1988 *Japan. J. Appl. Phys.* **27** 2073–7

Hornreich R M and Shtrikman S 1968 *Phys. Rev.* **171** 1065

Hubert A and Traeger G 1993 *J. Magn. Magn. Mater.* **124** 185–202

Hübner W and Bennemann K H 1989 *Phys. Rev.* B **40** 5973–9

Idzerda Y U, Tjeng L H, Lin H J, Gutierrez C J, Meigs G and Chen C T 1993 *Phys. Rev.* B **48** 4144

Iijima T, Ishii O and Hatakeyama I 1989 *Appl. Phys. Lett.* **54** 2376–7

Imada S and Jo T 1990 *J. Phys. Soc. Japan* **59** 3358

Imamura M, Nakahara M and Sasaki T 1987 *J. Magn. Soc. Japan Suppl.* 1 **11** 373–6

Imamura M, Ogata K, Tokubachi M and Nakahara M 1994 *IEEE Trans. Magn.* **30** 4936–8

Inoue F, Mutoh H, Itoh A and Kawanishi K 1983 *J. Magn. Magn. Mater.* **35** 17

Inukai T, Sugimoto N, Matsuoka M and Ono K 1987 *J. Magn. Soc. Japan Suppl.* 1 **11** 217–20

Ito H, Yamaguchi M and Naoe M 1990 *J. Appl. Phys.* **67** 5307–9

Itoh A and Nakagawa K 1992 *Japan. J. Appl. Phys.* **31** 790–2

Itoh A, Unozawa K, Shinohara T, Nakada M, Inoue F and Kawanishi K 1985 *IEEE Trans. Magn.* **21** 1672–4

Iwamura H, Hayashi S and Iwasaki H 1978 *Opt. Quantum Electron.* **10** 393

Iwata S, Parkin S S P, Suzuki T and Weller D 1992 *IEEE Trans. Magn.* **28** 3231–3

Jérôme B and Shen Y R 1993 *Phys. Rev.* E **48** 4556–73

Jiang Z S, Ji J T, Jin G J, Sang H, Guo G, Zhang S Y and Du Y W 1995a *J. Magn. Magn. Mater.* **140** 469–70

Jiang Z S, Jin G J, Sang H, Ji J T, Du Y W, Zhou S M, Wang Y D and Chen L Y 1995b *J. Appl. Phys.* **78** 439–41

Jin L, Ying N, Tingjun M and Ruiyi F 1995 *J. Appl. Phys.* **78** 2697–9

Jin Q Y, Xu Y B, Shen Y H and Zhai H G 1994 *Proc. SPIE* **2364** 198

Jo T 1992 *Synchrotron Radiat. News* **5** 21

Jo T and Imada S 1989 *J. Phys. Soc. Japan* **58** 1922

Jo T and Sawatzky G A 1991 *Phys. Rev.* B **43** 8771

Johansen T R, Norman D I and Torok E J 1971 *J. Appl. Phys.* **42** 1715

Johnson P B, Karnezos M and Henry R D 1980 *Mater. Res. Bull.* **15** 1669

Jones R C 1941a *J. Opt. Soc. Am.* **31** 488

——1941b *J. Opt. Soc. Am.* **32** 486

Jones R V 1976 *Proc. R. Soc.* A **349** 423–39

Judd B R and Pooler D R 1982 *J. Phys. C: Solid State Phys.* **15** 591

Kahn F J, Pershan P S and Remeika J P 1969 *Phys. Rev.* **186** 891–918

Kamada O, Minemoto H and Ishizuka S 1987 *J. Magn. Soc. Japan Suppl.* 1 **11** 401–4

Kamada O, Minemoto H and Itoh N 1994 *J. Appl. Phys.* **75** 6801–3

Kanaizuka T, Ohwada T, Morihara Y, Katayama T and Suzuki T 1987 *J. Magn. Soc. Japan Suppl.* 1 **11** 333–5

Kaneko M 1992 *IEEE Trans. Magn.* **28** 2494–9

Kaneko M, Okamoto T, Tamada H and Sato K 1987 *IEEE Trans. Magn.* **23** 3482–4

Kaneko M, Okamoto T, Tamada H and Yamada T 1983 *IEEE Trans. Magn.* **19** 1763

Kaneko M, Sabi Y, Ichimura I and Hashimoto S 1993 *IEEE Trans. Magn.* **29** 3766–71

Kapitulnik A, Dodge J S and Fejer M M 1994 *J. Appl. Phys.* **75** 6872–7

Kappert R J H, Vogel J, Sacchi M and Fuggle J C 1993 *Phys. Rev.* B **48** 2711

Katayama T, Awano H and Nishihara Y 1986a *J. Phys. Soc. Japan* **55** 2539

——1986b *IEEE Trans. Magn.* **23** 2949–51

Katayama T and Hasegawa K 1982 *Rapidly Quenched Metals* vol IV, ed T Masumoto and K Suzuki (Sendai: Japan Institute of Metals) pp 915–8

Katayama T, Suzuki Y, Awano H, Nishihara Y and Koshizuka N 1988 *Phys. Rev. Lett.* **60** 1426

Katayama T, Suzuki Y, Hayashi M and Geerts W 1994 *J. Appl. Phys.* **75** 6360–2

Kato T, Iwata S, Tsunashima S and Uchiyama S 1995a *J. Magn. Soc. Japan* **9** 205

Kato T, Kikuzawa H, Iwata S, Tsunashima S and Uchiyama S 1995b *J. Magn. Magn. Mater.* **140–144** 713–4

Katsui A 1976a *J. Appl. Phys.* **47** 3609–11

——1976b *J. Appl. Phys.* **47** 4663–5

Kawamura N, Sato R and Tamaki T 1995 *J. Magn. Magn. Mater.* **140** 2201–2

Kawase T, Ishida M, Hoshina S, Takakuwa A, Nebashi and Shimoda T 1994 *IEEE Trans. Magn.* **30** 4392–4

Kharchenko N F and Belii L I 1980 *Izv. Akad. Nauk SSSR, Ser. Fiz.* **44** 1451

Kharchenko N F, Bibik A V and Eremenko V V 1985 *Pis. Zh. Eksp. Teor. Fiz.* **42** 553

Kharchenko N F, Eremenko V V and Belii L I 1978 *Pis. Zh. Eksp. Teor. Fiz.* **28** 351–5

Kharchenko N F and Gnatchenko S L 1981 *Fiz. Nizhk. Temp.* **7** 475–93

Kielich S and Zawodny R 1973 *Opt. Acta* **20** 867

Kirillova M M, Makhnev A A, Shreder E I, Dyakina V P and Gorina N B 1995 *Phys. Status Solidi* b **187** 231–40

Kisielewski M, Maziewski A and Desvignes J M 1995 *J. Magn. Magn. Mater.* **140** 1923–4

Ko M and Garmire E 1995 *Appl. Opt.* **34** 1692

Kobayashi K and Seki M 1980 *IEEE J. Quantum Electron.* **16** 11–22

Kolmakova N P, Levitin R Z, Popov A I, Vedernikov N F, Zvezdin A K and Nekvasil V 1990 *Phys. Rev.* B **41** 6170

Komarov A V, Ryabchenko S M, Terletskii O V, Zheru I I and Ivanchuk D 1977 *Sov. Phys.–JETP* **46** 318

Kooy C and Enz U 1960 *Philips Res. Rep.* **15** 7

Kosobukin V A 1996 *J. Magn. Magn. Mater.* **153** 397–411

Kotov V A, Antonov A V, Shaburnikov A V, Neustroev N S and Yudichev A I 1981a *Proc. Conf. on Domain and Magnetooptical Memory (Batumi)* (Tbilisi: Institute Kibernetiki Press) p 45

Kotov V A, Antonov A V, Shaburnikov A V, Neustroev N S and Yudichev A I 1981b *Proc. Conf. on Domain and Magnetooptical Memory (Batumi)* (Tbilisi: Institute Kibernetiki Press) p 57

——1982 *Elektron. Tekh. Ser.: Mater.* **11** 64–5

Kotov V A, Antonov A V, Shaburnikov A V and Yudichev A I 1981c *Bubble Magnetic Domains: Physical Properties and Technical Applications* vol 6 (Moscow: Institute Problem Upravlenia) pp 13–4

Kotov V A, Balabanov D E, Grigorovich S M, Kozlov V I and Nevolin V K 1986a *Zh. Tekh. Fiz.* **56** 897–903

Kotov V A, Il'in V Yu and Rybak V I 1979 *Elektron. Tekh. Ser.: Mikroelektron.* **5** 30–3

Kotov V A, Nevolin V K, Shermergor T D and Balabanov D E 1986b *Mikroelektronika* **15** 338–43

Koyanagi T, Nakamura K, Yamano K, Sota T and Matsubara K 1988 *J. Magn. Soc. Japan* **12** 187

——1989 *Japan. J. Appl. Phys.* **28** L699

Koyanagi T, Watanabe T and Matsubara K 1987 *IEEE Trans. Magn.* **23** 3214–6

Krawczak J A and Torok E J 1980 *IEEE Trans. Magn.* **16** 1200

Krebs J J, Maisch W G, Prinz G A and Forester D W 1980 *IEEE Trans. Magn.* **16** 1179–84

Krenn H, Zawadzki W and Bauer G 1985 *Phys. Rev. Lett.* **55** 1510

Krichevtsov B B, Pavlov V V and Pisarev R V 1986 *JETP Lett.* **44** 607–10

——1988 *Sov. Phys.–JETP* **67** 378–84

——1989a *JETP Lett.* **49** 535

——1989b *Sov. Phys.–Solid State* **31** 1142

Krichevtsov B B, Pavlov V V, Pisarev R V and Gridnev V N 1993 *J. Phys.: Condens. Matter.* **5** 8233–44

Krinchik G S 1985 *Physics of Magnetic Phenomena* (Moscow: Moscow State University Press) p 336

Krinchik G S and Artem'ev V A 1968 *Zh. Eksp. Teor. Fiz.* **26** 1080

Krinchik G S and Chepurova E E 1973 *Proc. ICM-73 (Moscow, 1973)* (Moscow: Moscow State University Press) p 139

Krinchik G S and Chetkin M V 1960 *Sov. Phys.–JETP* **11** 1184L

——1961 *Sov. Phys.–JETP* **13** 509

Krinchik G S and Gan'shina E A 1973 *Zh. Eksp. Teor. Fiz.* **65** 1970

Krinchik G S and Gushchin V S 1969 *Sov. Phys.–JETP* **29** 984

Krinchik G S, Gushchin V S and Tsidaeva N I 1984 *Sov. Phys.–JETP* **59** 410–4

Krinchik G S and Kosturin 1982 *Pis. Zh. Eksp. Teor. Fiz.* **35** 295–7

Krishnan R, Kolobanov V N, Mikhailin V V, Orekhanov P A, Siroky P, Sramek J and Višňovský Š 1990a *J. Appl. Phys.* **67** 4803–5

Krishnan R, Nyvlt M, Smetana Z and Višňovský Š 1995 *J. Magn. Magn. Mater.* **140** 605–6

Krishnan R, Porte M and Tessier M 1990b *IEEE Trans. Magn.* **26** 2727

Krishnan R, Sikora T and Višňovský Š 1993 *J. Magn. Magn. Mater.* **118** 1–5

Krishnan R and Tessier M 1990 *J. Appl. Phys.* **67** 5391–3

Krumme J-P, Doorman V, Hansen P, Baumgart H, Petruzzello J and Viegers M P A 1989 *J. Appl. Phys.* **66** 4393–407

Krumme J-P and Hansen P 1973 *Appl. Phys. Lett.* **23** 576

Krumme J-P, Heitmann H, Mateika D and Witter K 1977 *Appl. Phys. Lett.* **48** 366

Krumme J-P and Schmitt H J 1975 *IEEE Trans. Magn.* **11** 1097

Krumme J-P, Verweel J, Haberkamp J, Tolksdorf W, Bartels G and Espinosa G P 1972 *Appl. Phys. Lett.* **20** 451

Kryder M H 1985 *J. Appl. Phys.* **57** 3913–8

Kubler J 1995 *J. Phys. Chem. Solids* **56** 1529

Kudo T, Johbetto H and Ichiji K 1990 *J. Appl. Phys.* **67** 4778–80

Kudryavtsev Y, Dubowik J and Stobiecki F 1995 *Thin Solid Films* **256** 171–5

Kuiper P, Searle B G, Rudolf P, Tjeng L H and Chen C T 1993 *Phys. Rev. Lett.* **70** 1549

Kulatov E T, Uspenskii Yu A and Halilov S V 1995 *J. Magn. Magn. Mater.* **145** 395–7

——1996 *J. Magn. Magn. Mater.* **163** 331–8

Kurtzig A J and Guggenheim H J 1970 *Appl. Phys. Lett.* **16** 43

Kurtzig A J, Wolfe R, LeCraw R C and Nielsen J W 1969 *Appl. Phys. Lett.* **14** 350

Lacklison D E, Scott G B, Giles A D, Klarke J A, Pearson R F and Page J L 1977 *IEEE Trans. Magn.* **13** 973–81

Lacklison D E, Scott G B, Pearson R F and Page J L 1975 *IEEE Trans. Magn.* **11** 1118

Landau L D, Lifshitz E M and Pitaevskii L P 1984 *The Electrodynamics of Continuous Media* (London: Pergamon)

Landsberg G S 1976 *Optics* (Moscow: Nauka) p 928

Larsen P K and Robertson J M 1974 *J. Appl. Phys.* **45** 2867

Lawler J F, Lunney J G and Coey J M D 1994 *Appl. Phys. Lett.* **65** 3017–8

LeCraw R C 1966 *IEEE Trans. Magn.* **2** 304

Lewicki G W 1969 *IEEE Trans. Magn.* **5** 298

Leycuras C, Le Gall H, Minella D, Rudashevskii E G and Merkulov V S 1977 *Physica* B **89** 43–6

Lin C-J and Do H V M 1990 *IEEE Trans. Magn.* **26** 1799–3

Loh E 1966 *Phys. Rev.* **147** 332–5

Lucari F, Mastrogiuseppe C, Terrenzio E and Tomassetti G 1980 *J. Magn. Magn. Mater.* **20** 84–6

Lucari F, Mastrogiuseppe C and Tomassetti G 1977 *J. Phys. C: Solid State Phys.* **10** 4869–75

Lundqvist S and March N M (ed) 1983 *Theory of the Inhomogeneous Electron Gas* (New York: Plenum)

Lyubimov V N 1969 *Sov. Phys.–Crystallogr.* **14** 168

MacNeal B E, Pulliam G R, Fernandez de Castro J J and Warren D M 1983 *IEEE Trans. Magn.* **19** 1766

Madison M R, Makansi T and McDaniel T W 1990 *J. Appl. Phys.* **67** 5331–3

Madison M R and McDaniel T W 1990 *J. Appl. Phys.* **67** 5325

Maillot C, Blanchard N, Valet T and Lehureau J C 1994 *IEEE Trans. Magn.* **30** 326–30

Malozemoff A P and Slonczewski 1979 *Magnetic Domain Walls in Bubble Materials* (New York: Academic)

Mansuripur M 1990 *J. Appl. Phys.* **67** 6466–75

——1994 *The Physical Principles of Magneto-optical Recording* (London: Cambridge University Press)

Martens J W D and Peeters W L 1983 *Proc. SPIE* **420** 231–5

Masterson H J, Lunney J G, Coey J M D, Atkinson R, Salter I W and Papakonstantinou P 1993 *J. Appl. Phys.* **73** 3917–21

Masterson H J, Lunney J G, Ravinder D and Coey J M D 1995 *J. Magn. Magn. Mater.* **140** 2081

Matsubara K, Koyama M, Koyanagi T, Watanabe Y and Yoshitomi T 1987 *J. Magn. Soc. Japan Suppl.* 1 **11** 213–6

Matsuda K, Minemoto H, Toda K, Kamada O and Ishizuka S 1987 *Electron. Lett.* **23** 203–5

Matsumoto K, Kitade Y, Mihara M and Nanba Y 1994 *Trans. Mater. Res. Soc. Japan* **15** 991–4

Matsumoto K, Sasaki S, Haraga K, Yamaguchi K and Fujii T 1992 *IEEE Trans. Magn.* **28** 2985–7

Matyushev V V, Kostylev M P, Stashkevich A A and Desvignes J M 1995 *J. Appl. Phys.* **77** 2087–9

Mayer L 1958 *J. Appl. Phys.* **29** 1003

Mayevskii V M 1985 *Phys Met. Metalloved.* **59** 1–7

Mayevskii V M and Bolotin G A 1973 *Phys. Met. Metalloved.* **36** 11–21

McElfresh M W, Plaskett T S, Gambino R J and McGuire T R 1990 *Appl. Phys. Lett.* **57** 730–2

McGuire T R, Gambino R J, McElfresh M W and Plaskett T S 1990 *IEEE Trans. Magn.* **26** 1349–51

Mee C D 1967 *Contemp. Phys.* **8** 385

Meiklejohn W H 1986 *Proc. IEEE* **74** 1570–81

Meilman M L, Kolomiizev A I, Kevorkov A N and Bagdasarov H S 1984 *Opt. Spectrosk.* **57** 239–41

Merkulov V S, Rudashevskii E G, Le Gall H and Leycuras K 1981 *Sov. Phys.–JETP* **54** 81–5

Meunier P-L, Castera J-P and Friez P 1987 *J. Magn. Soc. Japan Suppl.* 1 **11** 199–201

Miyazawa H, Ide T, Hoshina S and Ichikawa M 1993 *IEEE Trans. Magn.* **29** 3781–3

Mizumoto T, Mashimo S, Ida T and Naito Y 1993 *IEEE Trans. Magn.* **29** 3417–9

Moog E R, Bader S D and Zak J 1990 *Appl. Phys. Lett.* **56** 2687–9

Moser F, Ahrenkiel R K, Carnall E, Coburn T, Lyu S L, Lee T H, Martin T and Pearlman D 1971 *J. Appl. Phys.* **42** 1449–51

Moskvin A S, Zenkov A V, Gan'shina E A and Krinchik G S 1993 *J. Phys. Chem. Solids* **53** 101–5

Nakagawa K, Kurashina S and Itoh A 1994 *J. Appl. Phys.* **75** 7096–8

Nakaki Y, Fukami T, Tokunaga T, Kawano Y and Tsutsimi K 1992 *IEEE Trans. Magn.* **28** 2509–11

Nakamura H, Ohmi F, Kaneko Y, Sawada Y, Watada A and Machida H 1987 *J. Appl. Phys.* **61** 3346–8

Nakano Y and Miyazaki Y 1987 *J. Magn. Soc. Japan Suppl.* 1 **11** 393–6

Nakano T, Yuri H and Kihara U 1984 *IEEE Trans. Magn.* **20** 986–8

Nassau K 1968 *J. Cryst. Growth* **2** 215–21

Navas E, Starke K, Laubschat C, Weschke E and Kaindl G 1993 *Phys. Rev.* B **48** 14 753

Nersisyan S R, Sarkisyan V A and Tabiryan N V 1983 *Fiz. Tverd. Tela* **25** 2556–60

Nikitin P I and Savchuk A I 1990 *Sov. Phys.–Usp.* **33** 974–89

Nomura T 1985 *IEEE Trans. Magn.* **21** 1544–5

Numata T, Ihbuchi Y and Sakurai Y 1980 *IEEE Trans. Magn.* **16** 1197

Ohkoshi M, Utsunomiya K and Tsushima K 1995 *J. Magn. Magn. Mater.* **140** 1845–6

Okuda T, Katayama T, Kobayashi H and Kobayashi N 1990 *J. Appl. Phys.* **67** 4944–6

Okuda T, Koshizuka, Hayashi K, Takahashi T, Kotani H and Yamamoto H 1987 *J. Magn. Soc. Japan Suppl.* 1 **11** 179

Oppeneer P M, Antonov V N, Kraft T, Eschrig H, Yaresko A N and Perlov A Y 1995 *Solid State Commun.* **94** 255–9

Oppeneer P M, Maurer T, Sticht J and Kubler J 1992 *Phys. Rev.* B **45** 10 924–33

Ostrovskii V S and Loktev B M 1977 *Pis. Zh. Eksp. Teor. Fiz.* **26** 134–41

Pan Ru-Pin, Wei H W and Shen Y R 1989 *Phys. Rev.* B **39** 1229

Paroli P 1984 *Thin Solid Films* **114** 187

Pavlovskii A I, Druzhinin V V and Tatsenko O N 1980 *Pis. Zh. Eksp. Teor. Fiz.* **31** 659–63

Pershan P S 1963 *Phys. Rev.* **130** 919

Petrocelli G, Martelucci S and Richetta M 1993 *Appl. Phys. Lett.* **63** 3402

Petrovskii G T, Edelman I S, Stepanov S A, Zarubina T V and Kim T A 1994 *Glass Phys. Chem.* **20** 515–26

Pisarev R V, Krichevtsov B B, Gridnev V N, Klin V P, Frolich D and Pahlke-Lerch C 1993 *J. Phys. C: Solid State Phys.* **5** 8621–6

Pisarev R V, Krichevtsov B B and Pavlov V V 1991 *Phase Transitions* **37** 63–72

Pisarev R V, Sinii I G and Smolenskii G A 1969 *Pis. Zh. Eksp. Teor. Fiz.* **9** 112

Polanskii A A, Indenbom M V, Nikitenko V I, Osipian Yu A and Vlasko-Vlasov V K 1990 *IEEE Trans. Magn.* **26** 1445–7

Popkov A F 1977 *Sov. Phys.–Solid State* **19** 1288

Popma T J A and Kamminga M G J 1975 *Solid State Commun.* **17** 1073–5

Postava K, Bobo J F, Ortega M D, Raquet B, Jaffres H, Snoeck E, Goiran M, Fert A R, Redoules J P, Pistora J and Ousset J C 1996 *J. Magn. Magn. Mater.* **163** 8–20

Prokhorov A M, Smolenskii G A and Ageev A N 1984 *Usp. Fiz. Nauk* **134** 33–72

Pustogowa U, Hübner W and Bennemann K H 1994 *Phys. Rev.* B **49** 10 031

Raasch D 1993 *IEEE Trans. Magn.* **29** 34–40

Raether H 1988 *Springer Tracts in Modern Physics* vol 11 (Berlin: Springer)

Ray S and Tauc J 1980 *Solid State Commun.* **34** 769

Razeghi M, Meunier P-L and Maurel P 1986 *J. Appl. Phys.* **59** 2261–3

Red'ko V G 1989 *Microelectronics* **18** 72–7

Reif J, Zink J C, Schneider C-M and Kirschner J 1991 *Phys. Rev. Lett.* **67** 2878

Reim W, Husser O E, Schoenes J, Kaldis E and Wachter P 1984 *J. Appl. Phys.* **55** 2155–6

Reim W and Schoenes J 1984 *J. Appl. Phys.* **55** 2155–6

Reim W, Schoenes J, Hulliger F and Vogt O 1986 *J. Magn. Magn. Mater.* **54–57** 1401–2

Reim W and Weller D 1988 *Appl. Phys. Lett.* **53** 2453

Robson B A 1974 *The Theory of Polarization Phenomena* (Oxford: Clarendon)

Rosenberg G 1950 *Usp. Fiz. Nauk* **40** 328

Rugar D, Suits J C and Lin C-J 1988 *Appl. Phys. Lett.* **52** 1537–9

Rutkin O G, Kowshikov N G and Stashkevich A A 1985 *Pis. Zh. Tekh. Fiz.* **11** 386–7

Ryabchenko S M, Semenov Yu G and Terletskii O V 1982 *Sov. Phys.–JETP* **55** 557

Sacchi M, Sakho O and Rossi G 1991 *Phys. Rev.* B **43** 1276

Safarov V I, Kosobukin V A, Hermann C, Lampel G, Peretti J and Marlier C 1994 *Phys. Rev. Lett.* **73** 3584–7

Sainctavit P, Arrio M A and Brouder C 1995 *Phys. Rev.* B **52** 12 766

Saito J, Akasaka H, Birecki H and Perlov C 1992 *IEEE Trans. Magn.* **28** 2512–4

Saito J, Sato M, Matsumoto H and Akasaka H 1987a *Japan. J. Appl. Phys.* **26** 155–9

Saito T, Tamanoi K, Shinagawa K and Tsushima T 1987b *J. Magn. Soc. Japan Suppl.* 1 **11** 245–8

Santamato E, Daino B, Romagnoli M and Shen Y R 1986 *Phys. Rev. Lett.* **57** 2423

Santi K S, Mizumoto T and Naito Y 1987 *J. Magn. Soc. Japan Suppl.* 1 **11** 207–10

Sato K, Kida H and Kamimura T 1987 *J. Magn. Soc. Japan Suppl.* 1 **11** 113–6

Sato R, Saito N and Morishita T 1988 *IEEE Trans. Magn.* **24** 2458–60

Sauter G F, Hanson M M and Fleming D L 1977 *Appl. Phys. Lett.* **30** 11–3

Sauter G F, Honebrink R W and Krawczac J A 1981 *Appl. Opt.* **20** 3566

Sawatzky E 1971 *J. Appl. Phys.* **42** 1706–7

Sawatzky E and Street G B 1971 *IEEE Trans. Magn.* **7** 377–80

——1973 *J. Appl. Phys.* **44** 1789–92

Schiff L I 1955 *Quantum Mechanics* (New York: McGraw-Hill)

Schoenes J 1987 *J. Magn. Soc. Japan Suppl.* 1 **11** 99–105

Schoenes J and Brandle H 1991 *J. Magn. Soc. Japan Suppl.* 1 **15** 213–8

Schultz M D, Kryder M H, Sekiya M and Chiba K 1993 *IEEE Trans. Magn.* **29** 3772–7

Schütz G, Wagner W, Wilhelm W, Kienle P, Zeller R, Frahm R and Materlik G 1987 *Phys. Rev. Lett.* **58** 737

Scott G B and Lacklison D E 1976 *IEEE Trans. Magn.* **12** 292

Scott G B, Lacklison D E and Page J L 1974 *Phys. Rev.* B **10** 971–86

Scott G B, Lacklison D E, Ralph H I and Page J L 1975 *Phys. Rev.* B **12** 2562–71

Scott G B and Page J P 1977a *J. Appl. Phys.* **48** 1342

——1977b *Phys. Status Solidi* b **79** 203–13

Serber R 1932 *Phys. Rev.* **41** 486–506

Shen J X, Kirby R D and Sellmyer D J 1990 *J. Appl. Phys.* **67** 4929–31

Shen J X, Kirby R D, Wierman K W, Zhang Y B, Suzuki T and Sellmyer D J 1995 *J. Magn. Magn. Mater.* **140** 2139–40

Shen Y R 1984 *The Principles of Nonlinear Optics* (New York: Wiley)

——1989 *Nature* **337** 519

Shibukawa A, Katsui A and Egashira K T 1976 *Japan. J. Appl. Phys.* **15** 1912–20

Shieh H P D and Kryder M H 1986 *Appl. Phys. Lett.* **49** 473

——1987 *IEEE Trans. Magn.* **23** 171

Shimoda T, Shimokawato S, Funada S, Nebashi S, Aoyama A and Sugimoto M 1987 *J. Magn. Soc. Japan Suppl.* 1 **11** 337–40

Shinagawa K 1982 *J. Magn. Soc. Japan* **6** 247–53

Shiraishi K, Aizawa Y, Yanagi T and Kawakami S 1989 *Electron. Lett.* **25** 1335–6

Shirasaki M, Targaki N and Obokata T 1981 *Appl. Phys. Lett.* **38** 833

Siddons D P, Hart M, Amemiya Y and Hastings J B 1990 *Phys. Rev. Lett.* **64** 1967

Silin V P and Rukhadze A A 1961 *Physics of Plasma and Plasmalike Media* (Moscow: Fizmatgiz)

Šimša Z 1990 *J. Magn. Magn. Mater.* **83** 15–6

Šimša Z, Gornert P, Pointon A J and Gerber R 1992 *IEEE Trans. Magn.* **26** 2789–91

Šimša Z, Le Gall H, Šimšová J, Kolaček J and Le Paillier-Malécot A 1984 *IEEE Trans. Magn.* **20** 1001–3

Singh M, Wang C S and Callaway J 1975 *Phys. Rev.* B **11** 287

Smith D O 1965 *Opt. Acta* **12** 13

Smolenskii G A, Mironov S A and Ageev A N 1977 *Pis. Zh. Tekh. Fiz.* **3** 284

Smolenskii G A, Stinser E P and Ageev A N 1976a *Pis. Zh. Tekh. Fiz.* **2** 289–92

——1976b *Pis. Zh. Tekh. Fiz.* **2** 641–4

Sobel'man I I 1972 *Introduction in the Theory of Atomic Spectra* (Oxford: Pergamon)

Song K, Ito H and Naoe M 1993 *IEEE Trans. Magn.* **29** 3367–9

Song K and Naoe M 1994 *J. Appl. Phys.* **75** 6357–9

Spielman S, Fesler K, Eom C B, Geballe T H, Fejer M M and Kapitulnik A 1990 *Phys. Rev. Lett.* **65** 123–6

Spierings G, Koutsos V, Wierenga H A, Prins M W J, Abraham D and Rasing T 1993 *J. Magn. Magn. Mater.* **121** 109

Stähler S, Schütz G and Ebert H 1993 *Phys. Rev.* B **47** 818

Starostin Yu V and Kotov V A 1982 *Pis. Zh. Tekh. Fiz.* **8** 1518–22

Stephens P J 1970 *J. Chem. Phys.* **52** 3489–516

Stoffel A M 1968 *J. Appl. Phys.* **39** 563–5

Stoffel A M and Schneider J 1970 *J. Appl. Phys.* **41** 1405–7

Stöhr J and König H 1995 *Phys. Rev. Lett.* **75** 3748

Stoll M P 1972 *Solid State Commun.* **11** 437

Strange P and Gyorffy B L 1995 *Phys. Rev.* B **52** R13 019

Sugawara E, Katayama T and Masumoto T 1987 *J. Magn. Soc. Japan Suppl.* 1 **11** 277–80

Sugimoto T, Katayama T, Suzuki Y and Nishihara Y 1989 *Japan. J. Appl. Phys.* **28** L2333–5

Suits J C 1972 *IEEE Trans. Magn.* **8** 95–105

Suzuki K, Namikawa T and Yamazaki Y 1988 *Japan. J. Appl. Phys.* **27** 361

Suzuki T, Iwata S, Brandle H and Weller D 1994 *IEEE Trans. Magn.* **30** 4455–7

Suzuki T, Kawai H and Umezawa H 1996 *Japan. J. Appl. Phys.* B **35** L224–6

Tabor W J, Anderson A W and Van Uitert L G 1970 *J. Appl. Phys.* **41** 3018–21

Tabor W J and Chen F S 1969 *J. Appl. Phys.* **40** 2760–5

Takahashi A, Nakajima J, Murakami Y, Ohta K and Ishikawa T 1994 *IEEE Trans. Magn.* **30** 232–6

Takanashi K, Fujimori S, Shoji M and Nagai A 1987 *Japan. J. Appl. Phys.* **26** L1317

Taki K and Miyazaki Y 1987 *J. Magn. Soc. Japan Suppl.* 1 **11** 369–72

Tamada H, Kaneko M and Okamoto T 1987 *J. Magn. Soc. Japan Suppl.* 1 **11** 397–400

——1988 *J. Appl. Phys.* **64** 554–9

Tamada H and Saitoh M 1990 *J. Appl. Phys.* **67** 949–54

Tamaki T and Tsushima K 1987 *J. Magn. Soc. Japan Suppl.* 1 **11** 173–7

Tanabe T, Tanaka Y and Arai R 1995 *Appl. Opt.* **34** 338–45

Thavendrarajiah A, Pardavi-Horvath M and Wigen P E 1990 *J. Appl. Phys.* **67** 4941–3

Thole B T, Carra P, Sette F and van der Laan G 1992 *Phys. Rev. Lett.* **68** 1943

Thole B T, van der Laan G, Fuggle J C, Sawatzky G A, Karnatak R C and Esteva J M 1985a *Phys. Rev.* B **32** 5107

Thole B T, van der Laan G and Sawatzky G A 1985b *Phys. Rev. Lett.* **55** 2086

Thuy N P, Višňovský Š, Prosser V, Krishnan R and Vien T K 1981 *J. Appl. Phys.* **52** 2292–4

Tien P K, Martin R J, Blank S L, Wemple S H and Varnerin L J 1972a *Appl. Phys. Lett.* **21** 207–9

Tien P K, Martin R J, Wolfe R, LeCraw R C and Blank S L 1972b *Appl. Phys. Lett.* **21** 394–6

Tien P K, Schinke D P and Blank S L 1974 *J. Appl. Phys.* **45** 3059–68

Tobin J G, Waddill G D and Pappas D P 1992 *Phys. Rev. Lett.* **68** 3642

Toki K, Kariyada E, Shimada M, Yumoto S and Okada O 1990 *IEEE Trans. Magn.* **26** 1709–11

Tokumaru H and Nomura T 1986 *J. Magn. Soc. Japan* **10** 209

Tolksdorf W, Dammann H and Pross E 1987 *J. Magn. Soc. Japan Suppl.* 1 **11** 341–5

Toriumi Y, Itoh A, Mizobuchi E, Katayama K and Inoue F 1987 *J. Magn. Soc. Japan Suppl.* 1 **11** 249–52

Traeger G, Wenzel L and Hubert A 1992 *Phys. Status Solidi* a **131** 201–7

Traeger G, Wenzel L, Hubert A and Kambersky V 1993 *IEEE Trans. Magn.* **29** 3408–10

Treves D, Jacobs J T and Sawatzky E 1975 *J. Appl. Phys.* **46** 2760–5

Tsai C S and Young D 1990 *IEEE Trans. Magn.* **26** 560–70

Tsai C S, Young D, Chen W, Adkins L, Lee C C and Glass H 1985 *Appl. Phys. Lett.* **47** 651–4

Tseng S C C, Reisinger A R, Giess E A, Powell C G and Glass H 1974 *Appl. Phys. Lett.* **24** 265

Tsutsumi K, Fujii Y, Komori M, Numata T and Sakurai Y 1983 *IEEE Trans. Magn.* **19** 1760–2

Tsutsumi K, Nakaki Y, Tokunaga T, Fukami T and Fujii Y 1993 *IEEE Trans. Magn.* **29** 3760–5

Tsuya N Y, Saitoh Y and Saito N 1986 *J. Magn. Magn. Mater.* **54–57** 1681

Turner A E, Gunshor R L and Datta S 1983 *Appl. Opt.* **22** 3152

Uba S, Uba L, Gontarz R, Antonov V N, Perlov A Y and Yaresko A N 1995 *J. Magn. Magn. Mater.* **140** 575–6

Ueki M and Miyazaki Y 1987 *J. Magn. Soc. Japan Suppl.* 1 **11** 365–8

Umeda N and Eguchi T 1988 *Japan. J. Appl. Phys.* **27** 2173–4

Umegaki S, Inoue H and Yoshino T 1981 *Appl. Phys. Lett.* **38** 752–4

Uspenskii Yu A and Halilov S V 1989 *Sov. Phys.–JETP* **68** 588

Uspenskii Yu A, Kulatov E T and Halilov S V 1995 *Sov. Phys.–JETP* **80** 952–9

——1996 *Phys. Rev.* B **54** 474

Urner-Wille M, Hansen P and Witter K 1980 *IEEE Trans. Magn.* **16** 1888

Valiev U V, Klochkov A A, Lukina M M and Turganov M M 1987a *Opt. Spectrosk.* **63** 543–6

Valiev U V, Klochkov A A, Sokolov B Yu, Tugushev R I and Hasanov E G 1987b *Opt. Spectrosk.* **64** 1192–5

Valiev U V, Krinchik G S, Levitin R Z and Sokolov B Yu 1985a *Fiz. Tverd. Tela* **27** 233–5

Valiev U V, Krinchik G S, Levitin R Z, Sokolov B Yu and Turganov M M 1985b *Opt. Spectrosk.* **58** 1375–88

Valiev U V and Popov A I 1985 *Fiz. Tverd. Tela* **27** 2729–32

Valiev U V, Zvezdin A K, Krinchik G S, Levitin R Z, Mukimov K M and Popov A I 1983 *Sov. Phys.–JETP* **58** 181 van der Heide P A M, Baelde W, de Groot R A, Vromen A R, van Engen P G and Buschow K H J 1985 *J. Phys. F: Met. Phys.* **15** L75–L80

van der Laan G 1987 *Giant Resonances in Atoms, Molecules and Solids (NATO Advanced Study Institute, Series B: Physics, vol 151)* ed J P Connerade (New York: Plenum) p 447

——1990a *Proc. 2nd European Conf. on Progress in X-Ray Synchrotron Radiation Research* vol 25, ed A Balerna, E Bernieri and S Mobilio (Bolognia: SIF) p 243

——1990b *Phys. Scr.* **41** 574

van der Laan G, Hoyland M A, Surman M, Flipse C F J and Thole B T 1992 *Phys. Rev. Lett.* **69** 3827

van der Laan G and Thole B T 1990 *Phys. Rev.* B **42** 6670

——1991 *Phys. Rev.* B **43** 13 401

——1992 *J. Phys.: Condens. Matter.* **4** 4181

van der Laan G, Thole B T, Sawatzky G A, Goedkoop J B, Fuggle J C, Esteva J M, Karnatak R, Remeika J P and Dabkowska H A 1986 *Phys. Rev.* B **34** 6529

van der Ziel J P, Pershan P S and Malstrom L D 1965 *Phys. Rev. Lett.* **15** 190

van Engelen P P J and Buschow K H J 1986 *J. Magn. Magn. Mater.* **66** 291

van Engelen P P J, de Mooij D and Buschow K H J 1988 *IEEE Trans. Magn.* **24** 1728

van Engen P G 1978 *J. Appl. Phys.* **49** 4620

——1983 *PhD Thesis* Technical University Delft[1]

van Engen P G, Buschow K H J and Erman M 1983a *J. Magn. Magn. Mater.* **30** 374–82

van Engen P G, Buschow K H J, Jongenbreur R and Erman M 1983b *Appl. Phys. Lett.* **42** 202

Vedernikov N F, Zvezdin A K, Levitin R Z and Popov A I 1987 *Sov. Phys.–JETP* **60** 1232–42

Veselago V G, Rudov S G and Chernikov M A 1984 *Pis. Zh. Eksp. Teor. Fiz.* **40** 181–3

Vetoshko P M, Volkovoy V B, Zalogin V N and Toporov A Yu 1991 *J. Appl. Phys.* **70** 6298–300

Višňovský Š 1986 *Czech. J. Phys.* B **36** 625

——1991a *Czech. J. Phys.* B **41** 663–94

——1991b *J. Magn. Soc. Japan* **15** 67

Višňovský Š, Knappe B, Prosser V and Muller H-R 1976 *Phys. Status Solidi* a **38** K53–6

Višňovský Š, Nyvlt M, Prosser V, Atkinson R, Hendren W R, Salter I W and Walker M J 1994 *J. Appl. Phys.* **75** 6783–6

Višňovský Š, Prosser V, Krishnan R, Parizek V, Nitsch K and Svobodova L 1981 *IEEE Trans. Magn.* **17** 3205–10

Višňovský Š, Siroky P and Krishnan R 1986 *Czech. J. Phys.* B **36** 1434

Vollmer R, Kirilyuk A, Schwabe H, Kirschner J, Wierenga H A, de Jong W and Rasing T 1995 *J. Magn. Magn. Mater.* **148** 295–7

Voloshinskaya N M and Bolotin G A 1974 *Fiz. Met. Metalloved.* **36** 68

Voloshinskaya N M and Fedorov G V 1973 *Fiz. Met. Metalloved.* **36** 946

Voloshinskaya N M and Ponomariova V I 1969 *Opt. Spectrosk.* **27** 674–81

Voloshinskaya N M, Sasovskaya I I and Noskov M M 1974 *Fiz. Met. Metalloved.* **38** 1134

Walecki W and Twardowski A 1989 *Acta Phys. Pol.* A **75** 313

[1] The results of this work can also be found in Buschow (1988).

Wallenhorst M, Niemoller M, Dotsch H, Hertel P, Gerhardt R and Gather B 1995 *J. Appl. Phys.* **77** 2902–5

Wang C S and Callaway J 1974 *Phys. Rev.* B **9** 4897

Wang H Y, Schoenes J and Kaldis E 1986 *Helv. Phys. Acta* **59** 102

Wang S, Shah M and Crow J D 1972 *J. Appl. Phys.* **43** 1861–75

Wang X, Antropov V P and Harmon B N 1994 *IEEE Trans. Magn.* **30** 4458–60

Weller D, Stöhr J, Nakajima R, Carl A, Samant M G, Chappert C, Mégy R, Beauvillain P, Veillet P and Held G A 1995 *Phys. Rev. Lett.* **75** 3752

Wemple S H, Blank S L and Seman J A 1974 *Phys. Rev.* B **9** 2134

Wemple S H, Dillon J F Jr, van Uitert L G and Grodkiewicz W H 1973 *Appl. Phys. Lett.* **22** 331

Wenzel L, Kambersky and Hubert A 1995 *Phys. Status Solidi* a **151** 449–66

Wettling W, Andlauer B, Koidl P, Schneider J and Tolksdorf W 1973 *Phys. Status Solidi* b **59** 63

Whitcomb E C and Henry R D 1978 *J. Appl. Phys.* **49** 1803

Wickersheim K A, Lefever R A and Hanking B M 1960 *J. Chem. Phys.* **32** 271

Wijngaarden R J, Koblischka M R and Griessen R 1995 *Physica* C **235** 2699–700

Winkler H P, Dotsch H, Luhrmann B and Sure S 1994 *J. Appl. Phys.* **76** 3272–8

Wittekoek S, Popma T J A, Robertson J M and Bongers P F 1975 *Phys. Rev.* **12** 2777

Wolfe R, Fratello V J and McGlashan-Powell M 1987 *Appl. Phys. Lett.* **51** 1221–3

Wolfe R, Hegarty J, Dillon J F Jr, Luther L C, Celler G K and Trimble L E 1985 *IEEE Trans. Magn.* **21** 1647–50

Wolfe R, Kurtzig A J and LeCraw R C 1970 *J. Appl. Phys.* **41** 1218–24

Wood D L, Ferguson J, Knox K, and Dillon J F Jr 1963 *J. Chem. Phys.* **39** 890

Wood D L and Remeika J P 1967 *J. Appl. Phys.* **38** 1038

Wood D L, Remeika J P and Kolb E D 1970 *J. Appl. Phys.* **41** 5315–22

Wu Y, Stöhr J, Hermsmeier B D, Samant M G and Weller D 1992 *Phys. Rev. Lett.* **69** 207

Xu Y B, Lu M, Bie Q S, Zhai Y, Jin Q Y, Zhu X B and Zhai H R 1995 *J. Magn. Magn. Mater.* **140** 581–2

Yamamoto M, Nakanishi H and Hara S 1987 *IEEE Trans. Magn.* **23** 2689–91

Yamamoto S and Makimoto T 1974 *J. Appl. Phys.* **45** 882

——1977 *J. Appl. Phys.* **48** 1680–2

Yan X and Xiao S 1989 *J. Appl. Phys.* **65** 1664–5

Yanata K and Oka Y 1994 *Superlatt. Microstruct.* **15** 233–6

Yariv A 1975 *Quantum Electronics* 2nd edn (New York: Wiley)

Yeh P 1980 *Surf. Sci.* **96** 41

Yokoyama Y, Tsukahara S and Tanaka T 1983 *J. Magn. Magn. Mater.* **35** 175

Yokoyama Y, Umezawa H, Takahashi T, Okumura T and Koshizuka N 1987 *J. Magn. Soc. Japan Suppl.* 1 **11** 203–6

Yuen S Y, Wolff P A, Becla P and Nelson D 1987 *J. Vac. Sci. Technol.* A **5** 3040

Zak J, Moog E R, Liu C and Bader S D 1990 *J. Appl. Phys.* **68** 4203–7

Zapasskii V S and Feofilov P P 1975 *Sov. Phys.–Usp.* **18** 323

Zenkov A V 1990 *Magnetooptical properties of ferrites* Thesis Ural University, Sverdlovsk

Zeper W B, Jongenelis A P J, Jacobs B A J and van Kesteren H W 1992 *IEEE Trans. Magn.* **28** 2503–5

Zhang Y B, Woollam J A, Shan Z S, Shen J X and Sellmyer D J 1994 *IEEE Trans. Magn.* **30** 4440–2

Zhou S M, Lu M, Zhai H R, Miao Y Z, Tian P B, Wang H and Xu Y B 1990 *Solid State Commun.* **76** 1305–7

Zvezdin A K, Koptsik S V, Krinchik G S, Levitin R Z and Liskov V A 1983 *Pis. Zh. Eksp. Teor. Fiz.* **37** 331–4

Zvezdin A K and Kotov V A 1976a *Fiz. Tverd. Tela* **18** 967–70

——1976b *Proc. Moscow Physico-Technical Institute* (Moscow: Moscow Physico-Technical Institute Press) pp 201–6

——1977 *Mikroelectronika* **6** 320–6

——1988 *Magnetooptics of Thin Films* (Moscow: Nauka) p 192

Zvezdin A K, Popov A I and Turkmenov H I 1986 *Fiz. Tverd. Tela* **28** 1760

Index

Milton Keynes UK
Ingram Content Group UK Ltd.
UKHW031126141024
449569UK00006B/422